Disaster Risk, Resilience, Reconstruction and Recovery

Series Editor

Anna Lukasiewicz, Disaster Risk Science Institute, Australian National University, Canberra, ACT, Australia

This Palgrave Macmillan book series takes an interdisciplinary integrated approach to disaster risk, resilience and reconstruction. It brings together the multi-disciplinary expertise in the physical sciences and social sciences necessary for effective disaster risk reduction and recovery. With a broad focus on disaster risk science and sustainability, the series affords significant attention to those areas of the world which experience large scale societal and economic losses from disaster events. It includes disasters arising from climate change impacts and those from geological processes and tectonic movements, poor governance and emergency management practices. It takes into account both rural and urban environments in developing and developed countries and the ecological impacts of disasters.

By taking an integrated approach to the four research communities – poverty alleviation, development, climate change and disasters – this series brings new perspectives to the global disaster policy and practice communities. The series connects people and technology and examines the roles and contributions of multiple stakeholders across scale and their interactions with policy development and implementation. It provides significant inclusion of and acknowledgment of the importance of indigenous knowledge and cultural practices in shaping disaster risk reduction from numerous countries. By bringing together the multiple facets of disaster preparedness, response, recovery and reconstruction, the series opens new vistas on achieving reduced societal and economic losses from disasters around the world.

The series was founded by Prof. Helen James of the Disaster Risk Science Institute, at the Australian National University.

Anna Lukasiewicz · Tayanah O'Donnell
Editors

Complex Disasters

Compounding, Cascading, and Protracted

palgrave
macmillan

Editors
Anna Lukasiewicz
Institute for Climate, Energy &
Disaster Solutions
Australian National University
Canberra, ACT, Australia

Tayanah O'Donnell
Fenner School of Environment
and Society
Australian National University
Canberra, ACT, Australia

ISSN 2662-5660 ISSN 2662-5679 (electronic)
Disaster Risk, Resilience, Reconstruction and Recovery
ISBN 978-981-19-2427-9 ISBN 978-981-19-2428-6 (eBook)
https://doi.org/10.1007/978-981-19-2428-6

Cover illustration: @FC_Italy/Alamy Stock Photo

This Palgrave Macmillan imprint is published by the registered company Springer Nature Singapore Pte Ltd.
The registered company address is: 152 Beach Road, #21-01/04 Gateway East, Singapore 189721, Singapore

Series Dedication

The *Disaster Risk, Resilience, Reconstruction and Recovery* book series is dedicated to the late Prof. Helen James, who is its founding editor. Helen was an internationally ranked expert on Southeast Asia, especially Thailand and Myanmar. She was a true polymath, blending knowledge from a wide range of disciplines and always seeking practical solutions. She believed that collaboration—between disciplines, nationalities, and institutions—was the only way to advance knowledge and she built an international community of scholars and practitioners who are dedicated to preventing and mitigating disasters.

This series takes an interdisciplinary, integrated approach to disaster risk, resilience, and reconstruction. It brings together the multi-disciplinary expertise in the physical sciences and social sciences necessary for effective disaster recovery. With a broad focus on disaster risk science and sustainability, the series is global in scope and includes disasters arising from climate change impacts and those from geological processes and tectonic movements, poor governance, and emergency management practices. It provides significant inclusion of and acknowledgment of the importance of Indigenous knowledge and cultural practices in shaping disaster risk reduction from numerous countries. By bringing together the multiple facets of disaster preparedness response, recovery, and reconstruction, the series will open new vistas on achieving reduced societal and economic losses from disasters around the world.

Acknowledgements

The conceiving of the origins of this book, our first substantive collaboration, began just prior to COVID-19 really beginning to bite, in early 2020.

As the pandemic unfolded, during which time the world lost Helen James, we along with the rest of the world watched and wondered when things would 'return to normal'. As the pandemic prolonged, we held an online workshop for all authors within this volume, most of whom were able to attend. There, we shared ideas about the theoretical and practical underpinnings of compound, cascading, and protracted disasters. We discussed at length the conceptualisation of lingering, then in place of protracted, comparing it to our 2020 pandemic experience, to climate change, and a range of other events that had or were shaping our lives at that time. As newer scholars in the field of disasters, we explored at length the time element of disasters and settled on 'protracted' as an appropriate framing.

We wanted to highlight the complexity of disasters, and this was the perfect time to do so, as multiple and ongoing disasters were playing out all around us.

There are many people to whom we express our sincere thanks and appreciation. In no special order, and to acknowledge a long history of support: Karen Hussey, Steve Dovers, John Handmer, John Finnegan, Mark Crosweller, Jon Barnett, Dorean Erhart, Karyn Bosomworth, Tim

Smith, Mark Stafford-Smith, and many other colleagues—you know who you are. We'd also like to thank our families and support networks.

The reviewers who donated their time to this project are just as diverse as the chapters: Andrew Gissing, Amanda Lamont, David Karoly, Alessandro Antonello, Claudia Baldwin, Melissa Parsons, Gemma Sou, Nicholas Osborne, Harriot Beazley, Rikki Kersten, Chris Ballard, Enrique Oracion, Judy Rogers, Robert Glasser, Iain Mackenzie. We are very grateful for your time in undertaking peer review; it is an important and often under-acknowledged aspect of our work as scientists and academics.

What follows is a lengthy volume of case studies that offer theoretical insights into the compound, cascading, and protracted conceptualisation of disasters. Critically, we offer these chapters in this format so as to appeal to a broader audience than academics: we want to reach those with 'boots on the ground'. Practitioners, policy-makers, emergency services folks, volunteers, community members, and others. This book is mostly for you.

December 2021

Anna Lukasiewicz
Tayanah O'Donnell

PRAISE FOR *COMPLEX DISASTERS*

"This is an important, timely and provocative book! The authors explore the contested terrain of risk and disaster, challenging the reader through diverse, and at times disruptive, perspectives and analysis. Unusually for material on this subject, I found the book very accessible. It deserves to be widely read and I expect it to have significant influence on thinking and policy. The volume is also a wonderful tribute to Professor Helen James."

—Emeritus Professor John Handmer, *FASSA, Senior Research Scholar, International Institute for Applied Systems Analysis (IIASA), Vienna*

"Disasters can compound, cascade and linger. They are increasingly complex as networks grow in their interdependencies, settlements expand, inequality worsens, and climate change intensifies severe weather. Such complexities pose additional challenges to communities as they attempt to prevent, prepare, respond and recover. Exploration of the complexity of disasters in a multi-hazard context is crucial to improving future disaster resilience. The book provides an important contribution to an area of growing academic interest. Insights address both theoretical and practical aspects of disaster complexity. Its many case studies provide an opportunity for readers to gain an appreciation of the challenges that increasing complexity poses and how they are being addressed. For policymakers the book provides evidence of the need to invest in disaster resilience and adaptation to address the growing threats posed by complex disasters. It also provides relevant arguments as to how disaster risk is shaped and how future consideration of complex disasters at local to global levels is vital to ensuring safe and sustainable communities."

—Andrew Gissing, *Fellow at Macquarie University and General Manager, Risk Frontiers, Sydney, Australia*

"The authoring of complex disasters: compound, cascading, and protracted is both timely and essential. Timely because Australia as well as many other parts of the world are having lived experiences of complex disasters, and essential because nations can no longer ignore their compounding, cascading, and protracted effects on their communities and the systems that help both create and support them. The authors of the various chapters have done an excellent job in bringing together existing academic and practical insights about systems thinking in disasters as well as introducing us to new insights about how to prepare for a future of greater ambiguity, complexity, and uncertainty. With the real potential for exponentially increasing levels of disaster impact driven by climate change versus only incremental levels of capability improvement to manage their effects, it is critical that governments, the private and non-government sectors, and communities embrace a more holistic, long-term, and systemic approach to all aspects of disaster management. As recent lived experience demonstrates, treating disasters as one-off events and seeking to manage their effects through the traditional approach of prevention, preparedness, response and recovery no longer works and cannot continue to be the basis in which we address complex disasters into the future. This book makes a significant contribution in helping policy makers, academics, strategists, operational leaders as well as anyone else who is concerned about the current and future challenges in disaster risk management to think differently about disasters. I commend the authors for their dedication and commitment to this important topic and encourage all who read this book to incorporate its insights into all aspects of disaster management practice. In doing so, we can achieve real and sustainable change and help secure a safer and more sustainable world."

—Mark Crosweller, *AFMS, former Director-General Emergency Management Australia and Director at Ethical Intelligence, Canberra, Australia*

Contents

Notes on Contributors

Prof. Arbon Paul is a Matthew Flinders Distinguished Professor, Director of the Initiative and Professor of Nursing (Population Health) of Flinders University, Adelaide, Australia. The Torrens Resilience Initiative was established in 2009 to improve the capacity of organisations and societies to respond to disruptive challenges that have the potential to overwhelm local disaster and crisis management capabilities and plans. Professor Arbon is Past President of the World Association for Disaster and Emergency Medicine, a member of the International Health Regulations Roster of Experts, Member of the World Health Organisation, Health Security Interface, Technical Advisory Group, and Fellow of the Australian College of Nursing and the American Academy of Nursing.

Prof. Baldwin Claudia has taught Urban Design and Town Planning and is Co-director of the Sustainability Research Centre, University of the Sunshine Coast. She researches institutional and social-environmental change on topics as diverse as water allocation, coastal planning, rural and regional land use, and climate change adaptation as well as affordable housing and ability and age-friendly communities. She has published over 60 journal articles and books, including *Integrated Water Planning: Achieving Sustainable Outcomes* (2014, translated into Chinese) and *Natural Hazards and Disaster Justice: Challenges for Australia and its Neighbours* (2020) co-edited with Dr. Anna Lukasiewicz.

Dr. Barnes Paul is the Judith Neilson Research Fellow in Disaster Resilience at the University of New South Wales in Sydney, and an Adjunct Associate Professor in the School of Engineering & Information Technology at UNSW/ADFA, Canberra. He was formerly Head of the Risk and Resilience Program at the Australian Strategic Policy Institute and has served in specialist roles in emergency and risk management at State and Federal government levels and as a tenured academic. Dr. Barnes currently serves as a Council Member of the International Military Council on Climate & Security (Washington DC) and a Member of the World Economic Forum Expert Network on Risk & Resilience.

Cajilig Pamela Gloria is a design anthropologist as well as a Ph.D. candidate at the RMIT School of Architecture and Urban Design in which she conducts research on how marginalised fisherfolk in the Philippines improvise upon their environment to overcome climate extremes. She is also the Founding President of Curiosity, a Manila-based research firm that works with the private sector, civic society, and government agencies to design useful and meaningful products and services.

Chantra Victoria is a social and environmental scientist specialising in climate change risk, adaptation and resilience. She has worked with a diverse range of public, private, and not-for-profit organisations across the Asia-Pacific to develop strategies and frameworks that support cities and communities to build resilience, adapt to climate change, reduce disaster risk, and explore nature-based solutions for thriving and sustainable environments.

Cook Margaret is a history lecturer at the University of the Sunshine Coast, Honorary Research Fellow (University of Queensland) and Adjunct Research Fellow (La Trobe University). Her research interests are 'natural' disasters and water histories. She is the author of *A River with a City Problem: A History of Brisbane Floods* (University of Queensland Press, 2019) and co-editor (with Scott McKinnon) of *Disasters in Australia and New Zealand: Historical Approaches to Understanding Catastrophe* (Palgrave, 2020). She was awarded the John and Ruth Kerr Medal of Distinction for excellence in historiography, historical research and writing in 2020.

D'Arcy Zoë is undertaking a Ph.D. at the School of Architecture and Design, RMIT University. Through her research, she's exploring how

Australian communities could take a more strategic, rather than reactive, approach to fire in the landscape. Previous research has examined community-led disaster recovery programs and builds on Zoë's long-term career in the government sector, which focused on delivering community programs.

Zoë is a volunteer firefighter; a chapter lead for the Australasian Women in Emergencies network; and sits on the Lake George district Bushfire Management Committee as a representative for the Nature Conservation Council of NSW.

Dawson Alistair APM is Inspector-General Emergency Management has over 40 years' policing experience and holds a Master of Business Administration, an Executive Master Public Administration, a Graduate Diploma of Executive Leadership, and a Diploma of Public Safety (Police Search and Rescue Coordination). Alistair is a graduate of the Australian Institute of Company Directors and a graduate of the Leadership in Counter-Terrorism (LinCT) Program. In 2010, he was awarded the Australian Police Medal. Alistair also has been awarded the National Emergency Medal for Queensland 2010–2011 and a bar for TC Debbie 2017. In 2021, Alistair was appointed as an Adjunct Professor within the Centre for Environment and Population Health, School of Medicine and Dentistry at Griffith University.

Dovers Stephen is an Emeritus Professor with the Fenner School of Environment and Society, Australian National University, a Fellow of the Academy of the Social Sciences in Australia, and a Senior Associate with the firm Aither.

Dyer Matthew works across planning, research, and climate change at Queensland Fire and Emergency Services. In previous front-line disaster management roles, he was prominent in the response and recovery associated with the record flooding and tornado outbreak that impacted the Wide Bay area, Queensland in 2013, as well as preparedness and response actions accompanying Cyclone Marcia in 2015 and Cyclone Debbie in 2017. He has been an enthusiastic collaborator with the tertiary sector and local and state governments to reduce climate and disaster risk.

Dr. Graham Sonia has been studying the social impacts of sea-level rise and associated adaptation policies for almost ten years. During her research at the University of Melbourne, Sonia and her colleagues developed a method that enables policy-makers to identify which social values are likely to be affected by climate change, and how to incorporate such values into adaptation policies. Since then, she has worked with others to apply the method to coastal communities in Sydney, Hobart, and Barcelona.

Dr. Green David received his Ph.D. from Northeastern University, Boston in 2012. He has been an Assistant and now Associate Professor of Political Science at Nagoya University since 2012. Professor Green's research focuses primarily on Japanese immigration issues, looking at various aspects including public opinion towards immigration, immigrant integration, and central–local differences in Japanese immigration policy.

Harrison Ella is a Senior Principal Engineer at WMA Water Pty Ltd and Senior Flood Risk Adviser at Queensland Reconstruction Authority. She has over 15 years experience specialising in floodplain management in both Australia and the UK. Ella has worked in both the public and private sectors on projects ranging from single-site developments to cross catchment studies and has particular expertise in catchment management strategies, economic assessment, land use planning for natural hazard risk management, and flood risk management policy. Ella has been involved in the Brisbane River Catchment Flood Studies since 2015 as technical lead for the delivery of the Strategic Floodplain Management Plan (Phase 3) and subsequent implementation of actions (Phase 4).

Hicks John is an economist who has held a number of university positions in Australia and New Zealand and who has worked as a senior economist in the private sector. He is currently Professor of Economics at Charles Sturt University, Bathurst, Australia. His expertise encompasses labour economics and macroeconomics and his current research focuses on issues relevant to regional economic and social development.

Ingham Valerie is an Associate Professor of Emergency Management at Charles Sturt University. She is a founding member of the Disaster and Community Resilience Research Group and her research interests include community resilience, disaster recovery, and the tertiary education of emergency managers and fire investigators. Her research highlights the importance of local community organisations in building community connections for disaster resilience, with a particular focus on Bangladeshi and Australian communities.

Dr. Islam Mir Rabiul received his Ph.D. in social psychology from the University of Bristol. His research interest centres around issues related to environmental social psychology. In particular, his research focuses on environmental factors and health impacts relating to climate change; coping, adaptation, resilience, and recovery in natural disasters; climate justice, human rights, and displaced population.

Kim Christopher is an associate member of the News and Media Research Centre, University of Canberra. He received a Ph.D. in Communication from the University of Canberra, Australia. His research interests are in the area of public sector management and community engagement. His expertise lies in research methodology, leadership, and communications.

Legg Rupert is a Ph.D. Candidate and Research Assistant at the Institute for Sustainable Futures, University of Technology Sydney. His thesis research, which is situated within geography and psychology, examines how living on contaminated land is related to residents' mental health and wellbeing.

Lukasiewicz Anna has a diverse education that includes international development, sustainability and water governance. Anna synthesises fragmented and interdisciplinary topics around the governance of environmental resources and hazards using a justice framework. These topics include natural hazard management (especially policy around disaster resilience), the distribution of natural resources (including how decision-making processes are structured and stakeholder–decision-maker interactions); and water governance (including water management, climate change adaptation, and stakeholder engagement).

MacKenzie Andrew is Associate Professor and Director of USP, Vanuatu. Andrew's research interests include urban greening, green infrastructure planning, urban open space management, urban landscape planning, and governance. He has published widely on this topic including 'Understanding Metropolitan Landscapes' published by Routledge in 2020. Andrew's most recent research investigates how urban greening is planned and implemented in Pacific small island developing states.

O'Donnell Tayanah is an honorary Associate Professor at the ANU. She specialises in climate risk and resilience, coastal governance, urban and legal geography, climate change adaptation, and implementing behavioural and institutional change and governance solutions in complex social-ecological-technological systems.

Ramírez-Lovering Diego is Professor of Architecture and Director of the Informal Cities Lab (ICL) at Monash University. The Lab undertakes design-based research exploring and speculating on the conditions of informality in developing cities in the Global South and through a lens of planetary health.

Rifai Allison leads and implements the Queensland Disaster Management Research Framework, building collaborations between the disaster management and tertiary research sectors. Allison has more than 15 years operational disaster management experience in roles across planning, preparedness, response and recovery in Queensland. She has worked operationally during disasters such as the 2011 floods and Cyclone Yasi, as well as deployments to support local governments to develop recovery plans and governance arrangements. Allison has a Ph.D. in geography from the Queensland University of Technology. Allison is an Adjunct Associate with the Cities Institute at Griffith University.

Samnakay Nadeem is a researcher with the Institute for Water Futures and a Fellow at the Centre for Entrepreneurial Agri-Technology at ANU. He is currently working on the subject matter of drought resilience and its implications for water and agricultural policy. Nadeem's interests are in the environment and natural resource policy design, implementation, and evaluation.

Steffen Will is an Earth System scientist. He is a Councillor on the publicly funded Climate Council of Australia that delivers independent expert information about climate change. He is also an Emeritus Professor at the Australian National University (ANU); Canberra, a Senior Fellow at the Stockholm Resilience Centre, Sweden; and a member of the Anthropocene Working Group. From 1998 to mid-2004, Steffen was Executive Director of the International Geosphere-Biosphere Programme, based in Stockholm. His research interests span a broad range within Earth System science, with an emphasis on sustainability and climate change.

Thom Bruce is a member of the Wentworth Group of Concerned Scientists and former Chair of the National Coasts and Climate Change Committee.

Dr. Thomson Jack is a disaster risk specialist with over 20 years of field experience delivering human security programming and emergency operations across the Indo-Pacific. He has been deployed by the ICRC, UNFAO, UNHCR, and numerous multilateral institutions across early warning, institutional preparedness, emergency response, landscape systems, and livelihood security. He is a Research Fellow at The Torrens Resilience Initiative, and Member of the IFRC Displacement-Migration, Emergency Shelter, and Forecast-based Financing Clusters. He currently serves and Strategy Lead—Emergency Services for the Australian Red Cross.

Dr. Thomson Nicholas is a public health epidemiologist with over 20 years experience working across the Indo-Pacific region in infectious disease interventions. During that time, he increasingly sought to understand the influence of security sector actors on environments that either promoted or disabled infectious disease. He began spending time working alongside multiple policing agencies to examine opportunities to align policies and practices that would better support public health outcomes and enhance trust between police, health and civil society sectors. He is currently undertaking work with the Global Initiative on Transnational Organised Crime looking at the flows of commodities associated with environmental crimes across borders in the Mekong Sub Region and the implications for biological threats. He is also a research fellow at the ANU and works on multi-agency biosecurity governance in several countries in the Indo-Pacific.

Turner-Walker Skye has worked in the Asia-Pacific on environment and development issues for more than a decade, with a focus on climate change, climate change adaptation (CCA), disaster risk management (DRM), and disaster risk reduction (DRR). Skye's work has concentrated on strengthening the mechanisms for responding to environmental and climate change risk and uncertainty. Skye is a doctoral researcher being hosted by the Centre for Asia-Pacific Studies at Universitas Gadjah Mada (UGM) from the Australian National University (ANU) Fenner School of Environment and Society, and ANU Institute for Climate, Energy and Disaster Solutions (ICEDS). Skye's research focuses on community involvement in climate change adaptation and development programming in Indonesia.

Dr. Whitney Justin is a hazard researcher in the emergency management sector in Australia and a research fellow at Nagoya University, Graduate School of Law. He received his Ph.D. from the Coral Bell School of Asia-Pacific Affairs' Department of Political and Social Change at Australian National University in 2018. Dr. Whitney's primary research focus is on crisis response and emergency management policy in Japan, with a particular focus on the role of the not-for-profit sector and neighbourhood associations.

Wolff Erich is an interdisciplinary researcher and educator with a civil engineering and architecture background. His research delves into the concepts of risk and disaster to interrogate how these frameworks inform the production of infrastructure and the built environment.

List of Figures

List of Tables

PART I

Introducing the Concepts

CHAPTER 1

The Evolution of Complex Disasters

Anna Lukasiewicz and Tayanah O'Donnell

Shifting the Way that We View Disasters and Their Management

Societies all over the world now face the ever-present and increasingly ruinous challenge of dealing with the impacts of the Industrial revolution. Living with complexity is exacerbated by the warming of the Earth's atmosphere, due to anthropogenic climate change, already locked into the Earth's systems (Steffen et al., 2018). In the past 200 or so years, humans have simultaneously evolved (technologically, politically, economically, socially and culturally) and rapaciously consumed the Earth's resources to

A. Lukasiewicz (✉)
Institute for Climate, Energy & Disaster Solutions, Australian National University, Canberra, Australian Capital Territory, Australia
e-mail: anna.lukasiewicz@anu.edu.au

T. O'Donnell
Fenner School of Environment and Society, Australian National University , Canberra, ACT, Australia
e-mail: tayanah.odonnell@anu.edu.au; taodonnell@deloitte.com.au

A. Lukasiewicz and T. O'Donnell (eds.), *Complex Disasters*, Disaster Risk, Resilience, Reconstruction and Recovery, https://doi.org/10.1007/978-981-19-2428-6_1

do so. Thus, climate change will see warming trends anywhere from 1.5 degrees warmer on average to 4 degrees or higher; the impacts that follow include higher frequency and severity of the fire, flood, and storm events. We also know that the risks of a changing climate are far-reaching, that the impacts will be particularly negative for vulnerable societies around the world, and that biodiversity impacts will be severe (Roe, 2019). Climate change will elevate the frequency and severity of weather events and disasters; the risk associated with these events will also evolve due to the iterative nature of climate change. Climate change, biodiversity loss, and evolving technologies, therefore, demand a more nuanced understanding of disasters. As with Simpson et al. (2021), we use the term complex to mean '*the diversity of interactions among sectors and systems*' (p. 489), noting '*the complexity of interactions among multiple drivers of climate change risk*' (p. 489).

As our understanding of natural hazards and risk evolves, we are faced with an uncomfortable reality: any time that any community is responding to a disaster, it will increasingly be likely to also recover from previous disasters. This is while simultaneously (trying to) make decisions that affect its capacity to deal with other past and current events, while (ideally) aiming to embed adaptive capacity to better prepare for and respond to future events. The commonly used prevention, preparedness, response, and recovery (PPRR) framework to capture this iterativeness must also accommodate the increasing complexity that climate change and biodiversity loss bring to the disaster landscape.

There are numerous examples of multiple hazards occurring either simultaneously or in short succession, preventing a return to a normal or previous state of being. Internationally, the impacts of hurricanes Sandy in 2012 and Katrina in 2005 in the United States, along with the entirety of the 2017 US hurricane season, and the triple disaster of the Great East Japan Earthquake, tsunami and Fukushima Daiichi nuclear accident (2011) are well known complex disasters that created both immediate and prolonged impacts (UNDRR, 2019). Closer to home, in Australia, the 2019–2020 period has been marked by the Black Summer bushfires in which hundreds of catastrophic fires burned for weeks and months on end encompassing at one point much of the East Coast of Australia and resulting in loss of life and property: the biodiversity impact alone in the millions of deceased animals; rapidly followed by the impacts of a global pandemic and now flood events in a La Nina year. Indeed, the contributions to this book were written in 2020 and 2021—a time of unprecedented disruptions to livelihoods all over the world as a result of the COVID-19 pandemic (Mofijur et al., 2021). We now know that far from being singular disruptive events, climate risk, hazards, and complex disasters—and our response(s) to them—are forever part of the fabric of life.

This book represents a collection of case studies and analyses, highlighting that complex, multiple disasters occur simultaneously, or in close succession and that these are the norm, rather than the exception (Gissing et al., 2021). Moreover, this book illustrates that complex disasters are a more preferred framing than natural disasters and that we ought to define complex disasters as compound, cascading, and protracted.

In writing about lessons learned from the global response to the COVID-19 pandemic, Few et al. (2020, p. 5) state that:

> *There is an entrenched tendency amongst governments and aid agencies to regard hazards as singular events with clear start and end dates. This tends to arise from a deeply embedded project mentality that determines fund flows, reporting structures, and ultimately, governance. In the case of disasters however, such mentality translates into a risk reduction strategy that focuses only on the tip of the problem, while disregarding longer-term challenges.*

In writing this book for both practitioners and academics, our aim is to encourage focus on the messiness of our now ever-present, both short- and longer-term challenges in the context of disasters. Nowhere is this more evident than for climate change impacts and for biodiversity loss.

Definitions and Terminology Used in This Book

This book talks about hazards (most of which are 'natural'), disasters (none of which are natural!), risks and disaster management. Some discussion of the definitions of these terms—and how they fit together—is required.

The Sendai Framework defines a hazard as 'a potentially damaging physical event, phenomenon or human activity that may cause the loss of life or injury, property damage, social and economic disruption or environmental degradation' (UNDRR, 2020). This definition of a hazard was adopted by the United Nations General Assembly (UNGA) in

February 2017. While many hazards are environmental phenomena—such as floods, hurricanes, earthquakes, etc. and can thus be labelled as 'natural', this definition includes human processes, such as environmental degradation that may lead to droughts or the development and use of technology that may not withstand a hydro-meteorological event (such as an industrial plant failing during a flood). A general acceptance that not all hazards are 'natural' is thus a relatively new idea.

An oft-used definition of a disaster was formulated by the Special Report of the Intergovernmental Panel on Climate Change in 2012 (Managing the Risks of Extreme Events and Disasters to Advance Climate Change Adaptation). This report defines a 'disaster' as:

> *severe alterations in the normal functioning of a community or a society due to hazardous physical events interacting with vulnerable social conditions, leading to widespread adverse human, material, economic, or environmental effects that require immediate emergency response to satisfy critical human needs and that may require external support for recovery.* (IPCC, 2021, p. 558)

This definition, went on to influence subsequent international frameworks for disaster risk reduction (the Hyogo Framework for Action and the Sendai Framework for Disaster Risk Reduction 2015–2030) highlights that a disaster is the interplay of physical hazards with vulnerable social conditions. It thus steps away from the historically used term of 'natural disasters' to acknowledge that disasters are the consequences of societal decisions. For example, where a storm surge hitting a wetland may be somewhat damaging and disruptive, the same storm surge hitting the same area may have devastating effects if that wetland were to be subsequently turned into a housing estate. A significant contribution to the literature by Handmer and Dovers (2013) links emergency management, public policy, and institutional frameworks by examining disasters as phenomena that can be usefully categorised as routine, non-routine, and unbounded. This contribution also connects disasters with climate change adaptation, recognising then that the increasing climate risks would have flow-on effects for both the ability to adapt and for the frequency and severity of disaster-related impacts.

Risk dominates disaster academic and practitioner discussions. The UNDRR offers two different definitions of 'risk' (UNDRR, 2019):

1. *a synonym for probability of a harmful effect occurring.*
2. *a synonym for the mathematical expectation of the magnitude of the undesirable consequence.*

Here, the term 'risk' is used for both the probability and for magnitude of an event. An oft-used calculation of risk (in both its meanings) as risk = hazard x exposure x vulnerability (UNDRR, 2019; consider also Kelman, 2018; Perry, 2018; Quarantelli, 1987). Risk as the probability of an event occurring is defined as the combination of hazard, exposure, and vulnerability. Risk as the magnitude of the undesirable consequences (i.e., the disaster) is the function of the context of hazard, exposure, and vulnerability (UNDRR, 2019). More recently, scholars frame risk as compounding and the complexity as arising from any given emergency (e.g., Kruczkiewicz et al., 2021), noting the increasing effect of the inter-relation between the event, the various risks, and the underlying hazards.

Whichever definition is preferred, the way that emergency or crisis management professionals deal with disasters is usually portrayed as a series of phases, commonly known in Australia as the PPRR cycle–Prevention–Preparedness–Response–Recovery (PPRR cycle, referred to earlier). However, there are some expansions and variations of this cycle (e.g., Pede, 2020) and the use of this cycle comes with an acknowledgement that the different phases are not distinct, but blend into each other and can happen simultaneously (for example, some Recovery actions start during the Response phase). Moreover, the definitions of each phase of the cycle can vary; time and spatiality also influence the definitional approaches and while definitions are important, what is critical now in late 2021 (the time of writing this chapter) and beyond is that complex disasters highlight and test the limits of framing how we deal with disasters as a simple cycle. The case studies offered in this book, of both the protracted and changing nature of disasters, and the presence of several connected and unconnected disasters, offer concrete examples and new ways to think about complex disasters.

What Are Compound, Cascading, and Protracted Disasters?

Complex and multiple disasters, and risks and hazards are described in various ways. Simpson et al. (2021) categorise different 'complex' risks as those with an IPCC definition, and those without. Compound risk has an IPCC definition: 'compound risks arise from the interaction of hazards, which can be characterized by single extreme events or multiple coincident or sequential events that interact with exposed systems or sectors' (Simpson et al., 2021). Cascading risk, on the other hand, does not have an official IPCC definition, but is described by Simpson et al. (2021) as 'one event or trend triggering others; interactions can be one way (e.g., domino or contagion effects) but can also have feedbacks'. Alexander (2018) explains:

> *Cascading effects are the dynamics present in disasters, in which the impact of a physical event or the development of an initial technological or human failure generates a sequence of events in human subsystems that result in physical, social or economic disruption.*

In this book we highlight, analyse and discuss three types of complex disasters—compound, cascading, and protracted. While the differences may seem obvious, in reality, it can be hard to draw clear distinctions between these types of complex disasters, especially once we start talking about hazards and risk (Gissing et al., 2021). Our introduction in this book of a protracted framing, aimed as it is to draw attention to the time and spatial elements of complex disasters and explicitly connects these lenses to the disaster literature (consider, for example, Sutton et al., 2021). The overlay of hazards and risk to these three disaster typologies adds further complexity to both how we think about, and how we treat and act, an evolving field.

We also choose not to define *all* types of complex disasters and risks—examples include consecutive, interacting, interconnected and systemic, and conjoint (Cutter, 2018, 2020; Gissing et al., 2021)—and that is a non-exhaustive list. In other words, we are not suggesting that this book alone is a panacea to all the complexity that is complex disasters! Rather, we are opening doors, old and new, to think and rethink how we collectively might (re)theorise disasters, and what this might mean and how it then flows into the practicalities of disaster risk reduction, disaster management and response, and resilience to disasters.

The following sections discuss some relevant literature on the three chosen foci of this book, along with explanations and examples of how they are used by authors in this volume. It is not an exhaustive literature review and while we expand some thinking on the compound and cascading disaster literature in Chapter 2, we acknowledge that we may raise more questions than answers. We see this as a good thing.

Compound

Compound disasters are variously described as (Gissing et al., 2021; Pescaroli & Alexander, 2018; Ridder et al., 2020):

1. *two or more extreme disaster events occurring simultaneously or successively*
2. *combinations of extreme events with underlying conditions that amplify their impact*; or
3. *extreme impacts that result from combinations of "average" events.*

In relation to climate change, compound disasters often relate to risks (in terms of probability of occurrence) or drivers of hazards (Hallegate, 2016; Pescaroli & Alexander, 2018), and tangible impacts (Gissing et al., 2021; Ridder et al., 2020; Zscheischler et al., 2018). Adding to the quagmire of definitions on the range and type of disasters, compound impacts are also slightly different and hold inherent subjectivities such that while individual disasters may strike different locations, the impact may still be compounded. For example, the United States was hit by three large hurricanes within a three-week period in 2017—Harvey, Maria, and Irma. These were swiftly followed by wildfires in California, posing challenges for disaster recovery at a national level (Gissing et al., 2021). These individual disasters were far from 'average' and though national governments play a major role in disaster response and relief and cover multiple jurisdictions; the complexity brought to bear as a result of our present-day experience of climate change impacts must not be understated.

Cascading

Cascading disasters have captured the public imagination with the idea that one major event directly causes other disaster events. A poignant recent example is the 2011 Tōhoku earthquake, which was the direct cause of a tsunami that inundated the north-eastern parts of Japan, causing the Fukushima Daiichi nuclear disaster. Cascading disasters are thus differentiated from compound disasters by this idea of a direct 'causal chain', where one event triggers subsequent events with their own impacts (Cutter, 2018). While the primary trigger or impact is usually a natural hazard (such as heavy rainfall, seismic activity, or unexpectedly rapid snow melt) (UNDRR, 2019), cascading disasters usually involve the failure of essential infrastructure and interdependent systems (Gissing et al., 2021). Interestingly, when such a hazard triggers technological emergencies, the cascading disaster is sometimes referred to as 'natech', a sub-category of cascading disasters such as that described by Cutter (2018), whereby the 2017 Arkema chemical plant explosion and fire was a consequence of Hurricane Harvey flooding in the Houston region of the United States.

The triggering event in a cascading disaster does not have to have a big impact in itself in order to cause the secondary events that do have large impacts (Alexander, 2018). These secondary effects can be especially significant in man-made infrastructure (or the 'anthropogenic domain', as referred to by Pescaroli and Alexander [2018]), and in the vulnerability domain (UNDRR, 2019). They are also exacerbated by existing, underlying inequities, and inequality (Cutter, 2020). The magnitude of secondary effects, especially in human-made systems:

> *"can be exacerbated by the failure of physical structures, and the social functions that depend on them, including critical facilities, or by the inadequacy of disaster mitigation strategies, such as evacuation procedures, land use planning and emergency management strategies. Cascading disasters tend to highlight unresolved vulnerabilities in human society"* (Pescaroli & Alexander, 2018, quoted in Cutter, 2018, p. 22).

As with compound events, there is a slight but important difference between cascading hazards and cascading disasters. Cascading hazards often have a linear causal pathway. Therefore, the risk of cascading hazards is thus mostly predictable and could be calculated. Of course, the increasing frequency and severity of climate change and related impacts may influence the granularity of such predictability, and indeed Cutter

(2020) analyses the systemic nature and complexity of disasters and the various interdependencies across systems (see also Boulter et al., 2018). Recognising these interdependencies is essential, given that the earth's systems are responding to a warming climate, which in turn has an effect on hazard events (IPCC ARC6).

Another important element of 'cascading' is the impact of a cascading hazard as dependent on the interaction of natural hazard events with man-made infrastructure and its vulnerabilities. For example, Gissing et al. (2021) point out that in 2016, a series of tornadoes damaged 23 transmission towers in South Australia causing a power blackout in the city of Adelaide for several days. The effects of this power blackout were unanticipated and wide-ranging, including disruptions to food access, public transport, finances, telecommunications, water, medications, and fuel. A central vulnerability in all these systems was the lack of planning for an extended power blackout. The impact of the triggering event (the tornadoes) cascaded both in space and time and caused unpredictable effects in vulnerable systems. Alexander (2018) suggests that the magnitude of disasters should be viewed not only in terms of damaging events, but also in terms of damaging interactions between vulnerabilities (see also Cutter, 2020). These vulnerabilities should include governance issues such as the structural weaknesses of economic systems, corruption, and negligence (Cutter, 2020; Pescaroli & Alexander, 2016). Cutter (2020) is particularly insightful in terms of how we might think about and treat temporal (time) and spatial scales in complex disaster thinking and practice.

Protracted

Unlike 'compound' or 'cascading', protracted disasters do not have an official definition and their recognition as a 'type' of disaster is still in its infancy. The phenomenon implied by the term 'protracted' is certainly mentioned, but not really studied either in Australia or internationally in disaster risk management literature and practice. The traditional view of disasters has been that they are sudden, extreme events, 'contained' events with definite start and end points (Morley et al., 2020). More recently, the phenomenon of 'slow-onset' events—such as droughts—has been recognised, where the starting point is difficult to determine. Despite recognition of this phenomenon, there is a dearth of research on the protracted nature of disasters, particularly in light of the complexity of climate change risks argued for in Simpson et al. (2021).

While 'protracted' disaster does not have an official definition, the term has been picked up in the emergency literature, particularly in the health and medical disciplines (for example, Lynn et al., 2019). We argue that there is utility in the protracted frame for disaster types, following Cutter's (2020) centring of temporal and spatial elements of disasters and the complex risk frame that is climate change, in relation to mental health impacts of disasters (Chen et al., 2020) and rising inequalities due to climate change induced disasters (Cappelli et al., 2021). For Staupe-Delgado (2019), there is a focus on the temporal scale: slow-onset, slowly emerging, slow violence, slow-burning, gradually occurring, gradually manifesting, creeping environmental problems, creeping crises, and disasters. Another way of conceptualising these is as 'everyday' disasters, that may not invoke a formal disaster response (Hsu, 2019), but are nonetheless disastrous in terms of their impacts (e.g., mental health).

The COVID-19 pandemic has thrust protracted disasters into the spotlight. Kruczkiewicz et al. (2021) define the pandemic as a 'protracted crisis' where waves of infectious outbreaks act as shocks, while the ongoing response becomes an ever-present stressor. Few et al. (2020) describe COVID-19 as a 'long-duration hazard event' where transition out of the crisis is likely to take time and vary across scales. The same could be said for a number of health events, e.g., HIV, mental health impacts of disaster events. Few et al. describe COVID-19 as a crisis where 'there are much greater numbers of people affected by the crisis than physically harmed by the hazard (in this case a biological hazard—the outbreak of a virus' (Few et al., 2020). Other similar long-duration hazards include prolonged volcanic activity, periods of drought, and biological infestations (Few et al., 2020) and other protracted social disadvantage, which in turn influence vulnerability and adaptive capacity (Boulter et al., 2018).

A distinguishing feature of protracted (long-duration) events is that the parameters of risk and exposure are 'in flux' rather than static or linear and thus the management of protracted disasters is not straightforward. Nor is it segmented into clear-cut 'response and recovery' phases, and the time-frame for the 'protracted' is unknown. The estimation of the impact of a protracted disaster is therefore difficult to quantify as impact reports rarely include the secondary effects of costs of, for example, cleanup, or the economic costs from loss of the resources (UNDRR, 2019). Additional considerations on the time/spatial axis are the benefits of having more time to adjust, vs the challenges of disaster fatigue (Aridi, 2020; Ingham,

2020; Koebele et al., 2020). Such impacts, whether positive or negative, can also be difficult to attribute to a specific event and so there is always a question of how long a timeframe is considered, what the start and end frames are, and which geographical or spatial scale belongs to what specific event. We posit that while specific attribution is important for unlocking relief funding and other response efforts, there is utility in stepping back and examining the 'whole of system' nature of protractedness. Mental health and growing inequities and inequality are two tangible reasons for this. But, even if only to better understand the flow-on effect of increased disaster risks as a result, this time/spatial element of disasters demands explicit attention from a range of scholars, beyond the health/medical field.

Protracted disasters suffer from the same ambiguity (as with compound and cascading) of being both about impacts and risks. For example, environmental pollution is an example of a protracted risk in a landscape, while the financial effects of COVID management are an example of a protracted impact. An example of a protracted disaster that represents both risks and impacts (albeit not one started by a natural hazard), is the 1984 Bhopal gas explosion in India—a sudden onset event, where the initial death toll is estimated to be just over two thousand people. However, the gas dispersed from its origin and remained in the landscape affecting and killing thousands of people for years to come—a long-term, ongoing impact for those affected and an ongoing risk affecting a much wider area than the initial industrial plant where the explosion occurred (Deb, 2021; Hsu, 2019).

So, the question then is, does the protracted nature of an event qualify it as a disaster? While the IPCC definition quoted above does not specify a timeframe, most disaster researchers have historically confined the term 'disaster' to sudden onset, contained events, with slow-onset events being mentioned only as a means of delimitation (Staupe-Delgado, 2019). Disaster researchers have argued that including slow-onset, protracted phenomena in the definition of disasters would expand the definition to such a point as to render it unusable (Hsu, 2019). However, given the recognition of the complexity and interconnectedness of hazards, risks, and impacts described in compound and cascading disasters above, and the increasingly known and felt risks and impacts of a changing climate, growing inequities, and inequality, increased severity and frequency of exposure and continued degradation of biodiversity and land systems around the world, the argument for excluding protracted disasters becomes less meaningful. Discussions of compound and cascading disasters summarised above are moving away from the view of disasters as discrete events, to considering the socio-economic processes, systemic risks, and vulnerabilities (Hsu, 2019) that operate on longer timescales (Cutter, 2020; Kruczkiewicz et al., 2021). We support this shift and hope that this book helps to elevate this evolution.

What Now?

In practice, all of these definitions are blurred and problematised by the subjectivities in which they arise, and the specific jurisdiction and scales at which they are deployed. Moreover, fussing about definitions does little to solve the challenges (also perhaps opportunities) facing society as we stare down the impacts of climate change. The time to act is now, and the need for action is urgent.

The Australian 2019–2020 catastrophic bushfires, in which hundreds of out-of-control fires burned across the landscape, resulting in particular devastation on biodiversity and human systems on the south-east of the continent, were caused by compounding drivers: widespread drought and heatwaves, disaster response systems that were not used to managing an event of that magnitude, a larger than anticipated geographical scale of fire spread, intensity, and volatility, stranding thousands on beaches as the country burned. These compounding drivers magnified the scale, and therefore the impact, of the bushfires. Adding to the protractedness and longevity of the catastrophic impacts, the fires were soon followed by severe storms and flash floods (in many cases affecting the same communities which were affected by fires), causing fire retardant and other contaminants to enter waterways, and further damaging an already fragile landscape. These rains were then closely followed by the COVID-19 pandemic which significantly slowed down the recovery of bushfire-affected communities. There were compounding climate drivers of both fires and floods, as well as three separate cascading hazard events (fires, floods and pandemic) occurring across varying and protracted temporal and spatial scales. In addition, the ongoing COVID-19 response has delayed and frustrated efforts to respond to and recover from disasters. COVID-19 became a compounding factor in the response delays and barriers to Tropical Cyclone Harold in April 2020, as emergency responders had to follow quarantine procedures (Kruczkiewicz et al., 2021). The pandemic also created a number of cascading effects, affecting shipping and transportation, which affected the delivery of emergency supplies to affected areas (Kruczkiewicz et al., 2021). Notwithstanding the potential for the distinctions between these disaster types to become blurred, there is a utility to setting out how we think about and how we define or otherwise frame disaster types. This utility ranges from the positionality of the research response, to the policy and response efforts that must be iterative and evolving in response to hazards, risks, and disasters as our global environment experiences rapid and in many instances catastrophic change. As with Boulter et al. (2018), the practice of how we understand complex disasters is more critical than agonising over definitions, particularly given the urgency of climate change impacts and biodiversity loss, and the need for climate change adaptation.

How Complex Disasters Are Explored in This Book

This book is divided into five parts. The first introduces some key concepts. Chapter 2 expands on this chapter, exploring further the theory of compound, cascading, and protracted, contextualising this analysis against the recent extreme fire and flood events in Australia 2019–2022. Chapter 3 deals with terminology, explaining crucial concepts of vulnerability, resilience and systemic risks and providing an overview of how complex disasters challenge traditional government responses. Chapter 4 is an overview of climate change and its impacts on the natural hazards that constitute the risk factors in complex disasters and magnify their impacts. Chapter 5 further explores the intersection of rising climate hazards and vulnerabilities in human systems through an exploration of Australian coastlines and coastal development—locations of significant value and vulnerability to increasing hazards and climate related risks.

Part 2 showcases a range of compounding risks and hazards in Australia and the Asia Pacific. Chapter 6 explores the present and future flood risk in the Brisbane River floodplain. It is an example of climate change as a compound risk (as opposed to an impact). The lack of consideration of climate change in flood planning and the belief in the use of dams as a flood mitigation measure create further compounding risk factors, on top of the physical aspects of climate change changing rainfall intensity. Chapter 7 is an interesting study of compound hazards, but from the viewpoint of an individual—rather than a place. The chapter explores the cumulative impacts of multiple hazards threatening multiple locations that an individual is connected to. It is a very different perspective (individual-oriented, rather than place-based) to vulnerability and exposure that deserves more consideration. Chapter 8 explores the persistent risk of environmental pollution that is compounded by extreme weather events. It focuses on the mental health impacts of residents dealing with long-term PFAS contamination, coupled with the effects of drought.

Part 3 illustrates adaptation to cascading disasters and risks. Chapter 9 explores the local perceptions of residents living in flood-prone Indonesian settlements of Makassar and how they adapt to flood effects, such as disruptions to water and electricity supplies. Chapter 10 discusses transboundary 'choke-points' and the need for early warning systems to prevent the cascading impacts of emergent biological hazards, such as the coronavirus. The chapter focuses on preventing the cascading effects of pandemics. Chapter 11 examines the role of gardening in adaptation to cascading disasters for the urban poor in Port Vila, the capital city of Vanuatu.

Part 4 focuses on protracted disasters. Chapter 12 is a study of the effectiveness of local government provision of disaster information in the wake of the COVID-19 pandemic. This chapter is particularly focused on the response phase in a protracted event, with implications for how policy might be adapted or evolve in such circumstances. Chapter 13 details the experience of Indonesian villagers dealing with repeated earthquakes, exploring the protractedness of impacts which continued for weeks after the initial earthquake. Chapter 14 examines citizen participation of fisherfolk who are facing increasing levels of tidal flooding, whilst at the same time struggling to rebuild their homes after a super typhoon. In this instance, the protracted impacts of the disaster are affecting the community capacity to respond to new risks. Chapter 15 explores the concept of shared responsibility in the preparation and response phases of disaster management focusing on the underlying risks of bushfires.

Part 5 steps away from focusing on the three different types of complex disasters and looks at the challenges posed by disaster complexity to other sectors. Chapter 16 looks at how other policy sectors affect the vulnerabilities in human systems that the emergency management sector then has to deal with. Chapter 17 considers the effect of complex disasters on local communities by establishing the concept of community disaster fatigue, while Chapter 18 explores the role of cooperation and collaboration in research and practice, and learning to tackle complex disasters. Chapter 19 concludes this edited volume by reaffirming the three different complex disaster typologies set out in this book. It also explores the interactions between damaging events and damaging interactions between vulnerabilities across social systems and offers concluding thoughts on future opportunities for the disaster research, policy, and practice field.

References

Alexander, D. (2018). A magnitude scale for cascading disasters. *International Journal of Disaster Risk Reduction, 30*, 180–185.

Aridi, R. (2020). 'For scientists studying 'disaster fatigue', 2019 has been a year like no other' Science. https://www.science.org/content/article/scientists-studying-disaster-fatigue-has-been-year-no-other (Accessed 18 December 2021).

Boulter, S., Palutikof, J., Karoly, D. J., & Guitart, D. (2018). *Natural disasters and adaptation to climate change*. Cambridge University Press.

Cappelli, S., et al. (2021). The trap of climate change-induced "natural" disasters and inequality. *Global Environmental Change, 70*, 102329.

Chen, S., et al. (2020). Anxiety and resilience in the face of natural disasters associated with climate change: A review and methodological critique. *Journal of Anxiety Disorders, 76*, 102297.

Cutter, S. L. (2018). Compound, cascading, or complex disasters: What's in a name? *Environment: Science and Policy for Sustainable Development, 60*(6), 16–25. https://doi.org/10.1080/00139157.2018.1517518

Cutter, S. L. (2020). The changing nature of hazard and disaster risk in the Anthropocene. *Annals of the American Association of Geographers, 111*(3), 819–827.

Deb, N. (2021). Slow violence and the gas Peedit in Neoliberal India. *Social Problems*. https://doi.org/10.1093/socpro/spab058

Few, R., Chhotray, V., Tebboth, M., Forster, J., White, C., Armijos, T., & Shelton, C. (2020). *COVID-19 Crisis: lessons for recovery. What can we learn from existing research on the long-term aspects of disaster risk and recovery?* London, UK.

Gissing, A., Timms, M., Browning, S., Coates, L., Crompton, R., & McAneney, J. (2021). Compound natural disasters in Australia: A historical analysis. *Environmental Hazards*. https://doi.org/10.1080/17477891.2021.1932405

Hallegate, S. (2016). *Natural disasters and climate change: An economic perspective*. Springer.

Hsu, E. L. (2019). Must disasters be rapidly occurring? The case for an expanded temporal typology of disasters. *Time & Society, 28*(3), 904–921.

Ingham, V. (2020). *Disaster fatigue: The implications of Black Summer + COVID19 for Australia. Seminar presented at Susan L Cutter Seminar Series*. https://iceds.anu.edu.au/news-events/events/disaster-fatigue-implications-black-summer-covid-19-australia

IPCC. (2021). Summary for Policymakers. In V. Masson Delmotte, P., A. Zhai, S. L. Pirani, C. Connors, S. Péan, N. Berger, Y. Caud, L. Chen, M. I. Goldfarb, M. Gomis, K. Huang, E. Leitzell, J. B. R. Lonnoy, T. K. Matthews, T. Maycock, O. Waterfield, R. Y. Yelekçi, & B. Zhou (Eds.), *Climate change 2021: The physical science basis. Contribution of working group I to the sixth assessment report of the Intergovernmental Panel on Climate Change*. Cambridge University Press.

Kelman, I. (2018). Lost for words among disaster risk science vocabulary? *International Journal of Disaster Risk Science, 9*, 281–291.

Koebele, E., Crow, D. A., & Albright, E. A. (2020). Building resilience during recovery: Lessons from Colorado's Watershed Resilience Pilot Program. *Environmental Management, 66*, 1–15.

Kruczkiewicz, A., Klopp, J., Fisher, J., Mason, S., McClain, S., Sheekh, N. M., ... Braneon, C. (2021). Opinion: Compound risks and complex emergencies require new approaches to preparedness. *PNAS*, 118(19), e2106795118.

Lynn, M., et al. (2019). *Disasters and Mass Casualty Incidents*. Springer.

Mofijur, M., Rizwanul Fattah, I. M., Asraful Alam, M., Saiful Islam, A. B. M., Chyuan Ong, H., S.M. Ashrafur Rahman, ... Mahlia, T. M. I. (2021). Impact of COVID-19 on the social, economic, environmental and energy domains: Lessons learnt from a global pandemic. *Sustainable Production and Consumption*, 26, 343–359.

Morley, P., Barclay, E., & Parsons, M. (2020). Barriers and enablers in the long-term recovery of communities affected by natural hazards: A review of the literature. Retrieved from https://www.bnhcrc.com.au/sites/default/files/managed/downloads/recovery_literature_review_final.pdf

Pede, E. (2020). Responding to disasters. In E. Pede (Ed.), *Planning for resilience new paths for managing uncertainty* (pp. 33–60). Springer Nature.

Perry, R.W. (2018). Defining disaster: an evolving concept. In Rodriguez, H., Donner, W. and Trainor, J.E. (Eds.), *Handbook of disaster research*. Springer.

Pescaroli, G., & Alexander, D. (2016). Critical infrastructure, panarchies and the vulnerability paths of cascading disasters. *Natural Hazards, 82*, 175–192. https://doi.org/10.1007/s11069-016-2186-3

Pescaroli, G., & Alexander, D. (2018). Understanding compound, interconnected, interacting, and cascading risks: A holistic framework. *Risk Analysis, 38*(11), 2245–2257. https://doi.org/10.1111/risa.13128

Ridder, N. N., Pitman, A. J., Westra, S., et al. (2020). Global hotspots for the occurrence of compound events. *Nature Communications, 11*, 5956. https://doi.org/10.1038/s41467-020-19639-3

Roe, D. (2019). Biodiversity loss – More than an environmental emergency. *The Lancet: Planetary Health, 3*(7), 287–289. https://doi.org/10.1016/S2542-5196(19)30113-5

Quarantelli, E. L. (1987). Disaster Studies: An analysis of the social historical factors affecting the development of research in the area. *International Journal of Mass Emergencies and Disasters, 5*(3), 285–310.

Simpson, N. P., Mach, K. J., Constable, A., Hess, J., Hogarth, R., Howden, M., ... Trisos, C. H. (2021). A framework for complex climate change risk assessment. *One Earth, 4*(4), 489–501. https://doi.org/10.1016/j.oneear.2021.03.005

Staupe-Delgado, R. (2019). Progress, traditions and future directions in research on disasters involving slow-onset hazards. *Disaster Prevention and Management, 28*(5), 623–635.

Steffen, W., Rockström, J., Richardson, K., Lenton, T. M., Folke, C., Liverman, D., ... Schellnhuber, H. J. (2018). Trajectories of the Earth system in the anthropocene. *Proceedings of the National Academy of Sciences, 115*(33), 8252–8259. https://doi.org/10.1073/pnas.1810141115

Sutton, J., et al. (2021). Longitudinal Risk Communication: A Research Agenda for Communicating in a Pandemic. *Health Security, 19*(4), 370–378. https://doi.org/10.1089/hs.2020.0161

UNDRR. (2019). *Global assessment report on disaster risk reduction*. Retrieved from https://gar.undrr.org/

UNDRR. (2020). *Hazard definition & classification review: Technical report*. Retrieved from https://www.undrr.org/publication/hazard-definition-and-classification-review

Zscheischler, J., Westra, S., van den Hurk, B. J. J. M., Seneviratne, S. I., Ward, P. J., Pitman, A., ... Zhang, X. (2018). Future climate risk from compound events. *Nature Climate Change*, 8, 469–477. https://doi.org/10.1038/s41558-018-0156-3

CHAPTER 2

Fire and Flood: Contextualising Compounding, Cascading, and Protracted Disasters

Tayanah O'Donnell, Victoria Chantra, and Anna Lukasiewicz

Introduction

Globally, the growing propensity of wildfire events is causing substantial social, economic, and environmental loss and damage (Filkov et al.,

T. O'Donnell · V. Chantra
Australian National University, Canberra, Australia
e-mail: tayanah.odonnell@anu.edu.au; taodonnell@deloitte.com.au

V. Chantra
e-mail: vchantra@deloitte.com.au

A. Lukasiewicz (✉)
Institute for Climate, Energy and Disaster Solutions, Australian National University, Canberra, Australian Capital Territory, Australia
e-mail: anna.lukasiewicz@anu.edu.au

A. Lukasiewicz and T. O'Donnell (eds.), *Complex Disasters*, Disaster Risk, Resilience, Reconstruction and Recovery,
https://doi.org/10.1007/978-981-19-2428-6_2

2020). Now etched into the Australian landscape and memory, the 2019–2020 East-Australian bushfires came on the back of a two-year drought in the hottest and driest year on record. Over 24 million hectares were burnt across three states and the Australian Capital Territory; at one point over 160 fires were burning simultaneously. Emergency services were stretched beyond capacity, roads and transport systems ground to a halt, telecommunications blacked out, and communities were left stranded (Adcock, 2021a, 2021b; Filkov et al., 2020). The financial costs alone are an estimated impact of over $10AUD billion (Binskin et al., 2020), with impacts spanning catastrophic biodiversity loss (Filkov et al., 2020) through to unprecedented air pollution (Graham et al., 2021; Kruczkiewicz et al., 2021; Rodney et al., 2021; Li et al., 2021). People lost their homes, some tragically, their lives. Many are still rebuilding and the environmental impacts will be lifelong. The adequacy or otherwise of Australia's disaster policy arrangement is discussed in Chapter 16 of this volume (Samnakay and Dovers), who find that as the nature of disasters in Australia evolve, so too must the supporting policy frameworks and policy leadership. More critically, the 2019–2020 fires shone a light on Australia's Indigenous knowledge systems and enduring care for the Country (Fletcher et al., 2021a, 2021b).

The 2019–2020 Australian bushfires are illustrative of a compound, cascading, and protracted disaster. Such systemic disasters have been predicted and well anticipated, and can be expected to continue into the future both in Australia and internationally (Steffen, Chapter 4 of this volume). With over 26 million hectares burnt, 5900 buildings destroyed, and over $100 billion in damage and economic costs, the Australian bushfire season of 2019–2020 is estimated to be Australia's costliest natural disaster (Deloitte Access Economic, 2021; Read & Denniss, 2020). While this in and of itself is significant, the fires also started in the middle of winter in the hottest and driest year on record and following a two-year drought that had affected a large part of the country. Shortly thereafter came a catastrophic hail storm for the nation's capital, Canberra; and before the last fire was extinguished, Australia announced its first case of COVID-19 (Binskin et al., 2020).

V. Chantra
Risk Advisory, Deloitte, Melbourne, Australia

It is not without irony that at the time of writing this chapter (late February-early March, 2022) much of the Australian east coast is in flood. Now near two years into a La Nina weather event, the disaster risks and impacts due to extensive rainfall and flood are cascading and compounding. Brisbane, Queensland is preparing to open Wivenhoe Dam, and eight lives have been lost to drowning incidents in fast-moving floodwaters with the weather system expected to last several days (Callinan, 2022; Hanna, 2022). In March 2021, Eastern Australia suffered $652AUD damage due to widespread flooding (Insurance Council of Australia, 2021). The economic costs are but one feature of the loss and damage suffered across our systems as climate change impacts make themselves known, and as the now increased frequency and severity of those impacts are clear. In light of our new shared reality, this chapter expands on the framing provided in Chapter 1 with how we ought to conceptualise compound, cascading, and protracted disasters.

Compounding, Cascading, and Protracted

In Chapter 1 we (Lukasiewicz and O'Donnell) traced both the history and current scholarly framing of risk as applied in the context of hazards and disasters. There we see that a significant proportion of the framing of complex disasters in the literature is as 'compounding' disaster, usually in response to an emergency event, though not always (Gissing et al., 2021; Kruczkiewicz et al., 2021); frequently, compounding disasters are often framed as two or more disaster events, a combinations of disaster events, or as cumulative disaster impacts that push the range of impact into the extreme (Gissing et al., 2021; Pescaroli & Alexander, 2018; Ridder et al., 2020). This can be contrasted with the linearity of cascading disasters, often conceptualised as quick succession triggered events each with their own impacts (cf Cutter, 2018) but expanding in impact as the systemic failure across systems is fully realised (Gissing et al., 2021), and often exacerbated by existing social and economic inequality (Cutter, 2020). Indeed, the systemic nature and complexity of climate change impacts, biodiversity loss, and the various interdependencies of increasing disaster events across whole-of-systems are well known (e.g. Boulter et al., 2018; Cappelli et al., 2021), with the window for incremental action to address the urgent need for adaptation that is cognisant of the compound and cascading nature of climate change impacts fast closing (IPCC WGII

ARC6). It is worth repeating in this chapter that the magnitude of secondary effects, especially in human-made systems:

> *can be exacerbated by the failure of physical structures, and the social functions that depend on them, including critical facilities, or by the inadequacy of disaster mitigation strategies, such as evacuation procedures, land use planning and emergency management strategies. Cascading disasters tend to highlight unresolved vulnerabilities in human society.* (Pescaroli & Alexander, 2018, quoted in Cutter, 2018, p. 22)

Recognition of the utility of linking compound and cascading concepts to the quantification of risk reduction was recently recognised by Madruga de Brito (2021) who argues that these dual conceptualisations are further influenced by time and space (sensu, protractedness). While there has been considered thinking in disaster risk reduction with respect to interconnected systems (e.g. Helbing et al., 2006; Pescaroli et al., 2018), scientific and scholarly literature exploring these interdependencies through the frame of compounding, cascading, and protracted disaster is scarce. Beyond our discussion in Chapter 1 and in this chapter, there remains scant scholarly literature that is specifically concerned with the linkages and interconnectedness of compound, cascading, and protracted disasters. That society would eventually be required, demanded even, to engage with disasters in this way was not completely overlooked. For example, a 2012 US National Science Committee—chaired by Susan Cutter—found that the social, economic, cultural, and environmental costs of disasters were expected to grow exponentially with attendant impacts on quality and security of life (National Research Council, 2012).

And yet we know, watching the repeated acceleration of climate impact-related disasters unfold worldwide and at a rapid pace since at least 2018 (perhaps earlier, given Hurricane Sandy in 2017) that this acceleration is two-fold: the time between emergency events is shortening, and there are simultaneous, cross-system disaster interventions happening all at once (e.g. flood and power loss and damage to evacuations routes). Population growth and expanding urbanisation into vulnerable environmental localities is a further compounding feature; as climate change continues to escalate, we see more vulnerable people in riskier locations facing more frequent and more severe weather events. According to Cutter (2020), any hazard can compound and cascade but what we are now seeing—weather events combined with a global health crisis combined with geopolitical tensions (eg Russian invasion of Ukraine)—illustrates the effect of disaster events piling on top of each other, in effect stacking and then having their multiple effects ripple across socio-economic systems. Indeed, Walton's recent (2021) analysis of recent US emergency response to a range of weather-related disasters occurring there found that the combination of vulnerable people building in vulnerable locations, with increasing exposure to hazards that were being augmented by climate change as risk creation faster than can be mitigated.

Lawrence et al. (2020) reflect on the mounting evidence that climate change impacts and implications will propagate as cascades (Hilly et al., 2018; Rocha et al., 2018). Though they note that the scope and scale of their effects are not yet well understood, there is agreement on the potential for the combined effect of different impacts compounding to reduce the ability of individuals, governments, and the private sector to adapt in time before loss and damage occur (Lawrence et al., 2020). As a result, there is an urgent need to better understand the spatial and temporal implications of these interactions.

Lawrence's et al. (2020) analysis of the cascades across three domains: water, urban systems, and infrastructure and financial services, found that the connectivity between domains through physical linkages such as built infrastructure, social systems, and goods and services, demonstrated multiple ways in which the impacts of climate change were felt beyond the immediate direct impact, complicating adaptation, and policy responses. Cascades may emerge continuously, such as in the case of sea-level rise or rising groundwater tables, and suddenly, due to extreme weather events such as flooding. Further, there is the potential for a vicious cycle of cascades over time and space as climate stressors weaken infrastructure with the flow on impacts for human health and environments and corresponding impacts on livelihoods, services, property, and overall quality of life (Lawrence et al., 2020). Given the widely recognised role that climate change plays in interdependencies between global risks (World Economic Forum, 2018), failing to adequately account for cascading impacts due to climate change may lead to 'blind spots in adaptation planning, with implications for adaptive capacity' (p. 2).

This potential for vicious cycles of cascades over time and space, raises the perennial challenge of betterment. 'Betterment' is 'the restoration or replacement of a damaged essential public asset to a significantly more disaster resilient standard than its pre-disaster standard' (Binskin et al., 2020). While the financial benefits for all levels of government of investing upfront in stronger and more resilient communities is widely acknowledged (Binskin et al., 2020) betterment is intrinsically tried to funding and hence is contentious and inconsistent. Central to the challenges betterment faces is that reconstruction does not just address the damage to assets resulting from a disaster, but also existing issues such as the asset's age or lack of maintenance (Palliyagaru & Amaratunga, 2011). Therefore, it is difficult to know how much funding is required to restore the assets to a reasonable condition without unduly subsidising the long-term asset renewal program of a local government or region and/or create disincentives for adequate asset management and maintenance (Binskin et al., 2020). Macaskill and Guthrie (2018) in their analysis of reconstruction following the Christchurch earthquake describe the tension present in the provision of national funds to support reconstruction that primarily benefits specific regions.

Under the Australia's Disaster Recovery Funding Arrangements (2018) funding for reconstruction is limited to that necessary to restore essential public assets to a 'pre-disaster function'. While there is also some allowance for betterment under the DRFA, this is discretionary and to be agreed upon in exceptional circumstances. So far, such funding has only been approved four times, each in Queensland. The Queensland Government advised the Royal Commission that 260 of the 370 betterment projects that were funded have now, subsequently been affected by 16 natural disaster events. 96 percent of these remained undamaged or were only superficially damaged, representing approximately $145 million in avoided reconstruction costs (Binskin et al., 2020). In the new reality of compounding, cascading, and protracted disasters, without betterment, the potential for vicious cycles of cascades is more likely to be realised, particularly when coupled with compounding events and impacts (Besser, 2013; Chester et al., 2021; Fleming et al., 2016; Palinkas et al., 2021).

Pescaroli et al. in a (2018) special issue introductory article focus on cascading risks in interconnected socio-economic systems to argue that integration of these key aspects is critical to ensure the adequate acknowledgement of 'non-linear escalations of secondary emergencies' (p. 160). They posit that a better understanding of what they term 'vulnerability paths' better enables a holistic understanding of the particular risk (as opposed to a preoccupation with trigger events). Thinking through then the compounding nature of drought, extensive and widespread bushfires, and over stretched and stressed emergency management and response teams combined with the cascading effect of the fires on road closures, communication system breakdown, destructive air quality, and biodiversity impact (among others), Adcock, 2021a, 2021b describes in her book, *Currowan,* the reality and true nature reverberation of compounding events, cascading across the whole of (economic, social, environmental) systems is such that catastrophic failure of human and infrastructure (sensu Gissing et al., 2021) in the face of increased climate impact risk and biodiversity loss was, at that time, beyond imagination.

The Currowan Fire

Adcock's account of the Currowan fire, one of many burning on the South Coast of New South Wales in late 2019 and into 2020, is a searing first-hand account of an unfolding compound, cascading, and protracted disaster. In her own (abridged) words, as shared in the Financial Review,

> I know this is the reason I am awake now. The day is forecast to hold dangerous fire weather, conditions that will almost certainly incite the Currowan fire, which is now more than 2700 square kilometres in size, poised just inland of dozens of towns and villages along the South Coast of New South Wales—anything could happen. The 2019 bushfire season began in the drought-stricken north of the state back in September. The landscape was so arid that fires sparked easily, spread quickly and proved impossible to extinguish. By October, the state's Rural Fire Service was warning the only way out was decent rainfall—yet none was forecast to arrive for another four months. In mid-November, the NSW Premier declared a seven-day state of emergency, before what the RFS called "the most dangerous bushfire week this nation has ever seen", with people warned "there are simply not enough fire trucks for every house. If you call for help, you may not get it." By now, nearly 1000 homes across the nation have been destroyed and nine people are dead. This dangerous day is coming as more than 80 fires still burn across NSW, with serious blazes also under way in Victoria, South Australia and Western Australia. In the dark I reach for my phone. This constant checking for fire updates: another new routine. I look first at the Fires Near Me app, and what I see is so shocking, I think I must be dreaming. ... It's not the Currowan fire on the move but another one, about 150 kilometres south. I checked the status of this fire before I went to sleep, and it was burning in remote mountain ranges well west of any populated areas—my younger brother lives in the vicinity, and we'd been speaking on the phone about how it could mean trouble for him in the coming days. But now the app is showing it's already escaped, with two enormous fingers of red reaching towards the villages

> of Cobargo and Quaama—one almost on top of the farmland where my brother and his family live. *Are they still asleep? Do they even know?* I text him and he responds within minutes: they're already in the car with kids and animals, racing towards a coastal headland. (Adcock, 2021b)

Adcock's account is supported by 25 years' experience in investigative journalism, backed by a rigorous approach to *Currowan* that seeks to uncover and understand the policy landscape, the inertia in government and community climate change mitigation and adaptation response, and the stark reality that important questions about Australia's preparedness for the dangerous and increasingly complex world we now find ourselves in, exacerbated by the fast pace and sheer scale of climate change risk and impacts, and biodiversity loss, is outstripping our capacity to even imagine let alone act on how we now manage disaster complexity. The Currowan fire burned for more than 70 days, killed three people and devastated many more homes and wildlife spanning near 5000 square kilometres. It was only one of many bushfires burning across Australia during that fateful summer.

That the fires were abnormal—compared to past knowledge—in that they spread quickly at night, often against the prevailing wind and spawning spot fires many kilometres ahead of the main fire front(s), as well as generating pyrocumulonimbus firestorms further fuels the compounding and cascading impacts felt across evacuated and stranded communities and environments that fateful New Year's Eve and day back in 2019–2020 (Canadell et al., 2021). On the back of record-breaking drought conditions, and record high temperatures, a disaster framing that considered the nature of the compound, cascading, and protractedness in current day disasters might have been a country better prepared. It is here that relational knowledge systems show their central importance in the adaptive capacity disaster preparedness and resilience (Fletcher et al., 2021a, 2021b).

Interactions and Trade-offs

As Cutter (2020) argues, there is an increasing need to now focus not just on the specificity of the event(s), but precisely on the interactions between various vulnerabilities across systems as those systems are or operate, during an event (or compounding events). Doing this anticipates where cascading impacts might be readily identifiable or where weaknesses and vulnerabilities lie, so as to be more effective across the PPRR—and its extended variations—cycle by enabling a preparatory response to the identification of these risk points. The idea of feedback loops and interdependencies is common in the climate change adaptation literature, whereby climate impacts are argued as cascades in their own right, compounding to form multiple impacts across sectors, scales, resulting from the interdependencies across and between systems and sub-systems (comprising natural and socio-economic systems), with the combined effect of enhancing stressors and limiting responsiveness (e.g. Davila et al., 2021; Lawrence et al., 2020; Williams et al., 2017). This circularity affects a range of actors—from individuals to governments to industry—to adapt in time. Specifically, stressors include:

- Capacity—of planners, responders, and the public before, during, and after the event
- Fatigue—for responders and for the public (see Chapter 17)
- Unpredictability regarding the rate and scale of impacts
- Complicated governance systems influencing a range of decisions, including funding.

What all this points to, in our view, is the urgency and perhaps permanency of the protracted disaster frame in which we all find ourselves living, given widespread climate change impacts with no real clear pathway to reduce emissions let alone adapt, and the concomitant biodiversity loss that is already evident across multiple geographies and societal systems (Simpson et al., 2021). As foreshadowed in Chapter 1, protracted is yet to emerge as a key pillar of disaster theory and practice, though scholars, notably geographers, have begun to make headway into relational (Fletcher et al., 2021a, 2021b) and temporal factors as the crucial structures around which we must orient new disaster theory and practice (Cutter, 2020; Staupe-Delgado, 2019).

While there are some differentiating features of the definition and application of compound, cascading, and protracted disaster theory, in practice all three share common elements: the risk and/or event is not static; there are not delineated PPRR type phases; the temporal scale is un- or under-defined. These complexities place added pressure on health and wellbeing given that previously well-defined boundaries such as the 'start and finish' of an event, are now so blurred (Aridi, 2020; Ingham, 2020).

Disaster management in Australia prior to and since the Australian Royal Commission into National Natural Disaster Arrangements ('Royal Commission'), highlighted a culture of reacting to climate events as and when they happen, and of treating climate *change* impacts as though each individual impact can be remedied by defending ourselves from nature's ongoing and now locked in changes. Between 20 February and 31 August 2020, the Royal Commission received extensive evidence, from more than 270 witnesses, almost 80,000 pages of tendered documents and more than 1750 public submissions (Binskin et al., 2020). It makes 80 recommendations, largely focused on operational contexts in relation to disaster arrangements, with coverage of broader policy process and design and a core recognition of the challenges of the compound and cascading events (cf Samnakay and Dovers, Chapter 16 of this volume). The relevance of these findings is underscored by the broader scientific literature. As underlined by Simpson et al. (2021) and IPCC WGII ARC6, the rate, scale, and urgency of climate change impacts and biodiversity loss simply do not hold for such incrementalism. Holistic, across-systems approaches are now demanded in order to effectively and efficiently address the complexity in which we now find ourselves in relation to the Earth's system. We therefore need to move past merely providing decision makers and citizens with (potentially—it is often unknown) useful data and move to think about the human element in the use of that data, the type of data captured, and the rate and scale of data uptake, acceptance, and reliance by decision-makers and the broader public (AghaKouchak et al., 2018). Responding to this requires recognition of the evolving and fast-moving context in which these changes are occurring; it also demands a complex analysis of who, where, when, and how in prioritising disaster risk preparedness and response activities. In other words, an explicit engagement with the complexity of disasters as compound, cascading, and protracted phenomena in their own right.

The pressures that force vulnerable people to settle in vulnerable locations will only continue to increase—climate change, population growth, costs of living. The possibility of armed conflict displacing large populations is also growing (foreshadowed and anticipated decades ago now, by leading adaptation scholars such as Barnett, e.g., 2021). While as a society we have the knowledge necessary to mitigate disaster risk, the costs of doing so frequently overwhelm individuals, businesses, communities and governments. While disaster resilience is a shared responsibility, it cannot be shared equally; for instance, homeless people suffer disproportionately during disasters, yet have little capacity to shoulder their 'share' of responsibility for disaster resilience (Every & Thompson, 2014). This inequity is particularly egregious for Indigenous communities (Chakraborty et al., 2021); indeed, in the current 2022 Australian extreme flood event, Aboriginal people have been significantly disaffected signalling the need for institutional support along with self-determination (Nursey-Bray et al., 2019; Williamson, 2022). Even for those fortunate enough to be home owners, dealing with the aftermath of a single, contained disaster can be a struggle (Financial Rights Legal Centre, 2021). Dealing with complex disasters can therefore quickly overwhelm even those who have the capacity and resources. This means we all must be more attuned to the impacts and needs of marginalised communities, and communities who are often not at the forefront of planning (i.e. Indigenous peoples). Addressing complex disasters—whether compound, cascading, or protracted, must be tackled at the societal level and the sheer complexity of the phenomena brings us back to the intractable problems of modern societies: social inequality; conflicting priorities at different levels of government; colonisation; and conflicting values and societal goals.

Pressures for development that increase vulnerability are often exerted at the local or municipal planning level, which has most to gain from approving economically stimulating development, and also the most to lose when disasters strike. Similarly, the roles and functions of emergency services, the ones that respond to the impacts of a disaster, are removed from those of social services that influence the vulnerabilities which determine the levels of disaster impacts. Ultimately, tackling disaster complexity will continue to elude societies that continue to assume 'normal' equals 'disaster-free'.

References

Adcock, B. (2021a). *Currowan*. Black Inc Books.

Adcock, B. (2021b). 'Fire is to our south, west and north, the sea's the only clear front.' *Australian Financial Review*. 17 September 2021b. https://www.afr.com/policy/energy-and-climate/fire-is-to-our-south-west-and-north-the-sea-is-the-only-clear-front-2021b0910-p58qnb. Accessed 26 February 2022.

AghaKouchak, A., Huning, L. S., Mazdiyasni, O., Mallakpour, I., Chiang, F., Vahedifard, F., & Moftakhari, H. (2018). How to natural hazards cascade to cause natural disasters? *Nature, 561*, 458–460.

Aridi, R. (2020). 'For scientists studying 'disaster fatigue', 2019 has been a year like no other' Science. https://www.science.org/content/article/scientists-studying-disaster-fatigue-has-been-year-no-other. Accessed 18 December 2021.

Barnett, J., Graham, S., Quinn, T., Adger, N. W., & Butler, C. (2021). Three ways social identity shapes climate change adaptation. *Environmental Research Letters*, 16, 124029.

Besser, T. (2013). Resilient small rural towns and community shocks. *Journal of Rural and Community Development, 8*(1).

Binskin, M., Bennett, A., & Macintosh, A. (2020). *Royal commission into national natural disaster arrangements*. Commonwealth of Australia.

Boulter, S., Palutikof, J., Karoly, D. J., & Guitart, D. (2018). *Natural disasters and adaptation to climate change*. Cambridge University Press.

Chakraborty, L., Thistlethwaite, J., Minano, A., Henstra, D., & Scott, D. (2021). Leveraging hazard, exposure, and social vulnerability data to assess flood risk to indigenous communities in Canada. *International Journal of Disaster Risk Science, 12*, 821–838.

Callinan, R. (2022, 27 February). 'BOM says south-east Queensland will be battered by severe weather as system moves over Brisbane'. *ABC News*.

Canadell, J. G., Meyer, C. P., Cook, G. D., Dowdy, A., Briggs, P. R., Knauer, J., Pepler, A., & Haverd, V. (2021). Multi-decadal increase of forest burned area in Australia is linked to climate change. *Nature Communications, 12*, 6921.

Cappelli et al. (2021). The trap of climate change-induced "natural" disasters and inequality. *Global Environmental Change, 70*, 102329.

Chester, M., El Asmar, M., Hayes, S., & Desha, C. (2021). Post-disaster infrastructure delivery for resilience. *Sustainability, 13*(6) 3458.

Cutter, S. L. (2018). Compound, cascading, or complex disasters: What's in a name? *Environment: Science and Policy for Sustainable Development*, *60*(6), 16–25. https://doi.org/10.1080/00139157.2018.1517518

Cutter, S. L. (2020). The changing nature of hazard and disaster risk in the Anthropocene. *Annals of the American Association of Geographers, 111*(3), 819–827.

Davila, F., Plant, R., & Jacobs, B. (2021). Biodiversity revisited through systems thinking. *Environmental Conservation, 48*(1).

Deloitte Access Economics. (2021). *Special report: Update to the economic costs of natural disasters in Australia*. Australian Business Roundtable for Disaster Resilience & Safer Communities.

Every, D., & Thompson, K. (2014). Disaster resilience: Can the homeless afford it? *Australian Journal of Emergency Management, 29*(3), 52–56.

Filkov, I., Ngo, T., Matthews, S., Telfer, S., & Penman, T. D. (2020). Impact of Australia's catastrophic 2019/20 bushfire season on communities and environment. Retrospective analysis and current trends. *Journal of Safety Science and Resilience, 1*(1), 44—56.

Financial Rights Legal Centre. (2021). Exposed: Insurance problems after extreme weather events. https://financialrights.org.au/wp-content/uploads/2021/07/Financial-Rights-Exposed-Report_FINAL.pdf

Fleming, C. M., Manning, M., & Smith, C. (2016). Evaluating betterment projects. *Disasters, 40*(2), 365–383.

Fletcher, M. S., Romano, A., Connor, S., Mariani, M., & Maezumi, S. Y. (2021a). Catastrophic bushfires, indigenous fire knowledge and reframing science in southeast Australia. *Fire, 4*(3), 61.

Fletcher, M. S., Hamilton, R., Dressler, W., & Palmer, L. (2021b). Indigenous knowledge and the shackles of wilderness. *Proceedings of the National Academy of Sciences* 118 (40), National Academy of Sciences.

Gissing, A., Timms, M., Browning, S., Coates, L., Crompton, R., & McAneney, J. (2021). Compound natural disasters in Australia: A historical analysis. *Environmental Hazards*. https://doi.org/10.1080/17477891.2021.1932405

Graham, A. M., Pringle, K. J., Pope, R. J., Arnold, S. R., Conibear, L. A., Burns, H., Rigby, R., Borchers-Arriagada, N., Butt, E. W., Kiely, L., Reddington, C., Spracklen, D. V., Woodhouse, M. T., Knote, C., & McQuaid, J. B. (2021). Impact of the 2019/2020 Australian megafires on air quality and health. *GeoHealth*, 5(10). https://doi.org/10.1029/2021GH000454

Hanna, C. (2022). Floods in Queensland and NSW: What's happened to areas affected, and what's likely to happen next? *The Guardian*. https://www.theguardian.com/australia-news/2022/feb/28/queensland-floods-2022-nsw-flood-weather-forecast-brisbane-lismore-grafton-qld-flooding-what-we-know

Hilly, G., Vojinovic, Z., Weesakul, S., Sanchez, A., Hoang, D. N., Djordjevic, S., Chen, A. S., & Evans, B. (2018). Methodological framework for analysing cascading effects from flood events: The case of Sukhumvit Area, Bangkok, Thailand. *Water*, 10(81). https://doi.org/10.3390/w10010081

Helbing, D., Ammoser, H., Kuhnert, C. (2006). 'Disasters as extreme events and the importance of network interactions for disaster response management'. In S. Albeverio, V. Jentsch, & H. Kantz, (Eds.). *The Unimaginable and Unpredictable: Extreme Events in Nature and Society* (pp. 319–348). Springer.

Ingham, V. (2020). *Disaster fatigue: The implications of Black Summer + COVID19 for Australia*. Seminar presented at Susan L Cutter Seminar Series. https://iceds.anu.edu.au/news-events/events/disaster-fatigue-implications-black-summer-covid-19-australia

Insurance Council of Australia. (2021). *Insurance Catastrophic Resilience Report: 2020–2021*. Insurance Council of Australia.

Kruczkiewicz, A., Klopp, J. M, Fisher, J., Mason, S., McClain, C., Sheekh, N. M., Moss, R., Parks, R. M., & Braneon, C. (2021). Compound risks and complex emergencies require new approaches to preparedness. *Proceedings of the National Academy of Sciences, 118*(19), https://doi.org/10.1073/pnas.2106795118

Lawrence, J., Blackett, P., & Cradock-Henry, N. A. (2020). Cascading climate change impacts and implications. *Climate Risk Management, 29*, 100234.

Li, M., Shen, F., & Sun, X. (2021). 2019–2020 Australian bushfire air particulate pollution and impact on the South Pacific Ocean. *Scientific Reports, 11*, 12288.

Macaskill, K., & Guthrie, P. (2018). Funding mechanisms for disaster recovery: Can we affort to build back better? *Procedia Engineering, 212*, 451–458.

Madruga de Brito, M. (2021). Compound and cascading drought impacts to do not by chance: A proposal to quantify their relationships. *Science of the Total Environment, 778*, 146236.

National Research Council. (2012). *Disaster resilience: A national imperative*. The National Academies Press.

Nursey-Bray, M., Palmer, R., Smith, T. F., & Rist, P. (2019). Old wats for new days: Australian Indigenous peoples and climate change. *The International Journal of Justice and Sustainability, 24*(5), 473–486.

Palinkas, L. A., Springgate, B. F., Sugarman, O. K., Hancock, J., Wennerstrom, A., Haywood, C., Meyers, D., Johnson, A., Polk, M., Pesson, C. L., Seay, J. E., Stallard, C. N., & Wells, K. B. (2021). A rapid assessment of disaster preparedness needs and resources during the COVID-19 pandemic. *International Journal of Environmental Research and Public Health, 18*(2), 425.

Palliyaguru, P., & Amaratunga, D. (2011). Linking reconstruction to sustainable socio-economic development. In D. Amaratunga, & R. Haigh (Eds.), *Post-disaster reconstruction of the built environment—rebuilding for resilience*. Wiley-Blackwell.

Pescaroli, G., & Alexander, D. (2018). Understanding compound, interconnected, interacting, and cascading risks: A holistic framework. *Risk Analysis, 38*(11), 2245–2257. https://doi.org/10.1111/risa.13128

Pescaroli, G., Nones, M., Galbusera, L., & Alexander, D. (2018). Understanding and mitigating cascading crises in the global interconnected system. *International Journal of Disaster Risk Reduction, 30*(B), 159–163.

Read, P., & Denniss, R. (2020). 'With costs approaching $100 billion, the fires are Australia's costliest natural disaster'. *The Conversation*. http://theconversation.com/with-costs-approaching-100-billion-the-fires-are-australias-costliest-natural-disaster-129433. Accessed 22 February 2022.

Ridder, N. N., Pitman, A. J., Westra, S., et al. (2020). Global hotspots for the occurrence of compound events. *Nature Communications, 11*, 5956. https://doi.org/10.1038/s41467-020-19639-3

Rocha, J. C., Peterson, G., Bodin, Ö., & Levin, S. (2018). Cascading regime shifts within and across scales. *Science, 362*, 1379–1383. https://doi.org/10.1126/science.aat7850

Rodney, R. M., Swaminathan, A., Calear, A. L., Christensen, B. K., Lal, A., Labne, J., Leviston, Z., Reynolds, J., Trevenar, S., Vardoulakis, S., & Walker, I. (2021). *Physical and Mental Health Effects of Bushfire and Smoke in the Australian Capital Territory 2019–20*. Frontiers in Public Health. https://doi.org/10.3389/fpubh.2021.682402

Simpson, N. P., Mach, K. J., Constable, C., Hess, J., Hogarth, R., Howden, S. M., Lawrence, J., Lempert, R., Muccione, V., Mackey, B., New, G. M.,O'Neill, B., Otto, F., Pörtner, H.-O., Reisinger, A., Roberts, C. A., Schmidt, D. N., Seneviratne, S., Strongin, S., van Aalst, M., Totin, M., & Trisos, C. H. (2021). A framework for complex climate change risk assessment. *One Earth, 4*(4) 489–501. https://doi.org/10.1016/j.oneear.2021.03.005

Staupe-Delgado, R. (2019). Progress, traditions and future directions in research on disasters involving slow-onset hazards. *Disaster Prevention and Management, 28*(5), 623–635.

Walton, B. (2021, 6 July). 'Constant, compounding disasters are exhausting emergency response'. *Circle of Blue*. https://www.circleofblue.org/2021/world/the-disasters-are-exhausting-emergency-response/. Accessed 27 February 2022.

Williams, A., Kennedy, S., Philipp, F., & Whiteman, G. (2017). Systems thinking: A review of sustainability management research. *Journal of Cleaner Production, 148*, 866–881.

Williamson, B. (2022, 4 March). 'Like many disasters in Australia, Aboriginal people are over-represented and under-resourced in the NSW floods'. *The Conversation*. https://theconversation.com/like-many-disasters-in-australia-aboriginal-people-are-over-represented-and-under-resourced-in-the-nsw-floods-178420

World Economic Forum. (2018). *The Global Risk Report 2018*. World Economic Forum, Geneva, Switzerland.

CHAPTER 3

What's in a Name? Deconstructing Risk and Resilience

Paul Barnes

Introduction

Australia, like most developed and developing economies, faces the challenges of anticipating the nature and scale of climate variability and planning for related effects including those associated with weather-related disasters. Natural hazards are a perennial problem regionally and locally. They trigger a significant change to biotic environments, infrastructure (both critical and important), local and regional economies as well as the health and wellbeing, and livelihoods of impacted communities.

Australia has been described, ironically, as a 'lucky country' with an abundance of resources, opportunities, and optimism. It has also been described poetically as—'sunburnt with ragged mountains and droughts and flooding rains'—but from September 2019 onwards, pundits may have considered that luck to have run its course. What was arguably the

P. Barnes (✉)
University of New South Wales, Sydney, NSW, Australia
e-mail: paul.barnes@unsw.edu.au

A. Lukasiewicz and T. O'Donnell (eds.), *Complex Disasters*, Disaster Risk, Resilience, Reconstruction and Recovery,
https://doi.org/10.1007/978-981-19-2428-6_3

most devastating bushfire season in modern Australian history emerged from a background of decades of drought, widespread practices of land clearance, different water management practices across states impacting the flow of river systems, demographic changes, and a warming climate. Then the COVID-19 pandemic and the subsequent socio-economic and health impacts here and globally increased disruption by orders of magnitude.

The cascading and cumulative impacts of these events are likely to be evidenced long into the future. As these challenges increase, we should not underestimate the complexity of the regulatory and governance responses needed and the agility required by the public, and both the public and private sectors to manage them.

Coping with disasters is a challenge for governments, businesses, and especially communities with a full range of consequences often difficult to anticipate let alone control. The cascading nature of impacts from natural hazards means that institutions are unlikely to face only simple disruptions, but rather a series of systemic failures and related damage within and across essential services and other functional areas.

Disaster risk reduction and resilience are core terms in many a global and national policy repertoire in addition to an increasing plethora of consultancy offerings. While there are obvious benefits from applying practical approaches using these concepts, whether they are used accurately and consistently to the depth needed to address the emerging existential threat landscape is critically important: particularly in support of community safety and resilience outcomes.

This chapter explores a selection of caveats to the use of these terms by considering philosophical, linguistic and operational factors that can influence their meaning and practical benefits. This is particularly relevant in instances where regulatory and professional approaches to dealing with natural hazards and related impacts may be construed to be at odds with the experience of the hazards at community levels.

Background

The devastation in communities and the environment across Australia during the 2019–2020 fires was significant. A key thematic focus of submissions to and findings of the Royal Commission into National Natural Disaster Arrangements dealt with community needs and enhancing resilience in such settings. Many of the goals of the federal

National Recovery and Resilience Agency (circa 2021) relate to bolstering community preparation and readiness for future disruptions as well as ensuring supported capability and capacity are in place for recovery efforts.

Significant funding has been released by state and federal governments within Australia to support local resilience building and disaster risk reduction efforts. Many of the efforts planned and program developments that have been underway since the 2019–2020 bushfires have been directly targeted at community and regional recovery. A core aspect of these efforts has been working alongside communities.

An enabler of success in delivering meaningful support to communities will be ensuring recovery and readiness activities, and wider resilience outcomes, match needs in the local communities and support deepening and sustaining related capabilities and capacities 'on the ground.' Other issues of importance are community perceptions of trust and credibility in government.

Two key factors are important in considering issues that enhance or detract from issues of trust in and credibility of government support for community-focused disaster risk reduction. The first is that in most representative democracies (and locations with other forms of state-based governance) it is a reality that many decisions with the potential to impact our lives are invariably made by others: people may choose to self-evacuate as a cyclone or bushfire approaches or choose to take part in certain hazardous activities, like driving through floodwater, but it can be difficult to avoid the results of institutional decisions, especially when made centrally rather than at an individual or community level (Clarke, 1988; MacLean, 1982).

A second factor is that over the past 200 years the regulation of societal safety and authority for decision-making about situations of harm has been elevated from the level of folk or community discourse to the realm of a professionalised bureaucracy and scientific and technically qualified specialists (Plough & Krimsky, 1987; Rip, 1991). Along with these regulatory and technical advances came a specialisation of language and terminology. Socially and culturally, these professional knowledge holders may be seen as different to a 'lay public' in significant ways. This creates the potential for a communication and knowledge gap between specialists and communities exacerbated by differences in understanding the nature of hazards and concern about the impacts that can flow from them.

Given a core aspect of the recent change in national and state-level disaster risk reduction and resilience building policies in Australia, as embodied in several actioned recommendations of the Bushfire Royal Commission (Commonwealth of Australia, 2020), is working with and for communities, it is important that terminology and language use by all involved in disaster management is clear and commonly understood. This is particularly important as some community resilience activity, particularly in disaster remediation settings, remains focused on institutional delivery of services and support rather than as a sustained collaborative effort driven by community needs and experience.

Context and Experience in Disaster Settings

Language and terminology are important: both in terms of usage and the meaning they can convey. Risk is one example of a much-used term that carries a myriad of applications and is subject to variable definitions and perception: particularly between specialists and non-specialists.

An anecdote is helpful here. It has been reported that in his keynote address at the 1990 Annual Meeting of the Society of Risk Analysis, Norman Rasmussen, a pioneer of probabilistic risk assessment, deliberated on the challenges of establishing suitable techniques and approaches for deciding the relative importance of 'risk' ensuring society addresses real 'risks' rather than ephemeral concerns (Fisher, 1991). A side question of relevance is who decides what the real risk issues are.

A key example referred to in Rasmussen's presentation was the high level of concern by residents living near the Three Mile Island nuclear plant at the time of the 1979 partial meltdown of one of the reactors and resultant radioactive waste water. As part of the site remediation post-incident, the preferred official solution for dealing with this waste was onsite boiling of the contaminated water: a process that was known to liberate allowable concentrations of radioactive tritium into the air. The Community preferred the removal of the contaminated water off-site by road transport for treatment elsewhere: thus, avoiding what they perceived to be a hazard in their locality.

Official (technically assessed) views about this community position were that it was both more expensive and created the potential for avoidable hazardous exposure to the public. The implication being that onsite remediation of the wastewater limited exposure to the hazard and thus less risk (potential harm) to the public.

This dichotomy of views on acceptable solutions for dealing with radioactive waste highlights two core issues that underpin achieving risk communication outcomes, community safety and wellbeing and, ultimately, resilience. These are: namely, trust and confidence in public and private sector decision-makers (Kasperson, 1987) and fundamental differences between the [many] public and government specialists regarding relative safety and danger.

A social constructivist view on the perception of risk would suggest that in situations of actual or perceived exposure to hazards or threats, the experience of stress or concern arises from within a social (community) context and can be intensified or attenuated within such settings (Beck, 1992; Clarke & Short, 1993; Giddens, 1990, 1991; Renn, 1992).

That people, in general, make sense of their world collaboratively with others in social settings should not be surprising. In this sense notions such as safe and hazardous and what constitutes these conditions are consensually determined meanings (Douglas, 1985). Thus, people reacting to a disaster are unlikely to respond to physical factors alone, but also to what those events mean and represent as socially constructed and experienced events (see Chapter 12 for an illustration of this point). A fuller appreciation of the intensity of experience forced on communities by natural hazards is arguably inherently linked to recognising that making sense of shared experience is also linked to shared (common) language.

Figure 3.1 (derived from Krimsky & Plough, 1988) conveys suggested contrasting notions (mental schemas) between generic technical/professionalised approaches to risk assessment (limiting the potential for hazards or disruptions to cause harm or loss) against those grounded in a social/experiential and cultural context.

The contrasting bases for knowledge and understanding described here are indicative but the tensions can be seen to surface regularly in public debate on regulatory safety and impact assessments. Even with a recognition that human responses to high-risk events and danger are socially created historically, institutional intent in such cases has often been to discover the scope of public misunderstanding or misperception of the 'real (technical) risk' and provide suitable information to fill a knowledge deficit.

Historically information is more likely to have been expressed in technical terms and the approach to imparting detail designed to assist dealing with a presumed communication deficit: assuming an inability of the 'public' to understand technical knowledge (Bradbury, 1989) rather than

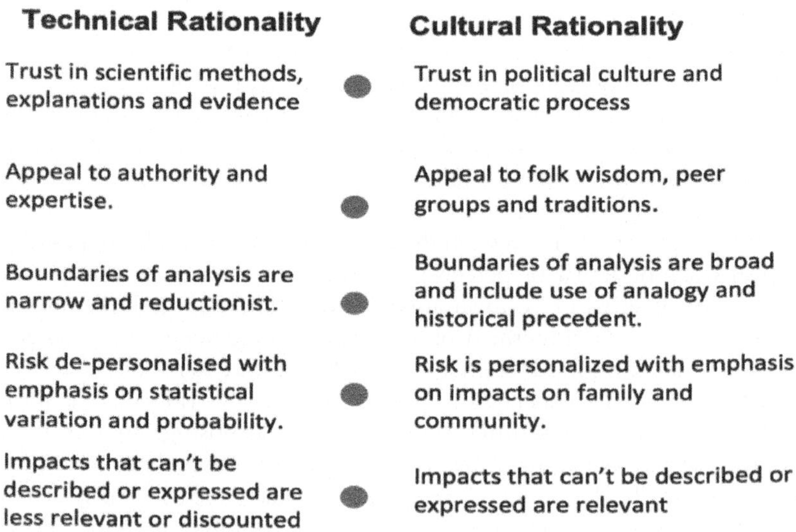

Fig. 3.1 Contrasting approaches to threat and harm (*Source* Author construct)

an absence of social agency and participation in decisions or explanations of the nature of potential harm that they faced. From a wider policy perspective. Two key factors seem relevant: the actual experience of danger or a threatening situation and concern about degrees of control/choice about (living through) the experience (Kasperson & Pijawka, 1985).

This reality was evident in the aftermath of the 2019/2020 Australian bushfires where communities have been in recovery and rebuilding for an extended period. Equally poignant were instances of visits by government or elected officials to disaster zones, such as the visit by the [then] Australian Prime Minister Scott Morrison to the bushfire-stricken town of Cobargo where he was harangued by locals (Hall & Chang, 2020). Responses to the Prime Minister's visit to Cobargo might be seen as indicative of both a difference in a direct lived experience of a disaster and local perceptions of a shallow fly-in fly-out photo opportunity leading to what may be seen as an adding to an already eroded sense of trust and credibility and perceptions of a lack of real sincerity. The Prime Minister later acknowledged that 'people were angry' with him in a post-visit radio interview. While institutional sensitivity to the experience of risk (as danger) remains an issue of great importance, an added frame of reference is the embedded technical meaning and context often applied in professional recovery and risk reduction activities.

Terminology and Professional Usage

Beyond the issue of power differentials embodied in the 'distance' between those providing governance and society in general, further insight can be derived from considering the application of the concept of risk in particular professional knowledge frames. These frames add difficulty to the formation of common policy and decision-making space for regulatory institutions and government.

Figure 3.2 shows analytical detail on applications of risk from actuarial science (insurance) to sociology and anthropology. The manner of usage of the term in professional settings differs in both applications and on an epistemological basis across this continuum.

From the left Fig. 3.2 depicts positivist applications such as those used in the insurance industry, toxicology, and engineering. The actuarial usage of the term 'risk' is a calculated expected value of loss-causing events that occur at a place, time and circumstance leading to the calculation of a premium changed for a particular insurance cover. The notion of actual loss 'events' underpins the calculation of insurance premiums matched

	Actuarial	Toxicology Epidemiology	Probabilistic Assessment	Economics	Psychology	Social Theories	Cultural Theories
Base Unit	Expected Value	Modelled Value	Synthesised Expected Value	Expected Utility	Subjective Expected Value	Perceived fairness & Competence	Shared Value
Prime Method	Extrapolation	Experiments Statistics	Event & Fault Tree Analysis	Risk Benefit Analysis	Psycho-metrics	Surveys - Structured Analysis	In-depth Interpretive
Scope of Concept	Universal	Health & Environment	Safety, Construction Engineering	Universal	Individual Perceptions	Social Interests	Cultural Clusters
Major Application	Insurance	Health & Protection of Environment	Safety Engineering	Decision Making	• Policy Making & Regulations • Conflict Resolution (Mediation) • Risk Communication		
General Function	Risk Sharing	Early Warning Standard Setting	Improving Systems	Resource Allocation	Individual Assessment	Fairness, equity, societal acceptance	Cultural Identity
Social Function	Assessment	Risk Reduction & Policy Selection (Coping with Uncertainty)					Political Legitimisation

Fig. 3.2 Applications of the concept of risk (*Source* Author construct, derived from Renn, 1992)

against the potential payments of insurance claims. Instances of motor vehicle accidents for example provide data at a certain location, time, and age group of drivers all contribute to variable costs of coverage and to determine the costs of premiums.

A common measure of 'risk' in toxicological studies is often expressed in the form of a 'dose–response curve' which express in simple terms the concentration of a toxicant against physiological effect. Epidemiological applications also offer variations in usage: notably 'relative risk.' Relative risk is not expressed in absolute terms such as a '30% chance of disease occurring'. Rather a relative risk of 1.3 for a particular lifestyle disease suggests that the potential for instance of the disease is 30% higher in groups with specifically identified lifestyle habits as compared to a similar matched group that does not have these habits.

The concept from economic and psychometric perspectives are less positivist in context, yet they retain degrees of rigidity in their practical applications. Such specialisations cannot capture the personal experiences of those who suffer losses from an accident or wide-area disaster. Risk perception studies have of course provided useful explanatory frameworks about personal assessments about or contact with a range of natural or technical hazards, but they cannot comprehensively uncover why such associations exist or the rich social contexts within which they are created.

Furthermore, while these behavioural factors are obviously useful, the influence of information on personal decision-making (including sources and contexts) remains unaddressed in many approaches to risk perception (Dunwoody & Neuwirth, 1991, p. 18). An overall critique of psychometric approaches to risk perception is the absence of a credible coverage of the experience of danger in the real world.

While the knowledge classifications detailed in Fig. 3.2, are scaled at a generic level they are important place holders for establishing different institutional and professional applications of the concept of 'risk'. This frame further establishes differences in language and terminology used between professional groups themselves in addition to differences between professional usage and that of a lay public.

More appreciation of the variability of usage and application of 'risk' can be seen in official standards documentation. An example of the complexity of some risk modelling can be found in a definition from Australian Standard HB 167–2006 Security Risk Management (Standards Australia, 2006). One of the definitions suggests that Risk—as a likelihood of disruption in some system—can be defined as the result of a combination of a 'threat likelihood' (T_L), a 'vulnerability' (V), and a 'criticality factor' (C_r). The model is complicated by a number of nested analytical sub-sections for each variable: namely,

> $\mathbf{T_L}$ is threat likelihood (comprising a general plus a specific threat likelihood)
> **V** is vulnerability comprising: (VG + VS + VR) where $\mathbf{V_G}$ (vulnerability to the general threat), $\mathbf{V_S}$ (vulnerability to the specific threat), & $\mathbf{V_R}$ (vulnerability to the specific and general threats)
> $\mathbf{C_r}$ is Criticality comprising: (CS + CP + CO) where $\mathbf{C_S}$ (social criticality), $\mathbf{C_P}$ (personnel criticality), & $\mathbf{C_O}$ (organisational criticality).

While this standard was published in 2006 and may be pending a revision it is indicative of an approach to calculating 'risk' with many assumptions about the veracity of the causal relationships between very complex attributes and access to numerical (probabilistic) estimates of the sub-components and the usefulness/functionality of resulting estimates of the term.

As a marked contrast, the current international Standard on Risk Management (ISO 31000, 2018) defines 'risk' as '[The] *effect of uncertainty on objectives*'. Subsequent text in the standard progresses to clarify this definition by adding several clarifying notes, that:

- An effect is a deviation from the expected positive and/or negative and may itself create in opportunities or new threats.
- Objectives can have different aspects (such as financial, health and safety, and environmental goals) and can apply at different levels (such as strategic, organisation-wide, project, product, and process).
- Risk is often expressed in terms of a combination of the consequences of an event (including changes in circumstances) and the associated likelihood of occurrence.
- Uncertainty is the state, even partial, of deficiency of information related to, understanding or knowledge of an event, its consequence, or likelihood.

Notwithstanding the fact that standards often constitute complex advice, there seems to be a degree of incompatibility between the head lining statement focus on uncertainty and clarification 'notes': which do seem to be of a different order.

Figure 3.3 depicts a framework that may unify the contrasts within the 'standard' by integrating knowledge about future outcomes (positive or negative) and their likelihoods—couched in terms of ambiguity, uncertainty, and ignorance.

It represents both a practical approach to dealing with data-poor analytical choices and the importance of institutional awareness in cases where ignorance is in abundance.

While the value of clarity in the technical and professional usage of risk and natural hazard-related terminology is recognised internationally (UNDRR, 2021) it is important also to examine aspects of the relevant language in everyday use. For this, we can look to the etymology of key terms. An analysis of the lexical roots of hazard and risk is a useful next step that highlights wide semantic variation in language development over time.

In a detailed consideration of English word sources, Ingles (1991) suggests that derivation of the term 'hazard' can attributed to an Arabic word *al-zahr* meaning 'the dice'. From this, the word *khatar*, meaning to gamble, influenced the Spanish word *azar* (a game of chance) in use circa 1100 AD. Ingles further noted linguistic diffusion from Spanish across to French as *hasard* and then into English with similar meaning. Thus

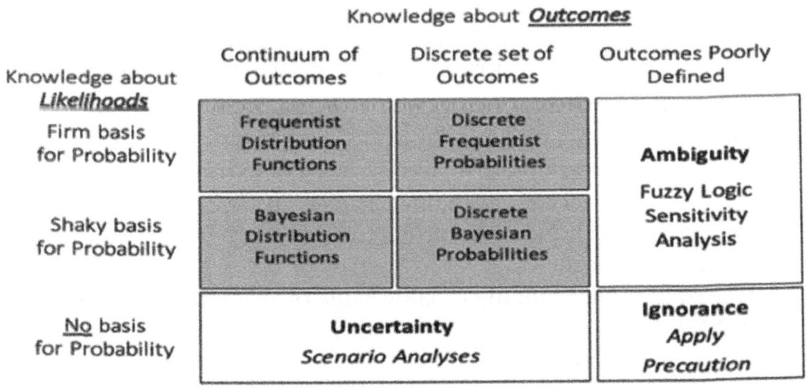

Fig. 3.3 Addressing uncertainty in risk analyses (*Source* Author construct, after Stirling, 2001)

hazard, historically, has been linked with games of chance and outside direct human control and associated with notions of involuntariness and un-foreseeability.

Ingles considers the word 'risk' to be a relatively recent addition to English sourcing from the low Latin *resecare*, alleged to mean 'to being shipwrecked' or the Greek *rhiza*, (reef). He further links its history to Greek *riskos*, (a money chest) and then into Arabic as *rizq* (wealth) and as *al-rizq*, (a fateful windfall). Extension from Arabic via Moorish influence to Spanish as arrisco is noted during the fourteenth century. From around the year 1477, 'risk' can be found in High German as *arreschq* and by 1518, as *risigio*, a precursor to the contemporary Germanic usage, *risiko*.

Figure 3.4 depicts a basic etymology of key semantic differences between risk and hazard with the former holding connotations of voluntary behaviour and the latter, involuntary behaviour.

At a meta-level further differentiation can be made into notions such as predictability, emphasis on wealth or life and whether humans are deemed masters of their own destiny or if life is fated by some extra-human influence. If 'voluntary human control' was aligned to adventurousness we may, as seen in modern English usage note the phrase, 'to take a risk'. Further complexity in usage may be seen in the 'foreseeability' of events and the nuance of precision.

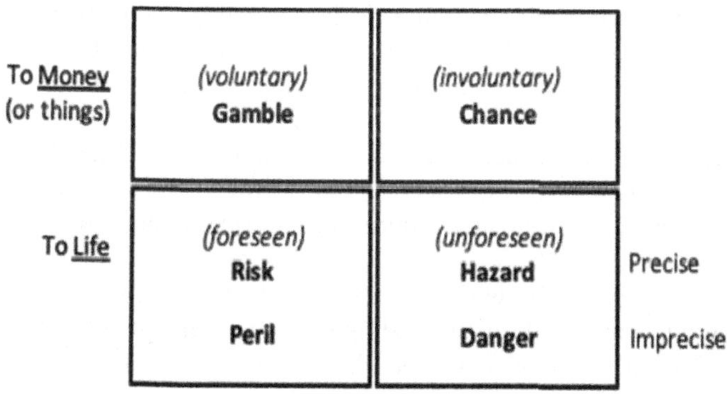

Fig. 3.4 A basic etymology of terminology (*Source* Author construct, after Ingles, 1991)

Resilience

Compared with the many interpretive options for risk, resilience—in the context of impacts from disasters—carries a different dilemma. Arguably there is a lack of clarity in the usage of the terminology in disaster settings.

The (Australian) Productivity Commission inquiry into natural disaster funding arrangements in 2014 defined resilience as '*the ability of communities to continue to function when exposed to hazards and to adapt to changes rather than returning to the original pre-disaster state*' (Productivity Commission, 2014).

The UN Sendai Framework for Disaster Risk Reduction (2015–2030) defines resilience as the '*ability of a system, community or society exposed to hazards to resist, absorb, accommodate, and recover from the effects of a hazard in a timely and efficient manner, including through the preservation and restoration of its essential basic structures and functions*' (UNDRR, 2017).

The US Department of Homeland Security defines resilience as '*the ability to prepare for and adapt to changing conditions and withstand and recover rapidly from disruptions; includes the ability to withstand and recover*' (Fisher et al., 2018).

A key to establishing clarity for resilience may be to accept that it is a systems phenomenon: an outcome of many moving parts. Dekker et al. (2008) interpreted systemic resilience as the ability to recognise, absorb and adapt to disruptions that fall outside a system's design base. A similar definition by Cox (2008), suggests that a resilient system has high adaptive capacity and can withstand disturbances without a decline in crucial functions.

Figure 3.5 depicts a meta-analysis of 'resilience, in engineering, organisational/social systems and ecological settings drawing on selected seminal references (Hollnagel, 2006; Fiksel, 2006; Folke, 2006; Holling, 1996).

Specialists historically working in the field of (disaster) resilience may have limited insight and understanding of how to engage with the variable epistemological bases of 'risk' and the etymological assessment detailed above. Figure 3.5 steps a distance along the path of providing a deeper insight into 'resilience' as an applied concept. Three thematic foci are relevant to deeper consideration: an efficiency of function (engineering systems resisting change), flexibility of function (adaptive and recovery in organisational and social settings) and persistence of function (ecological

Resilience	**Resistance** *(Withstand)*	**Absorption + Restoration** *(Adapt) + (Recover)*	
	Engineering	**Organisational/Social Science**	**Ecology**
	***Efficiency** of Function*	***Flexibility** of Function*	***Persistence** of Function*
	efficiency, constancy, predictability	resistance, mindfulness, uncertainty	persistence, change, unpredictability
	Ability to Maintain Stability Near Equilibrium	Ability to Cope with & Respond to the Unexpected	Ability to Absorb Shocks & Adapt to Change
	Single Operating State to be Maintained other operating states should be avoided by applying safeguards & optimal designs (search for equilibrium)	As a Universal Strategy for dealing with Uncertainty to persist (act reliably) in the face of change by having appropriate institutional structures and resources	Multiple Stability Domains multiple regimes of behaviour for survival
	Rigidity	*Flexibility*	

Fig. 3.5 A theoretical typology of resilience (*Source* Author construct, after Sinclair, 2009)

systems). All themes depict regularity of function or stability but describe differences in how they may respond to change or disturbances.

Importance of Meaning: What is Needed

This chapter has explored some of the differential approaches to the meaning and usage of risk and related terms across a range of professional settings, detailing what 'we think we know' about their use and what might still need clarification for effective risk governance in the face of cascading impacts and cumulative effects.

We no longer inhabit a world of once familiar global security tensions, instabilities in financial markets and systems, imbalances in trading relationships or even seasonal variations in natural hazards driven by warming effected climate/weather patterns. Collectively humanity faces an evolving threat landscape categorised by multi-sourced hazards, impacting (directly and/or indirectly) previously unknown, under-examined or ignored vulnerabilities requiring a paradigm shift in national capabilities.

Societal recovery after crises is often complicated and inevitably slow. A viable intent for governmental support should be returning communities and local economies to a semblance of pre-disturbance normality as efficiently as possible. Physical damage to essential lifelines, loss of housing stock, and the difficulties caused by emergency evacuations will significantly add to community-wide and regional impacts. However, the experience of living through a disaster can challenge the wellbeing and sense of safety of all those involved and is particularly disruptive to community cohesion and viability.

Disasters and their aftermath arguably constitute what Wahl (2017) calls wicked problems that maybe generically characterised as:

- Individually unique.
- consistently multi-causal, multi-scalar and interconnected.
- involving multiple stakeholders often with conflicting agendas.
- straddling organisational and disciplinary boundaries.
- having candidate solutions that are not right or wrong but merely better or worse.
- requiring a long time to evaluate solutions.
- being never completely solved.

Across the long tail of disaster recovery people living in impacted communities will be spoken with and to by many response and recovery specialists from across the private and public sectors. The language (and embedded meaning) used in these conversations may need attention. Real and authentic engagement with disaster-impacted communities is similarly a 'wicked' challenge.

Appreciating that lay communities may speak a different language to that of regulatory and policy professionals is critical. This is particularly important if an issue of differences in valuing what is lost in disasters is factored into policy debates and decision-making.

One significant challenge for Australia involves not having a nationally coordinated multi-threat, risk management strategy. It is important to note, though, that Australia is not bereft of strategic efforts on systemic risk reduction.

The development of the National Disaster Risk Reduction Framework (Department of Home Affairs, 2018) along with companion reports on vulnerability assessment have been produced but the comprehensive implementation of these guidance documents is still ongoing. The recently announced National Emergency Management, Resilience and Recovery Agency is likely to continue this work.

Our future is likely to be populated by a convergence of natural and socio-technically sourced disruptions: some can and should be anticipated, and their impacts reduced. Some will be unexpected, and some may result from neglect or loss of control or failures of imagination. How well Australia copes with these stresses and the wicked problems they cause will benefit from greater clarity and common understanding and language and terminology of risk and resilience.

Several steps might be considered to bridge these issues of credibility gaps and the lay and professionalisation of risk and resilience practices. These include establishing a national strategy for risk management that clearly establishes a common knowledge base for terminology and use in regulatory, policy and decision-making settings.

Supporting this nationwide thinking should be a comprehensive threat taxonomy that incorporates both known and possible threats in both domestic and international contexts: in other words what are the most likely sources of disruption Australia should be planning for. This would need to be supported by an effective technical lexicon to ensure that private and public sector agencies 'speak the same language' and, importantly, provide degrees of clarity for communities across the country.

This national strategy could benefit also from approaches employed in the United Kingdom National Risk Register but updated and adapted to Australian needs. The UK Register, while an ever-adapting tool with room for improvement (Raine, 2021), applies clear definitions of disturbances that can cause significant harm to human welfare, public health and safety, and continuity of essential supplies of fuel, food, water, and energy.

Australian endeavours would not be starting from a zero base. Existing work, such as the 2018 Department of Home Affairs reports profiling vulnerability and disaster risk reduction frameworks specific to Australia's national resilience needs, should be a key starting point (Commonwealth of Australia, 2018).

Establishing a definition of 'resilience', particularly as it applies to communities impacted by natural hazards, is not easy. The existing National Strategy for Disaster Resilience (Commonwealth of Australia, 2011) sets out a mandate for coordinating national thinking on readiness and resilience needs, but it doesn't establish adequate clarity on the underlying principles of community resilience in the context of recovery from the impacts of significant crises.

The US National Institute of Standards and Technology has suggested three key gaps in official knowledge that are relevant to community resilience outcomes (Flynn, 2015). These are the provision of:

- clear and measurable standards for community resilience and for essential infrastructure and buildings

- pathways to adopting and implementing standards for resilience in communities, infrastructure, and buildings
- the identification of the political, economic, and social incentives to promote investment by the public and private sectors and by individuals in supporting resilience.

Current State and federal policies and practices in Australia give considerable attention to enhancing the response and recovery aspect of communities in need, but a deeper understanding of the underlying principles that make up a truly resilient community are needed. Effective disaster mitigation includes improving the safety of community members, reducing damage to property, rapid recovery, and a reduction in overall costs to national, state, and regional economies.

As Australia progresses into a problematical and more disaster-prone future, common bases for sustained dialogue and engagement through a common understanding of the meaning of the terminology used in regulatory and technical assessment will benefit all. Trust and credibility are central issues in the 'wicked space' between those who provide societal regulatory governance and risk reduction services and those who benefit from them. Clarity and understanding of the language of risk and resilience will assist those who serve and those served.

References

Beck, U. (1992). *Risk society – Towards a new modernity*. Sage Publications.

Bradbury, J. A. (1989). The policy implications of differing concepts of risk. *Science Technology & Human Values*, *14*(4).

Clarke, L. (1988). Explaining choices among technological risks. *Social Problems*, *35*(1), 22–35.

Clarke, L., & Short, J. F. (1993). Social organisation and risk: Some current controversies. *The Annual Review of Sociology*, *19*, 375–399.

Commonwealth of Australia. (2020). *The Royal Commission into National Natural Disaster Arrangements Report*. Retrieved from https://naturaldisaster.royalcommission.gov.au

Commonwealth of Australia. (2011). *National Strategy for Disaster Resilience: building the resilience of our nation to disasters*. Retrieved from https://www.homeaffairs.gov.au/emergency/files/national-strategy-disaster-resilience.pdf

Commonwealth of Australia. (2018). *Profiling Australia's vulnerability: The interconnected causes and cascading effects of systemic disaster risk*, Department of Home Affairs. Retrieved from https://www.aidr.org.au/media/6682/national-resilience-taskforce-profiling-australias-vulnerability.pdf

Dekker, S., Hollnagel, E., Woods, D., & Cook, R. (2008). *Resilience engineering: New directions for measuring and maintaining safety in complex systems*. Lund University School of Aviation.

Douglas, M. (1985). *Risk acceptability according to the social sciences*. Russell Sage Foundation.

Dunwoody, S., & Neuwirth, K. (1991). Coming to terms with the impact of communication on scientific and technological risk judgments. In L. Wilkins & P. Patterson (Eds.), *Risky business: Communicating issues of science, risk and public policy* (pp. 11–30). Greenwood Press.

Fiksel, J. (2006). Sustainability and resilience: Towards a systems approach. *Sustainability: Science, Practice and Policy*, *2*(2), 14–21.

Fisher, A. (1991). Risk communication challenges. *Risk Analysis, 11*(2), 173–179. https://doi.org/10.1111/j.1539-6924.1991.tb00590.x

Fisher, R., Norman, M., & Peerenboom, J. (2018). *Resilience History and Focus in the United States*. January, Idaho National Laboratory (INL/MIS-16-40758-Revision-0).

Flynn, S. E. (2015). International Resilience Symposium: Understanding standards for communities and built infrastructure resilience, Engineering Laboratory, National Institute of Standards and Technology, US Department of Commerce, Gaithersburg, Maryland, June. Retrieved from https://nvlpubs.nist.gov/nistpubs/gcr/2015/NIST.GCR.15-917-37.pdf

Folke, C. (2006). Resilience: The emergence of a perspective for social-ecological systems analyses. *Global Environmental Change, 16*(3), 253–267.

Giddens, A. (1991). *Modernity and self-identity (self society in the late modern age)*. Polity Press.

Giddens, A. (1990). *The consequences of modernity*. Polity Press.

Hall, J., & Chang, C. (2020). Prime Minister Scott Morrison harangued by locals in bushfire-stricken town of Cobargo. News.com.au. Retrieved from https://www.news.com.au/technology/environment/prime-minister-scott-morrison-harangued-by-locals-in-bushfire-stricken-town-of-cobargo/news-story/a2b40b36bb6400ad3d71dac053ad0f3a

Holling, C. S. (1996). Engineering resilience versus ecological resilience. In P. Schulze (Ed.), *Engineering within ecological constraints* (pp. 31–44). National Academy Press.

Hollnagel, E. (2006). Resilience – the challenge of the unstable. In E. Hollnagel, D. D. Woods, & N. Leveson (Eds.), *Resilience Engineering: Concepts and precepts* (pp. 9–17). Ashgate.

Ingles O. (1991). A linguistic approach to hazard, risk and error. In Handmer, J. et al. (Eds.), *New perspectives on uncertainty and risk* (pp. 66–78). Centre for Resource and Environmental Studies, Australian National University, and Australian Counter Disaster College Natural Disasters Organisation

ISO 2018. *Risk management – Guidelines* (31000:2018), International Standards Organisation.

Kasperson, R. E., & Pijawka, D. (1985). Societal response to hazards and major hazard events: Comparing natural and technological hazards. *Public Administration Review, 45*, 7–18.

Kasperson, R. E. (1987). Trust and credibility: The central issue? In J. C. Davies, V. T. Covello, & F. W. Allen (Eds.), *Risk Communication* (pp. 43–46). The Conservation Foundation.

Krimsky, S., & Plough, A. (1988). *Environmental Hazards: Communicating Risks as a Social Process*, Auburn House, Massachusetts.

MacLean, D. (1982). Risk and consent: Philosophical issues for centralized decisions. *Risk Analysis, 2*(2), 59–67.

Plough, A., & Krimsky, S. (1987). The emergence of risk communication studies: Social and political context. *Science Technology and Human Values, 12*(3&4), 5.

Productivity Commission. (2014). *Natural Disaster Funding Arrangements, Inquiry Report no. 74*, Canberra. Retrieved from https://www.pc.gov.au/inquiries/completed/disaster-funding/report

Raine, S. (2021). Half of the National Risk Register is Missing, *RUSI Newsbrief*, January/February 2021, *41*(1).

Renn, O. (1992). Concepts of risk: A classification. In S. Krimsky, & D. Golding (Eds.), Ch.3 in *Social Theories of Risk* (pp. 53–79), Praeger.

Rip, A. (1991). The danger culture of industrial society. In R. E. Kasperson & P. J. M. Stallen (Eds.), *Communicating risks to the public-international perspectives* (p. 351). Kluwer.

Sinclair, N. (2009). *Resilience in critical infrastructures: the case of the Queensland electricity industry*. Master of Science (Research) thesis, Queensland University of Technology. Retrieved from https://eprints.qut.edu.au/35667/1/Natalie_Sinclair_Thesis.pdf

Standards Australia. (2006). *Australian Standard HB 167–2006 Security risk management*, Handbook. Retrieved from https://www.standards.org.au/standards-catalogue/sa-snz/publicsafety/ob-007/hb-167-2006

Stirling, A. (Ed.). (2001). *On Science and Precaution in the Management of Technological Risk*. An ESTO Project Report Prepared for the European Commission – JRC Institute Prospective Technological Studies Seville. Retrieved from https://citeseerx.ist.psu.edu/viewdoc/download?doi=10.1.1.470.2712&rep=rep1&type=pdf

UNDRR. (2017). *Disaster Resilience Scorecard for Cities: Preliminary level assessment*, United Nations Office for Disaster Risk Reduction (with the support of USAID, European Commission, IBM and AECOM). Retrieved from https://www.unisdr.org/campaign/resilientcities/assets/toolkit/Scorecard/UNDRR_Disaster%20resilience%20%20scorecard%20for%20cities_Preliminary_English.pdf

UNDRR. (2021). *Hazard Information Profiles: Supplement to UNDRR-ISC Hazard Definition & Classification Review* – Technical Report Retrieved from https://www.undrr.org/publication/hazard-information-profiles-supplement-undrr-isc-hazard-definition-classification

Wahl, D. C. (2017). Facing Complexity: Wicked Design Problems. Retrieved from https://medium.com/age-of-awareness/facing-complexity-wicked-design-problems-ee8c71618966

CHAPTER 4

The Nature of Climate-Related Disasters in Australia

Will Steffen

Introduction

Australia is known as the land of drought and flooding rains, a poetic reminder that the country has always had more than its fair share of climate- and weather-related extreme events. The Federation drought in the early 1900s and the 1939 bushfires still resonate with Australians as the risks with which we must live in our naturally chaotic climate. But recently, over the last two decades in particular, the number of record-breaking extreme weather events is accelerating strongly.

In this chapter I review the scientific background to climate change and the way that it is changing the nature of extreme weather events across the continent. I focus on the fundamental physics of the climate system,

W. Steffen (✉)
Australian National University, Canberra, ACT, Australia
e-mail: will.steffen@anu.edu.au

A. Lukasiewicz and T. O'Donnell (eds.), *Complex Disasters*, Disaster Risk, Resilience, Reconstruction and Recovery,
https://doi.org/10.1007/978-981-19-2428-6_4

the ways in which humans are changing the climate and thus the nature of extreme weather events, the dangers of compound extreme weather events, and what we might face in the future as the planet continues to warm.

What is Climate Change?

Climate change is normally defined as a long-term change in the fundamental characteristics of the climate system—changes in temperature, rainfall patterns, atmospheric circulation patterns and so on. These changes are measured at time scales of many decades (a minimum of three) or, more appropriately, centuries or longer. Climate change is different from short-term patterns of natural variability in the system, such as heatwaves, heavy rainfall events, storms, droughts and so on. In a stable climate system, these weather patterns operate on timescales of days and weeks, or sometimes longer in terms of drought. Importantly, they operate within well-defined limits that are stable over time scales of many decades or centuries. Because climate change means that the fundamental climate system itself is changing over these long time frames—currently the climate system is becoming warmer and more energetic—the normal patterns and limits of extreme weather events also change, leading to unprecedented events that are more frequent and/or more intense than those that occurred when the climate system was stable.

How is Climate Change Influencing Extreme Weather in Australia?

Because climate change is ultimately driven by the increasing energy content of the lower atmosphere, it is changing the background conditions in which extreme weather events occur. There are three fundamental ways in which climate scientists can link the changes in the frequency and severity of extreme weather events to the background changes in the climate system.

Fundamental Physics

All extreme weather is now occurring in an atmosphere that is warmer, therefore containing more energy, and also holding more water vapour. This means that when extreme weather events do occur, they have more energy and water vapour to draw on, thus becoming more intense and destructive.

Observations

The Australian Bureau of Meteorology (BoM) has long-term observational records, beginning from 1910 for most indicators, that can show us how the nature and frequency of extreme weather is changing. By comparing trends in the frequency and intensity of extreme weather events, such as heatwaves and daily rainfall extremes, in the recent past—over a minimum of 30 years—with the long-term observational record, significant changes in the characteristics of extreme weather can be determined. For other extreme weather events, such as extreme winds, tropical cyclones, thunderstorms and extreme sea levels, the observational data quality is not good or long enough to identify clear trends.

Attribution Studies

These are model-based simulations of the occurrence of particular extreme weather events, with and without the increase in greenhouse gases. That is, for the extreme weather event, climate models can be run thousands of times with the pre-industrial concentrations of greenhouse gases and thousands of times with the current level of greenhouse gases and compare how often the extreme event occurs in each case. For example, such an attribution study was carried out for the (then) record hot year of 2013 in Australia, when average temperature was 1.2 °C above the 1961–1990 average (BoM, 2016). The study found that, without the additional greenhouse gases, the record temperature would occur only once every 12,300 years (Knutson et al., 2014; Lewis & Karoly, 2014), compared to frequent occurrences in the model run that included the additional greenhouse gases from human activities.

Major Types of Extreme Weather Events in Australia and How They Are Changing

Extreme Heat

The increase in extreme heat is linked to the underlying increase in global temperature, as shown in Fig. 4.1. A small increase in the average temperature drives a disproportionately large increase in the frequency of extreme heatwaves as well as in their intensity. That is, heatwaves are occurring more often and they are getting hotter. Ultimately, this trend is being driven by accelerating global heating: the 2010–2019 period was 0.2 °C hotter than the 2000–2009 period, fueling more extreme heat around the world and here in Australia.

Globally, heatwaves and extremely hot days are increasing as a result of a warming climate. In Australia, 33 days above 39 °C—the temperature averaged over the entire continent—were recorded in 2019, more

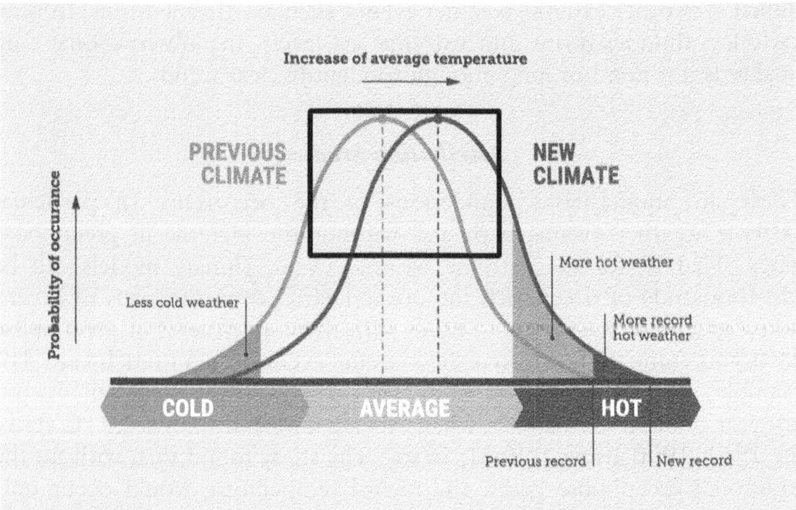

Fig. 4.1 When average temperatures increase and the curve showing the distribution of temperatures moves to the right, there is a significantly greater probability of experiencing very hot, and record hot, weather (*Source* Climate Council, 2021)

than all such days recorded over the previous 60 years (CSIRO & BoM, 2020). A single-day record was set on 4 January 2020, when Penrith, in western Sydney, recorded a maximum temperature of 48.0 °C, making it the hottest place on Earth on that day (Thompson, 2020). More generally, extreme heat is occurring more frequently in central Australia than in coastal regions (IPCC, 2021). Heatwaves are also afflicting the coastal seas around Australia, as more than 90% of the heat trapped by greenhouse gases is absorbed by the oceans. The rate of ocean warming has more than doubled since 1993 (IPCC, 2019).

The impacts of such extreme heat are serious, both for humans and our native wildlife. Heatwaves kill more Australians than all other natural disasters combined—bushfires, cyclones, earthquakes, floods and severe storms (DIT, 2013). More than a third of Australia's population of spectacled flying foxes were killed by a single heatwave in 2018 (Welbergen, 2019). Marine heatwaves have been deadly for Australia's iconic Great Barrier Reef. Consecutive heatwaves in 2015 and 2016, followed by a 2020 heatwave that affected the entire 2300 km length of the reef, have together killed about half of the hard corals (Dietzel et al., 2020; GBRMPA, 2020).

Intense Rainfall

Heavy rainfall events, typically associated with thunderstorms and east coast low-pressure systems, are becoming even more severe at the climate continues to warm. The amount of water vapour that the atmosphere can hold rises by about 7% for every degree C of warming, providing the fundamental physical basis for an increase in the intensity of rainfall (Fig. 4.2).

The multiplying effect is impressive. The most intense rainfall events that we experience today are projected to nearly double in frequency with every degree of warming (Myhre et al., 2019).

Observations across Australia provide solid evidence for the increase in intense rainfall (IPCC, 2021), typically associated with thunderstorms and east coast lows. Daily rainfall totals from thunderstorms have increased since 1979, especially in northern Australia (IPCC, 2021), where the intensity of short-duration (e.g., hourly) extreme rainfall has increased by around 10 per cent, increasing the risk of flash flooding. For example, in north and far north Queensland heavy rainfall in early 2019 triggered widespread and severe flooding, with impacts on health and wellbeing,

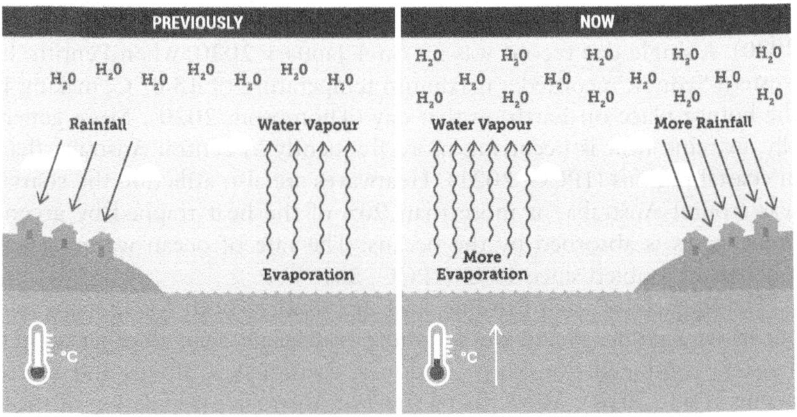

Fig. 4.2 The influence of climate change on the water cycle. Left: The pre-climate change water cycle. Right: The water cycle operating under higher surface and ocean air temperatures, leading to more water vapour (H_2O) in the atmosphere, and, in turn, more rainfall (*Source* Climate Commission, 2013)

homes and infrastructure, and agricultural systems. On the hand, the number of low-pressure systems that bring heavy rainfall to southern Australia has decreased in recent decades (CSIRO & BoM, 2020).

Drought

Drought is one of the most common extreme events in Australia, particularly in the southeast and southwest of the continent. Both agricultural and ecological droughts have increased in southern Australia (IPCC, 2021). Since the 1990s, cool season (April–October) rainfall, which is the important season for the agricultural sector, has decreased by about 12%, while in southwest Western Australia, the drying trend has been longer and more severe—about a 20% drop in rainfall since 1970 (CSIRO & BoM, 2020).

Extreme Sea-Level Events

Although sea-level rise is a much slower process than the other extreme weather events described in this section, extreme sea-level events can occur when storm surges, particularly at a high tide, leading to more

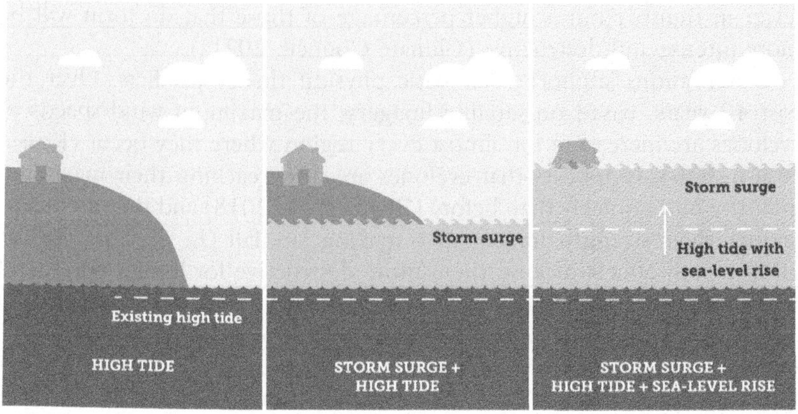

Fig. 4.3 Climate change increases the base sea level and thus exacerbates the effects of a storm surge on coastal flooding (*Source* Climate Commission, 2013)

extensive coastal inundation and more severe damages because of the higher base sea level on which they occur (Fig. 4.3). In addition, extreme sea-level events are often an important component of compound extreme events (next section).

Globally, sea levels have risen by around 20 cm between 1901 and 2018, but recently the rate has increased, from 1.3 mm per year in the 1901–1971 period to 1.9 mm per year in the 1971–2006 period, and to 3.7 mm per year in the 2006–2018 period. For Australia, relative sea-level rise has been higher than the global average in recent decades (IPCC, 2021).

Tropical Cyclones

With a very long coastline, much of which lies in the tropics and subtropics, changes in tropical cyclones can have devastating impacts on our infrastructure and health. Climate change is affecting tropical cyclones in a number of ways—how often they form, how intense they become (maximum windspeed), how much rain they drop, how fast they move across the ocean, how strong they remain over land, and their geographic range and the length of the cyclone season (Kossin et al., 2014; Li & Chakraborty, 2020). Clearly, the various ways that climate change is influencing tropical cyclones are complex, but in general they are likely to be fewer in number but a higher percentage of those that do form will be more intense and destructive (Climate Council, 2021a).

Observations support what basic physical theory predicts. Over the past 40 years, based on satellite imagery, the maximum wind speeds of cyclones are increasing for almost every region where they occur (Kossin et al., 2020). It appears that cyclones are now reaching their maximum intensity more quickly than before (Bhatia et al., 2018) and they are maintaining their strength longer after making landfall (Li & Chakraborty, 2020), both effects making them more destructive for longer periods of time. Tropical cyclones are likely to have higher rainfall amounts for the same intensity because of the higher water vapour content in the atmosphere. To make them even more deadly, there is a trend towards slower speeds as cyclones move across the ocean and land (Kossin, 2018), thus allowing more time for the storms to drop enormous amounts of water over a small area and sustaining high windspeeds for a long time (Resnick, 2020). For Australia, there is not enough data to show a clear trend in cyclone intensity or speed, but there has been a decrease in the number of tropical cyclones in the Australian region since 1982 (CSIRO & BoM, 2020; IPCC, 2021).

Compound Extreme Weather Events

As many extreme weather events become more damaging because of climate change, the risks of 'compound events' are even higher. Compound extreme events occur when multiple destructive extreme events occur simultaneously or in quick succession, often driving impacts that are worse than if the individual events occurred in isolation. Compound extreme events can also form when a number of events, each of which may not be extreme if it occurred in isolation, come together to drive impacts that become extreme.

Compound extreme events are more likely to occur when the climate change influence is magnified by patterns of natural variability. As an example, the long-term warming trends across Australia and the drying trends in the southeast can be amplified when the modes of El Niño–Southern Oscillation or the Indian Ocean Dipole also lead to warmer and dryer conditions, worsening the impacts of climate change alone.

Examples of these various types of compound extreme events are given below.

Bushfires

Perhaps the most damaging compound extreme event in Australian history was the unprecedented 2019–2020 bushfires that burned about 5.8 million hectares along the eastern coast. This massive event was also an example of an abrupt, nonlinear event where a critical tipping point was crossed (see Sect. Abrupt, Nonlinear Extreme Events: Tipping Points). Fire is an integral feature in eastern Australian broad-leafed (primarily eucalypt) forests; about 2–3% of the forests burn in a normal fire season. However, about 21% of the forests burned in 2019–2020, an order-of-magnitude increase in the area burned over any previous fire season. Climate change played the dominant role in driving this unprecedented disaster.

Two extreme weather events, both fueled by climate change, combined to form the ultimate driving force behind this compound extreme event. As shown in Fig. 4.4, the maximum temperature anomaly—that is, the

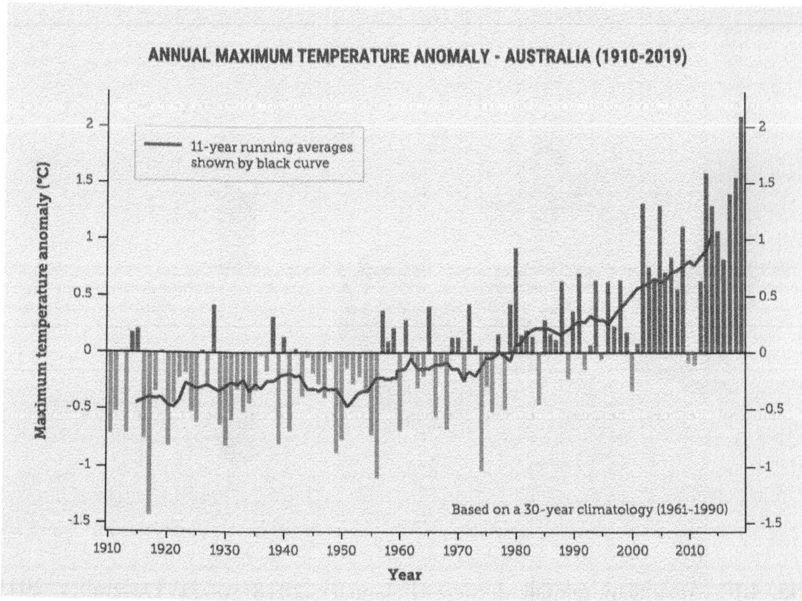

Fig. 4.4 Annual maximum temperature anomaly Australia (1910–2019) (*Source* BoM, 2020a)

maximum temperature recorded as an annual average across the continent—set a record level of over 2 °C above the 1961–1990 average maximum temperature (BoM, 2020a). The long-term, sharp increase in extreme heat across the continent is part of a global trend that is clearly driven by climate change (IPCC, 2021).

The second climate-related driver of the disaster was the exceptionally dry conditions in the two years leading up to the fires. Over that period much of eastern Australia suffered record-breaking dry conditions, as shown in Fig. 4.5. The two-year rainfall total over most of southeast Australia, covering virtually all of the area that burned, was the lowest on record (BoM, 2020a). Exceptionally strong dry phases of the Indian Ocean Dipole (IOD) and the Southern Annular Mode (SAM), two modes of natural climate variability, were major contributors to the very dry conditions in late 2019 (van Oldenborgh et al., 2021). The exceptionally

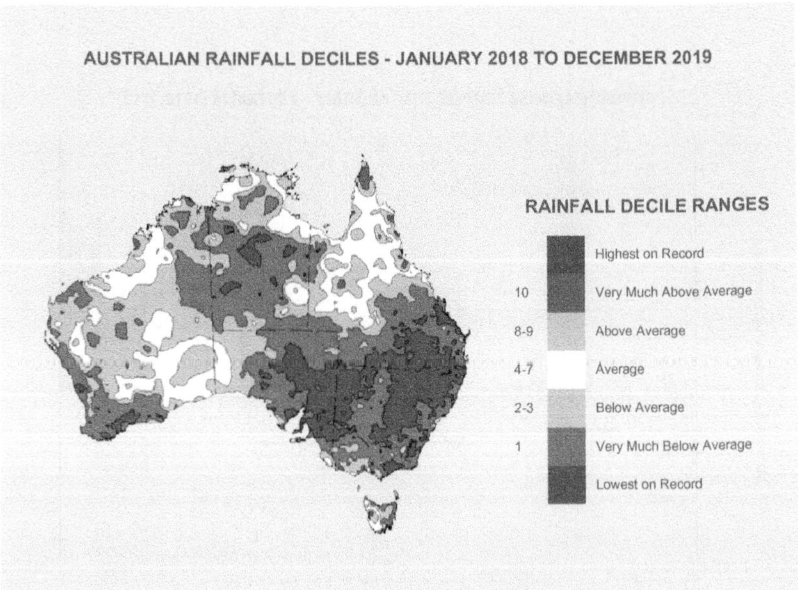

Fig. 4.5 Australian rainfall deciles—1 January 2018 to 31 December 2019 (*Source* BoM, 2020b)

low rainfall in 2018–2019 is consistent with a longer-term decrease in cool season rainfall over southeast Australia since the early 1990s, a trend that has been linked to climate change (CSIRO & BoM, 2015). Along with the extreme heat, the antecedent record low rainfall conditioned the forests to burn extensively and violently.

Figure 4.6, which shows the area of forest in a critically dry state, demonstrates the compound effect of these two climate drivers on the condition of the eastern Australian forests (Boer et al., 2020). The condition of forests vis-à-vis bushfires naturally varies widely from year-to-year, consistent with the high natural variability of Australia's climate. However, up until 2017, the area of forest that was critically dry varied between 3000 and 6000 million hectare-days. From about 2011, the critically dry area rose strongly and reached 7000 million hectare-days, more than two standard deviations above the average. This area of critically dry forest, coupled with the exceptionally hot, dry conditions, resulted in the

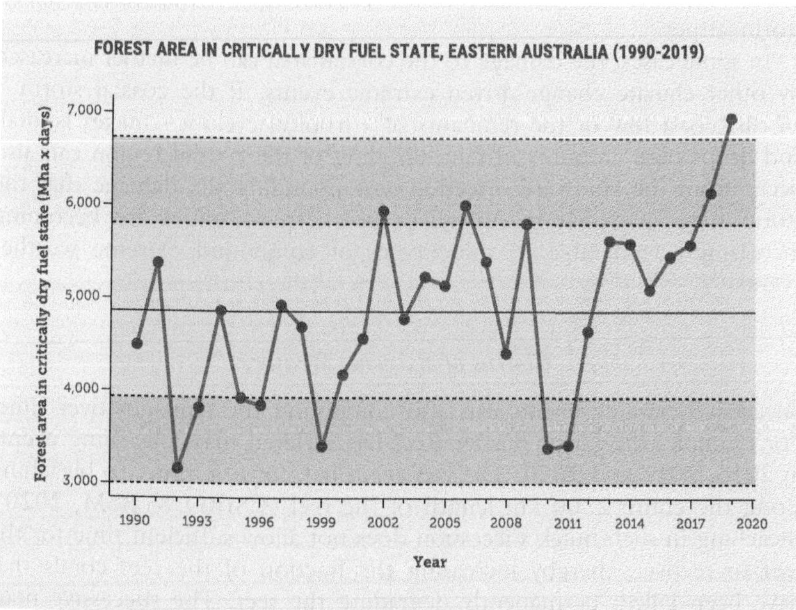

Fig. 4.6 Forest area in critically dry fuel state, eastern Australia (1990–2019) (*Source* Boer et al., 2020)

huge area burned and the exceptionally violent fires of 2019–2020. In essence, a critical tipping point was crossed in 2019 that triggered the massive fires (Climate Council, 2021a, 2021b).

Coastal Flooding

The increase and severity in inundation events along Australia's coastline is another example of a compound extreme event. The most damaging inundation events occur when a coastal storm coupled with a high tide and the long-term rise in sea level drive large storm surges that send massive amounts of water ashore (Fig. 4.3). Climate change plays at least two roles in the events. First, the long-term rise in sea levels is clearly driven by the warming climate, which is leading to the expansion of ocean water through heating of the water and is adding additional water to the ocean from melting ice sheets. In addition, although tropical cyclones are not increasing in number, the fraction of the more intense storms is increasing, which in turn increases the probability of exceptionally strong storm surges.

In some cases, the damage to the coastal area can be further increased by other climate change-driven extreme events. If the coastal storm—an east coast low or the remnants of a tropical cyclone—makes landfall and drops high amounts of rain, flooding of the coastal region can also occur from the landward direction, compounding the damage that the storm surge had driven. Australia's northern coastlines are becoming increasingly vulnerable to these types of compound extreme weather events.

Chains of Extreme Weather

Successive extreme events can also compound the damages over time. For example, the Great Barrier Reef has suffered mass bleaching events in 2016, 2017 and 2020. The last event led, for first time, to bleaching along the entire 2300 km length of the reef (CSIRO & BoM, 2020). Bleaching in such quick succession does not allow sufficient time for the reef to recover, thereby increasing the fraction of the reef corals that have been killed, permanently degrading the reef. The successive mass bleaching events over the past decade have led to the death of about 50% of the hard corals on the Great Barrier Reef (Climate Council, 2021a).

Eastern Australia experienced a dramatic chain of extreme weather in early 2020. Bushfires raged along the east coast through January 2020 (see Sect. Bushfires), but they were almost immediately followed in February 2020 by extreme rainfall along the east coast and by a devastating hail storm in Canberra (Climate Council, 2020).

Northeast New South Wales and southeast Queensland experienced exceptionally heavy rainfall. Further south over 390 mm of rain fell over Sydney in just four days, breaking a 30-year-old record for the total amount of rainfall in a four-day period. The city experienced flash flooding as well as damage to power supplies as gale-force winds brought down trees. Sections of beach in Collaroy and Narrabeen were washed away by storm surges. Following close behind the massive bushfires, the heavy rain drove extensive erosion in the burned-out forests, washing away valuable topsoil that is important in aiding the recovery of the forests.

In Canberra on 20 January 2020 an intense hail storm swept through the centre of the city, damaging several iconic sites—the National Museum of Australia, the Australian National Botanic Gardens and Australian Academy of Science's Heritage-listed Shine Dome. The Australian National University was directly in the storm's path and suffered extensive damage; about 80 buildings suffered damage, with many of them severely damaged. Local wildlife also suffered, with many birds killed or injured and 300 flying foxes killed.

Abrupt, Nonlinear Extreme Events: Tipping Points

Perhaps the most iconic Australian example of a tipping point being transgressed is the repeated bleaching of the Great Barrier Reef. Corals thrive in a rather narrow water temperature range, and so modest increases in surface ocean temperature can lead to transgressions of this temperature and thus to large areas of bleaching. The GBR was formed and has thrived through much of the Holocene, the 11,700-year warm period that we have now left because of climate change. However, the upward trajectory of surface temperature has increasingly driven the GBR across the threshold below which corals thrive.

There was some bleaching along the GBR in 1998 and 2002 as the rising temperatures approached the threshold level and transgressed in some regions (Hughes et al., 2018). However, much more serious and extensive bleaching occurred later as ocean temperature continued to rise and transgressed the threshold over much larger sections of the reef and for longer periods. In both 2016 and 2017, mass bleaching events occurred on large sections of 2300 km long reef. However, in 2020 another mass bleaching event hit, this time damaging and killing corals along the entire length of the reef. Together, these events have led to the death of about half of the hard corals of the GBR, a classic case of crossing a tipping point and triggering an abrupt change in a marine ecosystem (Climate Council, 2021a).

Climate-related tipping points can also occur in land-based ecosystems such as forests, particularly those that are already near temperature or precipitation limits. For example, alpine forests are becoming increasingly stressed as temperatures rise, resulting in dieback in the most vulnerable regions. Similarly, forests and woodlands at the edge of arid zones are vulnerable to dieback as temperatures rise and long-term rainfall declines in the southeast and southwest of the continent. Often the conversion to a new ecosystem, for example, a grassland, scrubland or savanna, occurs abruptly driven by a climate change-related extreme event or disturbance—an intense heatwave, large bushfires or widespread and severe drought—that tips the forest out of its environmental envelope and opens the landscape for the emergence of a new ecosystem, better suited to the changing climatic conditions.

Future Projections

Climate change-fueled extreme weather events are now part of Australia's everyday life, with their damaging impacts on our lives, our infrastructure and our natural ecosystems. The evidence that they are worsening is irrefutable—the Black Summer bushfires, three mass bleaching events of the Great Barrier Reef in the last six years, massive flooding in North Queensland, record-breaking drought in the southeast and rising sea levels driving increased erosion and flooding. These events will worsen for at least the next two decades based on the further warming that is already locked into the climate system (CSIRO & BoM, 2020; IPCC, 2021), and will become even more severe and damaging if global greenhouse gas emissions are not now set onto a rapid, long-term downward trend to cap temperature rise to as close to 1.5 °C as possible (Fig. 4.7).

The current global average temperature is 1.1 °C above the pre-industrial level, and is accelerating. The global temperature has increased faster since 1970 than in any other 50-year period in at least the last 2000 years. All of the emissions scenarios for the future, including the

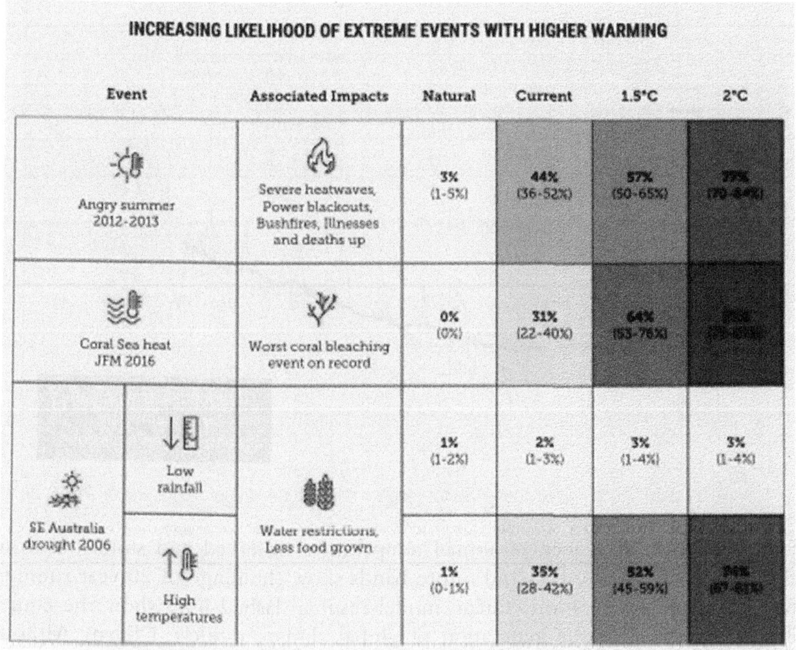

Fig. 4.7 The changing likelihood of Australian extreme events (*Source* King et al. [2017], reprinted from Climate Council [2021b])

most ambitious, project that global temperature will likely breach the lower Paris goal of 1.5 °C by mid-century (IPCC, 2021). Only the two low emissions scenarios achieve the upper Paris goal of limiting warming to 'well below 2 °C'.

The Australian average temperatures are rising at 1.4 times the global average and are rising at an increasing rate (Fig. 4.8). 2019 was Australia's hottest year on record, surpassing the Australian equivalent of 1.5 °C global warming above pre-industrial. Over the past five years, Australia's average annual temperature has been rising at the upper limit of the climate model projections.

The recent IPCC Sixth Assessment Report, especially its Regional Fact Sheet on Australasia (IPCC, 2021), the CSIRO and BoM State of the Climate 2020 report (CSIRO & BoM, 2020) and the Australian Academy of Science report on Australia in a Three-Degree World

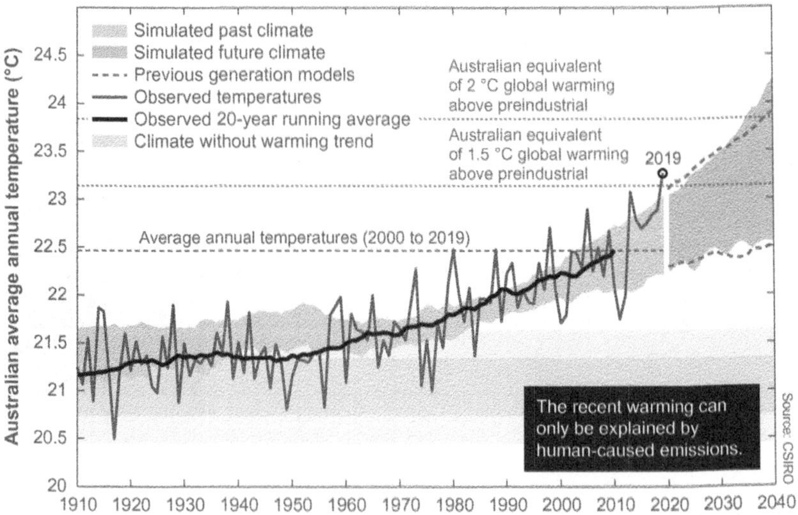

Fig. 4.8 Australian average annual temperature observed and simulated from global climate models. Past and future bands show the range of 20-year running average of new generation climate model results, dashed lines show the equivalent from the previous generation of global climate models. Climate without the warming trend shows both an inner and an outer band which are one and two standard deviations, respectively, from the 1850–1900 average, i.e., prior to the rapid growth in greenhouse gas emissions from human activities. The black dashed lines show the Australian equivalent of the global warming thresholds 1.5 °C and 2 °C above the pre-industrial baseline period (1850–1900), equating to warming levels of around +2.1 °C and +2.8 °C respectively, based on the observed ratio of Australian to global temperature of around 1.4 (*Source* CSIRO and BoM [2020])

(Hoegh-Guldberg et al., 2021), set out a grim picture of Australia's future if the climate system is not rapidly stabilized. In assessing the future trajectory of climate change and its consequences below, a risk-averse approach has been taken, as in Hoegh-Guldberg et al. (2021). That is, if there is scientific evidence for a link between the extreme event and climate change, even though the strength or probability of that link is still under scientific debate, the risk must nevertheless be considered. To put it simply, if a potentially damaging effect cannot be ruled out, it must be taken seriously.

Extreme Heat

Heatwaves will become more frequent and more intense across virtually all of the continent. Marine heatwaves will also increase in frequency and intensity, and, coupled with increasing ocean acidity, will drive more extensive destruction of the Great Barrier Reef. Land-based ecosystems will also be severely affected—large-scale mortality of trees, birds and tree-dwelling mammals; shifts in habitats at speeds with which many animals cannot keep up; and, under higher levels of heating, fundamental changes to the structure and functioning of ecosystems that would render them unrecognizable. Increasing extreme heat will damage our health and well-being, leading to increased mortality from heart and lung disease as well as increasing the risk from vector-borne diseases. Excessive heat would also stress our energy system and other infrastructure to levels at which they will increasingly fail; and drive increasingly severe impacts on the agricultural sector through crop failures and stress on livestock.

Heavy Rainfall

The incidence and intensity of heavy rainfall are likely to increase across many parts of the continent, primarily in the northern, central and eastern parts of Australia. As a result, severe river flooding events are likely to increase in number and extent. The frequency of tropical cyclones in the north is projected to decrease but the proportion of severe cyclones is expected to increase. Many of our cities and towns will become increasingly vulnerable to flooding events from heavy rainfall, leading to increasing financial risks and escalating insurance costs.

Drought

Drought is projected to increase in the southern and eastern parts of Australia, connected with a projected decrease in cool season rainfall. Drought is very likely to become even more severe and frequent in southwest Western Australia. The combination of more extreme heat and more drought will increase the intensity, frequency and duration of fire weather throughout Australia. Sand storms and dust storms are projected to increase across the continent. The combination of drought and more extreme heat will have major impacts on ecosystems in the continent's southeast, as well as exacerbating the already severe impacts on the agricultural sector via reductions in crop yields and increasing stress on livestock. Urban water supplies are also at rapidly increasing risk as longer and more severe droughts in the southeast and southwest dry out catchments and reduce runoff into dams. A severe lack of water resources has serious impacts on human health and wellbeing, and can lead to large-scale dislocations of rural populations.

Coastal Erosion and Flooding

Sea level will continue to increase for many centuries into the future. Uncertainties around polar ice sheet stability as well as the magnitude of greenhouse emissions generate a large range of potential global sea-level rises by the end of this century, from 0.28 to 1.01 m based on 2000 levels. With high emissions scenarios, sea-level rises of over 1.5 m by 2100 and up to 15 m by 2300 cannot be ruled out.

Globally, sea-level rise is accelerating, and around many of Australia's coasts, sea levels are rising faster than the global average. For most of the Australian coast, extreme sea levels that had a probability of occurring once in a hundred years are now projected to occur every year by the end of this century with lower emissions and by mid-century for higher emissions. This means that the frequency of coastal flooding will continue to increase, and coastal erosion and shoreline retreat will increase, driving escalating impacts on coastal infrastructure and property, and on Australia's economy more generally. An estimated 160,000 to 250,000 properties are at risk of coastal flooding if sea levels around Australia rise by about 1 m at the end of the century, a possibility under current emission trajectories. The combination of rising sea levels and more severe storms will also damage coastal ecosystems such as mangroves and wetlands, such as the World Heritage Kakadu National Park.

To minimise the severity of these impacts and stabilise the climate system within the Paris Accord temperature goals, greenhouse gas emissions must be rapidly and deeply reduced until 'net zero' emissions are achieved. 'Net zero' is achieved when the amount of CO_2 emissions of carbon dioxide (CO_2) is matched by the permanent removal of the same amount of CO_2 from the atmosphere. Of the five global emission reduction scenarios in Fig. 4.9a, only the two lowest emission scenarios, SSP1-1.0 and SSP1-2.6, are consistent with meeting the Paris Accord temperature goals. Both of these scenarios require significant drawdown

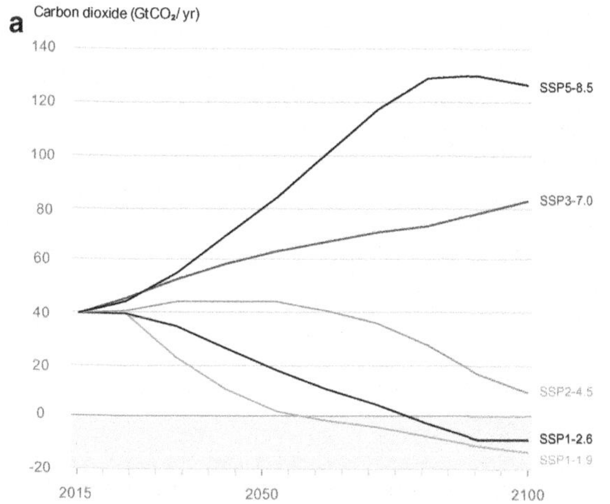

Fig. 4.9a Future global emission trajectories (Gt CO_2/yr) for the five IPCC AR6 scenarios. Note that the vertical axis is annual emissions (*Source* IPCC, 2021)

of CO_2 from the atmosphere in the second half of the century. Here 'drawdown' refers to the situation where our removals of CO_2 from the atmosphere are significantly larger than any residual emissions, thus leading to an overall reduction of the CO_2 in the atmosphere (Figs. 4.9b and 4.9c).

Australia's contribution to the global effort to reduce emissions was first estimated by the Federal Government's Climate Change Authority in 2014. Using the same methodology, the Climate Targets Panel recently updated the analysis to estimate the size of remaining budget consistent with the Paris Accord temperature goals. Figures 4.9a–4.9c shows that to do our fair share for a 67% probability of meeting a 2 °C goal, Australia must reduce its emissions by 50% on 2005 levels and achieve net-zero emissions by 2045. For a 50% probability of meeting a 1.5 °C goal, we must reduce our emissions by 74% by 2030 and achieve net-zero emissions by 2035. The remaining carbon budget for a 67% probability of meeting the 1.5 °C goal is vanishingly small.

These analyses convey a simple but powerful message—the risks of climate change are escalating rapidly and the time we have left to meet the climate change challenge effectively is rapidly running out. The choices

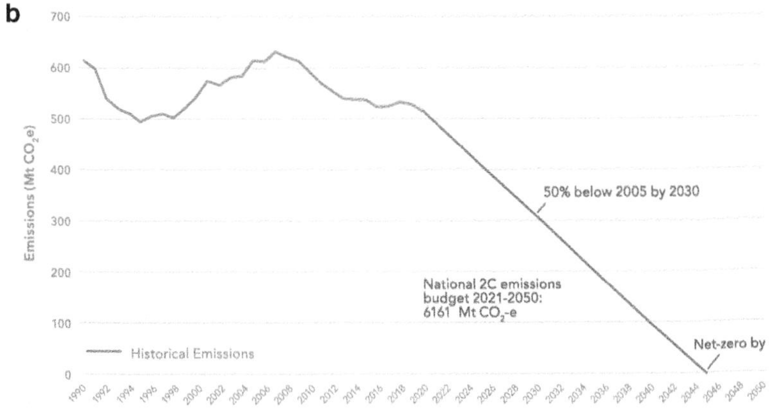

Fig. 4.9b Paris-compliant emissions trajectory for Australia for a 2 °C target. The vertical axis is annual emissions (*Source* Climate Targets Panel, 2021)

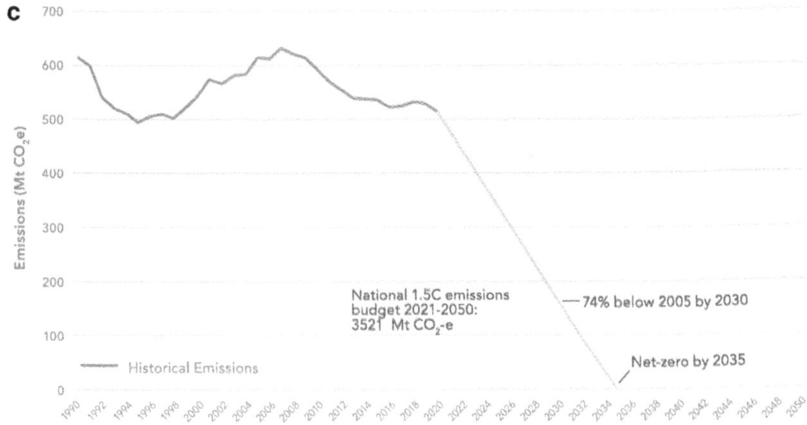

Fig. 4.9c Paris-compliant emissions trajectory for a 1.5 °C target. The vertical axis is annual emissions (*Source* Climate Targets Panel, 2021)

we make now here in Australia and around the world over the next few years may well be the difference between a planet on which our children and grandchildren can survive in a reasonable state, albeit with a much more difficult climate, or a planet whose environment will become exceptionally volatile and hostile, incompatible with well-functioning human societies.

References

Bhatia, K., Vecchi, G., Murakami, H., Underwood, S., Kossin, J. (2018). Projected response of tropical cyclone intensity and intensification in a global climate model. *Journal of Climate, 31*(20). https://doi.org/10.1175/JCLI-D-17-0898.1

Boer, M. M., Resco de Dios, V., & Bradstock, R. A. (2020). Unprecedented burn area of Australian mega forest fires. *Nature Climate Change, 10*. https://doi.org/10.1038/s41558-020-0716-1

BoM (Bureau of Meteorology). (2016). Australian climate variability and change—Time series graphs. http://www.bom.gov.au/climate/change/index.shtml

BoM (Bureau of Meteorology). (2020a). Climate change—trends and extremes. http://www.bom.gov.au/climate/change/index.shtml

BoM (Bureau of Meteorology). (2020b). Australian in December 2019. Accessed at: http://www.bom.gov.au/climate/current/month/aus/archive/201912.summary.shtml

Climate Commission. (2013). The critical decade: Extreme weather. https://www.climatecouncil.org.au/extreme-weather-report

Climate Council of Australia. (2020). *Summer of Crisis*, 28p.

Climate Council of Australia. (2021a). *Hitting home: The compounding costs of climate inaction*, 60p.

Climate Council of Australia. (2021b). *Aim high, go fast*, 75p.

Climate Targets Panel. (2021). Australia's Paris Agreement pathways: *Updating the Climate Change Authority's 2014 Emissions Reduction Targets*, 13p. (Panel members: John Hewson, Will Steffen, Lesley Hughes, Malte Meinshausen).

CSIRO and Bureau of Meteorology (BoM). (2015). *Climate Change in Australia—Technical Report*, CSIRO and Bureau of Meteorology, 216p.

CSIRO and Bureau of Meteorology (BoM). (2020). *State of the Climate 2020*. https://www.csiro.au/en/Showcase/state-of-the-climate

Dietzel, R., Bode, M., Connolly, S.R., & Hughes, T. P. (2020). Long-term shifts in the colony size structure of coral populations along the Great Barrier Reef. *Proceedings of the Royal Society B: Biological Sciences* 287 (1936). https://doi.org/10.1098/rspb.2020.1432

DIT (Australian Government Department of Infrastructure and Transport). (2013). *State of Australian Cities 2013*. https://infrastructure.gov.au/infrastructure/pab/soac/files/2013_00_infra1782_mcu_soac_full_web_fa.pdf

GBRMPA (Great Barrier Reef Marine Park Authority). (2020). Statement: Coral bleaching on the Great Barrier Reef. 26 March 2020, https://www.gbrmpa.gov.au/news-room/latest-news/latest-news/coral-bleaching/2020/statement-coral-bleaching-on-the-great-barrier-reef

Hoegh-Guldberg, O., Bowen, K., Capon, T., Church, J., Howden, M., Hughes, L., Jotzo, F., Karoly, D., Palutikof, J., Pearman, G., Quiggin, J., King, A., Lee, E., Nursey-Bray, M. & Steffen, W. 2021. *The Risks to Australia of a 3C Warmer World*. Australian Academy of Science, 97p.

Hughes, T. P., Anderson, K. D., Connolly, S. R., Heron, S. F., Kerry, J. T., Lough, J. M., et al. (2018). Spatial and temporal patterns of mass bleaching of corals in the Anthropocene". *Science 359*(6371). https://doi.org/10.1126/science.aan8048

IPCC (Intergovernmental Panel on Climate Change). (2018). *Special Report on Global Warming of 1.5°C*, http://ipcc.ch/report/sr15/

IPCC (Intergovernmental Panel on Climate Change). (2019). Summary for Policymakers. In: *IPCC Special Report on the Ocean and Cryosphere in a Changing Climate*, 42p. https://www.ipcc.ch/srocc

IPCC (Intergovernmental Panel on Climate Change). (2021). *Working Group I Contribution to the IPCC Sixth Assessment Report (AR6), Climate Change 2021: The Physical Science Basis. Summary for Policymakers*, 41p, including 'Regional fact sheet – Australasia'.

King, A. D., Karoly, D. J. & Benjamin, J. H. (2017). Australian climate extremes at 1.5 °C and 2 °C of global warming. *Nature Climate Change*. https://doi.org/10.1038/NCLIMATE3296

Knutson, T. R., Zeng, F. & Wittenberg, A. T. (2014). Seasonal and annual mean precipitation extremes occurring during 2013: A U.S focused analysis. In "Explaining extreme events of 2013 from a climate perspective". *Bulletin of the American Meteorological Society, 95*, S19–S23.

Kossin, J. P. (2018. A global slowdown of tropical-cyclone translation speed. *Nature, 558*. https://doi.org/10.1038/s41586-018-0158-3

Kossin, J. P., Emanuel, K. A., & Vecchi, G. A. (2014). The poleward migration of the location of tropical cyclone maximum intensity. *Nature, 509*. https://doi.org/10.1038/nature13278

Kossin, J. P., Knapp, K. R., Olander, T. L., & Velden, S. (2020). Global increase in major tropical cyclone exceedance probability over the past four decades. *Proceedings of the National Academy of Sciences, 117*(22). https://doi.org/10.1073/pnas.1920849117

Lewis, S. C., & Karoly, D. J. (2014). Are estimates of anthropogenic and natural influences on Australia's extreme 2010–2012 rainfall model-dependent? *Climate Dynamics, 45*, 679–695.

Li, L., & Chakraborty, P. (2020). Slower decay of landfalling hurricanes in a warming world. *Nature, 587*. https://doi.org/10.1038/s41586-020-2867-7

Myhre, G., Alterskjær, K., Stjern, C. W., Hodnebrog, Ø., Marelle, L., Samset, B. H., ... & Stohl, A. (2019). Frequency of extreme precipitation increases extensively with event rareness under global warming. *Scientific Reports*, *9*(1), 16063. https://doi.org/10.1038/s41598-019-52277-4

van Oldenborgh, G. J., Krikken, F., Lewis, S., Leach, N. J., Lehner, F., et al. (2021). Attribution of Australian bushfire risk to anthropogenic climate change. *Natural Hazards and Earth System Sciences, 21*, 941–960.

Resnick, B. (2020). *Why Hurricane Maria is such a nightmare for Puerto Rico.* Vox, 22 September 2017. https://www.vox.com/science-and-health/2017/9/21/16345176/hurricane-maria-2017-puerto-rico-san-juan-meteorology-wind-rain-power

Thompson, A. (2020). *The Sydney suburbs that hit 50C last summer*. The Sydney Morning Herald, 5 October 2020. https://www.smh.com.au/national/nsw/the-sydney-suburbs-that-hit-50c-last-summer-20201002-p561by.html

Welbergen. (2019, 15 January). Cited in: Mao 2019. How one heatwave killed "a third" of a bat species in Australia. *BBC*.

CHAPTER 5

Coasts: A Battleground in Disaster Preparedness, Response and Climate Change Adaptation

Tayanah O'Donnell and Bruce Thom

Introduction: Dynamic Forces Impact Coastal Regions

Coastal areas as the interface between land and sea are forever subject to dynamic processes. This interplay of solids like sand, mud and gravel with energy fluxes and biological activity has gone on over eons of geological time. But human intervention has changed that interplay in part directly through engineering works and indirectly through land use practices in catchments. In more recent times climate change attributed to

T. O'Donnell (✉)
Australian National University, Canberra, ACT, Australia
e-mail: tayanah.odonnell@anu.edu.au

B. Thom
University of Sydney, Sydney, NSW, Australia
e-mail: bruce.thom@sydney.edu.au

A. Lukasiewicz and T. O'Donnell (eds.), *Complex Disasters*, Disaster Risk, Resilience, Reconstruction and Recovery,
https://doi.org/10.1007/978-981-19-2428-6_5

greenhouse gas emissions has become a factor in driving changes to the character of coastal areas, both along open ocean shorelines and in estuaries.

Over the last five decades much has been written about shoreline change and increased inundation by the sea linked to the new world of the *Anthropocene*. This has been captured with increasing degrees of certainty in the six successive assessment reports of the *Intergovernmental Panel on Climate Change* (IPCC). These reports make clear that our understanding of biophysical processes under these emerging climatic conditions must converge with an integration of knowledge from multiple disciplines. In their introduction to their edited book on "Tomorrow's Coasts", Wright and colleagues make the point with a degree of urgency:

> *Over the next few years, advances in our ability to anticipate, plan for and mitigate the impacts of adverse changes in coastal processes and coastal communities will increasingly require not only continued refinements of natural science and social science models but also rely on development and application of complex systems models that account for a hierarchy of interconnections and non-linear feedbacks.* (Wright et al., 2019, p. 5)

Authors in that volume develop details on the nature of such complexities and the need to progressively test and calibrate with high resolution data linked to local and regional dynamics. One of us was fortunate to participate in the production of this volume and contributed to its analysis on adaptive management and in promoting resilience for tomorrow's impermanent coasts (Thom, 2019, Ch. 19; Wright & Thom, 2019, Ch. 21).

Recognition of the potential for acceleration in rates of changes to coastal conditions that will adversely impact through recurrent flooding, shoreline recession, salt intrusion into groundwaters and other effects, concerns the livelihoods of as many as two billion people (Dawson, 2017). Coastal development, both residential housing and state significant infrastructure, compound these pressures as the propensity for Australians to reside near coastlines and often in or near floodplains continues to grow. With this demand comes legal and political pressures to permit more development, larger holdings on smaller land sizes of a type and design not always suited to the locality (O'Donnell, 2018, 2019a, 2020b). People's attachments to these locations are enduring (Barnett et al., 2014) and often with a perverse sense of historical knowledge of weather, a reduced responsibility of risk, and a disembodied sense of who bears responsibility (O'Donnell, 2020a). There is therefore a melting pot of cascading, cumulative and protracted conditions of disaster in coastal locations. As demand for coastal infrastructure and housing continues to rise; climate conditions and impacts continue to adversely affect coastal ecosystems; peoples' attachments, values and behaviours with respect to their (real or perceived) property rights, place attachments, and avoidant responsibilities of risk; and legal and policy systems struggle to be proactive in providing frameworks to guide decision-making, coastlines are ground zero for where we can expect to see complex disasters to unfold.

As sea levels continue to rise, accelerated impacts will become more evident, stimulating all sorts of individual, community, commercial and government responses or lack thereof. All this calls for increased understanding of the nature of change at different spatial and temporal scales. In some cases, the changes may be rapid with dramatic effects (so-called "fast responders") compared to the more insidious "slow-responding" systems that at some stage reach an irreversible tipping point. Wright and Thom (2019, p. 341) pointed to three categories of actions that humans need to take in order to minimise detrimental impacts of climate change on tomorrow's coastal systems: (1) reduce emissions and other destructive actions such as overdeveloping in areas at risk, severing sediment supply to and along the coast; (2) promote the need to improve understanding of the complex interconnections of coastal natural and human processes to improve societies' capacity to predict and develop strategies that enable long-term adaptation strategies to become more acceptable; and (3) doing this ensures that quality coastal science informs relevant state and federal government policies and law.

Adaptive strategies that can be implemented are fundamental in addressing the disastrous impacts of climate change on coastal areas. However, it is necessary to recognise barriers to long-term planning at different scales across different nation states (Barnett et al., 2013). Furthermore, nation states, and other governments (such as constituted states in a federation) have to seriously address those barriers and in so doing provide the appropriate legal means to implement and enforce it. Without such legal powers embedded in legislation, any commitments made in the short term can easily be overlooked or discarded with political change. The nature of coastal change is such that future cannot be made to be easily sacrificed to the whims of the greedy present.

Differences exist within societies as to how to ensure coastal land use does not generate risk to economic, social and environmental values. Conflicts often exist between and within the landholder class and government authorities (Thom, 2020). Pressures to capitalise on property values, investment opportunities, alienation of public land and construction of major works are but some of the outcomes that local communities face over land adjoining waterways and the sea. This is why we term coastal lands ***contested spaces***, subject ***to death by a thousand cuts***.

Thompson (2007) and Stocker and Kennedy (2009) have eloquently described the various forms of coastal conflict in the USA and Australia, respectively. Vested interests seek benefits that may thwart the longer-term strategic efforts of those who seek to use scientific understanding of coastal change to reduce exposure of future generations to risks. It is vital to know up front who and why those involved in different coastal land uses would want to place barriers in the face of strategic planning. This knowledge would help combat those often using political and even legal means to allow for developments to occur in places that are or will be in harm's way (Harvey & Clarke, 2019; Harvey & Smithers, 2018; O'Donnell, 2016, 2019a).

Thom (2019, p. 306) identifies six barriers that make it difficult to develop adaptive strategies in coastal management:

- It is often very difficult for a community or even a government to reach agreement on a long-term vision for the future of any given coastal area. This difficulty is compounded by political opportunism leading to changes in power structures with government officials being forced to change directions at the behest of new leaders. This is manifest in changing priorities for investment or changes in the support for certain stakeholders at the expense of others. All this could and does change after elections as the power base shifts.
- Officials in government agencies are more inclined to follow land use planning procedures, engineering design and management practices that are based on historical evidence and avoid what may appear to be alarmist predictions from climate science. This innate conservatism can be easily justified to political masters who may otherwise be nervous about incorporating uncertain projections in planning when they know that this will more than likely add to development costs and be more readily challenged in courts. This short-term thinking is quite common and future risks associated with climate

change are seen as beyond decision-making horizons, such as next 4-year budget cycle.

- Many decision-makers are prepared to allow the future to take care of itself; here lies the moral hazard in coastal management. Those with property at risk may be content to accept that future governments or even insurance industry will support them after a disaster so that there is a disincentive to prepare for future risk of the scale and type forecast under climate change. This results in little local pressure to invest in protective works or other adaptive measures.
- Existing use rights and the pressure on governments to pay market prices for compensating those at risk to relocate often serves to discourage the introduction of policies that stop or force changes to development activity in areas that are or will soon be in harm's way. Long-term benefits are put to one side as landowners can appeal for retention of certain property rights under just terms or similar regulation to continue their existence and then demand help when inevitable disasters occur.
- Across institutions of government there are often jurisdictions with different legislative and budget responsibilities for coastal management. It is the so-called silo effect; agencies or entities within an agency may operate to an agenda defined historically or by the political/managerial authority of the day (see Chapter 15 of this volume for a general discussion of policy integration). As noted by Mumford and Harvey (2014) in their discussion of parliamentary inquiries in Australia, a major impediment to achieving integrated coastal zone management (ICZM) is "*disintegrated decision making that occurs between local, state, and the commonwealth government*". Even at the level of leadership, the role of short-term performance criteria in assessing outcomes within an entity can dominate decision-making so that integrating or coordinating planning for an adaptive future within government takes lower priority, or at worst is treated with disdain. This inhibits the mainstreaming of cost-effective, long-term focussed adaptation approaches.
- In many countries, the downsizing of governments within a neo-liberal context has resulted in reduced technical capacity across all levels of government. In many places this has resulted in outsourcing advice and practices to consultants operating within constrained budgets and poorly designed briefs. Lack of qualified staff within government agencies to interpret and implement actions has also

restricted capacity of officials to develop adaptive strategies for the future by being able to assess and reduce climate risk through frank, fearless, and well-informed advice based on best available science and research.

Australia is one of several countries where attempts are being made to address these barriers with varying degrees of success. With its highly urbanised society clustered in coastal cities, towns and villages, there is a strong argument that we are rapidly approaching "ground zero" in addressing these barriers and in securing more resilient and adaptive coastal societies. But this has not been without trying!

Adaptation Policy Framework Relevant to Coastal Australia

Australia, with its federated government system (local, state and commonwealth governments, each with differing responsibilities as set out in the Commonwealth and relevant State *Constitutions*), is an illustrative example of how coastal living explicates compounding, cascading and protracted disasters. Australia has a long history of conflicting policy arrangements as a result of its federated system; it is further problematised by its colonial heritage. This complexity is perhaps why Australia has been the site of many experiments concerned with streamlining decision-making in the disaster space. Indeed, we can trace the development of interest in climate change adaptation back to the large and interdisciplinary conference known as GREENHOUSE 87 held at Monash University in late November 1987. Its proceedings were published by CSIRO and subtitled "Planning for climate change" (Pearman, 1988). Many of the papers covered issues on potential impacts including those on coastal areas even exploring in an insightful way a range of scenarios (e.g., Stark, 1988). Following this conference, the Australian Labor Government established the *National Greenhouse Advisory Committee* (NGAC). This began a series of efforts over the next three decades by the federal government to look at ways to document coastal risk in association with state governments and research institutions. Over this lengthy period there have been many shifts in direction on both climate change mitigation and adaptation. All three levels of government, federal, state and local, have made attempts to develop and implement adaptation strategies but rarely in concert with one another (Harvey, 2016). The 2012 statement on roles and responsibilities for climate change adaptation in Australia agreed by the then *Select Council on Climate Change* is still extant. This statement gives the Australian Government a role in providing national science and information and leadership on national adaptation reform.

One recent outcome is the release on 29 October 2021, of the *National Climate Resilience and Adaptation Strategy* (NCRAS). This was presented to COP26 alongside Australia's first international *Adaptation Communication*. The Minister for Environment commented that "*working with business, government and the community I am confident that we can guide collective investment in adaptation to reduce the impacts of major weather events*". This is a huge challenge given the need to reinforce the "shared responsibilities" of local, state and federal governments with a defined program of action that recognises the fiscal imbalance between these levels of government and issues (Harvey, 2016). This is especially so with disaster prevention, preparedness, response and recovery (PPRR), where it is not only governments but also emergency services volunteers and other non-government actors who play critical roles (Lukasiewicz et al., 2017).

There is no doubt that Australian Government since the 1980s has invested in coastal science and research. This is seen in the work of those in CSIRO, AIMS, Bureau of Meteorology, Geoscience Australia, various Cooperative Research Centres, National Climate Change Adaptation Research Facility ("NCCARF"—itself globally recognised and highly regarded, focused as it was on applied, highly relevant research for which it partnered across sectors), and in universities. Successive governments may have approached such investments with different degrees of enthusiasm, but our national computational capacities have strengthened over the years in ways that benefited coastal managers. Of particular note are the sea-level studies by John Church and colleagues at CSIRO which have been incorporated into IPCC assessments. Kathy McInnes also in CSIRO and John Hunter in the Antarctic CRC have demonstrated ways of assessing impacts of sea-level rise and associated events on coastal communities. Much of this work was summarised by Woodroffe et al. (2012) in their comprehensive review of approaches to risk assessment on Australian coasts as part of NCCARF phase 1. During NCCARF phase 2, a specific program termed *CoastAdapt* allowed for a more detailed multidisciplinary analysis of the science and its links to the development of climate adaptation strategies. Such strategies are aimed to reduce the impact of compound, cascading and/or protracted disasters by enabling adaptive capacity to changing climate systems.

An enormous trove of valued intellectual material that requires continuous refinement as the science and policy of climate change evolves and communities become more receptive to action, is now buried in reports with little apparent interest in follow-up work. This applies to CoastAdapt too, designed as it was to directly assist local governments and communities. Work on "coastal sediment compartments" is one example (Thom et al., 2018). This national-scale work built upon by Short (2020) requires testing at what we called "tertiary scale" compartments or cells (Kinsela et al., 2017). The *NSW Coastal Management Act 2016* incorporated the concept into requirements of councils in developing certain *Coastal Management Programs*. However, there is no further national research guidance emerging to help with this and to help other states and local councils.

Perhaps the most direct engagement by a federal government agency in coastal adaptation was demonstrated by the release in 2009 and 2011 of two "first pass" risk assessments of Australia's coasts. One of us was directly involved with the then *Department of Climate Change* on this work (DCC, 2009), and the accompanying *National Climate Change Forum* on "Developing a national coastal adaptation agenda" (Adelaide, South Australia in 2010). These studies identified at a local government level where the coast was at high risk to climate change impacts; they also provided estimates of costs of adapting under certain assumptions. While courageous with both intent and output, they showed state and local governments (e.g., Gold Coast in the state of Queensland) that there was a need to undertake more detailed ("second pass") assessments. The state of New South Wales has pursued this challenge as seen in recent studies by Kinsela et al. (2017) and Hanslow et al. (2018). What has become clear is that risks to infrastructure and property around our estuaries over the next 50 or more years are an order of magnitude greater than on the open coast.

The Contested Nature of Coasts and the Disasters that Affect Them

As shown above, the compounding policy issue for how, when and by whom coastal management and governance has occurred has resulted in an uneven, often inconsistent plethora of responses. Laws, regulatory and policy frameworks, varied local responses, often-times ad hoc or incomplete data, and a failure to properly incorporate all of the complexity that arises due to social and cultural value systems have left coastlines as highly contested spaces. An additional, crucial, layer is that the voice for the natural environment is often missing. Coastal management ideas to manage competing property interests such as rolling easements (Kay, 2016; O'Donnell & Gates, 2013) run the risk of failing to give voice to broader public interest concerns such as the rights of nature (O'Donnell, 2018). A Public Trust Doctrine (PTD) that encapsulates all interests in coastal spaces, including that of the coast itself, would be an important tool that can be utilised by regulatory frameworks to help manage these competing interests (Thom, 2012).

Beaches have long been contested places where public interests clash with private "rights". Australian law provides a common law basis for enshrining beach use in the public interest, overcoming the claimed right of exclusive use by landowners. As discussed in Thom (2020), this ancient principle of a public right to access and to use the beach or foreshore has been extended over time from safe landing places for boats to bathing and to the need to protect both ecological values and serve as a "shock absorber" or buffer to shoreline erosion. But the natural beach is not a fixed structure in time and place. Any attempt to maintain/secure public "rights" must contend with dynamic movements of beaches both landwards into private property and seawards allowing private interests to acquire accreted land (doctrine of erosion and accretion). This becomes complicated when human activities intervene with natural processes often to the detriment of the public good (O'Donnell & Gates, 2013; Thom, 2020).

It is also important to note at this juncture that we recognise not all values can be saved. Climate change impacts demand that there will be changes to coastlines around the world. In some places, these will be catastrophic; in others, hard engineering solutions will forever change the ecosystem but will provide human security. For a place like Australia, with its vast coastline, the engineering adaptation options are varied. With

36,000 km of coastline, it will be an expensive exercise to provision hard engineering options to protect it all. So, as with elsewhere in the world, there are going to be trade-offs and for some locations, those trade-offs are going to involve relocating away from the vulnerable site.

The contestation about public interest values, private interests, and what to do about them has been evidenced in numerous Australian coastal locations over the past two decades. These threads together illuminate the compounding, cascading and protracted character of climate impacts on coastlines, a challenge for which there is no clear or single way to prepare and to respond. Along the Australian coastline, there have been numerous approaches to trying to thread the needs and desires of present-day coastal living and amenity with the ever-nearing coastal impacts expected under a changing environment. One obvious example is that of the very popular Collaroy-Narrabeen beach, on the New South Wales northern beaches just north of Sydney, Australia.

The Collaroy-Narrabeen waterfront has long been subject to adverse weather events, culminating in coastal disasters. With expansive homes built on soft shorelines right to the waters' edge, it is a fraught location and definitive example of how past planning decisions have created legacy issues for current and future homeowners/occupiers of these waterfront properties. For two decades property owners, local councils and state governments passed the proverbial buck in either acknowledging the high risk of coastal inundation or an erosion event that would cause the properties severe damage or to be lost to the sea, and subsequently in not acknowledging nor accepting who was responsible for managing the risk.

This all came to a head in June 2016, when a severe coastal storm, combined with a king tide event, hit the beach and took with it several metres of private land, the public beach, a private pool, a privately owned flagpole and one owner's dogs' cremated ashes (Cornwell, 2018; Coultan, 2016). Furious and distraught property owners blamed the municipal council: after all, they had rejected the erection of a sea wall that might have protected against such damage in 2002 (Houghton, 2016). A planned retreat strategy was rejected by Council (Northern Beaches Council, 2016), and the wall, now almost complete, has been the subject of much derision since. At a cost of $300,000 per resident, many of whom didn't really want a wall, the 1.3 kms long and 7 m high seawall is viewed as a "brutal engineering solution" that does not adequately balance the multiple values attached to this small strip of sand (Kurmelovs, 2021). The then New South Wales Minister for Transport and Roads further entrenched the elevation of private property rights, stating, "we can't just throw our hands up and walk away from people, their property rights have to be protected" (Cornwell, 2018). However, this perspective discounts other residents and beach users who place their values on having access to a beach. It also ignores the systemic complexities of a managed, and of a planned, retreat—both of which are an increasing reality for vulnerable communities around the world (O'Donnell, 2022).

Legal Interventions and Behavioural Change

There remain questions about trade-offs. Who decides, and how, in reconciling contested known values and desires with the known dynamic coastal environment, and now the ever-increasing risks of climate change impacts to such locations? This question is partly being explored by the scientific literature (for example, Wright & Reid Nichols, 2019), although how it will be answered in practice is likely to be more reactionary than the supporting science would prefer (O'Donnell, 2019b; Rush, 2018; Thom, 2020). This is because of the diversity of attendant values and identities that are attached to coastlines, and range of ideas of property and associated rights on coastlines and social-economic-cultural perceptions of coastal amenity (Barnett et al., 2021; Elrick-Barr & Smith, 2022; Frolich et al., 2019; O'Donnell, 2020a; Thompson, 2007). The case for both incremental and transformational responses to the expected increasing disasters for coastlines, and the necessary adaptation to climate change impacts, is clear.

Behavioural change to support the incremental steps can be well supported by legal frameworks. It is not however without its political challenges (Gibbs, 2016; Harvey, 2019; O'Donnell, 2019a). Identity too, and its linkages to climate risk perception, response and associated actions and behaviours, is increasingly gaining attention (Barnett et al., 2021). Further, place-specific responses are important for adaptation effectiveness (Adger, 2016), especially at the juncture of land use planning and expected sea-level rise (Hurlimann et al., 2014; O'Donnell, 2019b). The role of law in attempting to manage all the incremental and transformative changes ahead remains as complicated as ever:

> *Law is however as much about power as it is about an objective window toward justice. There is a clear need to open up the black box of law, to examine its pretensions of objectivity and while laws underpin sovereignty*

> *and geopolitics, defining boundaries, marking out ownership and territory in often unseen but nonetheless very powerful ways, law is not, in fact, always well-structured, nor wholly objective. In practice, law is shown to be open to interpretation, influenced by power and by politics, and co-constituted with local places and concerns—in other words, by legal geographies. There is some danger in this openness and there is a need to be cognisant of current populist realities while also alert to the rapid and cascading material changes impacting our environmental, social, economic, and cultural systems—and perceptions thereof—as a result of climate change. Climate change adaptation efforts will be multilayered in their complexity, and iterative in their design and implementation. We are, therefore, all in for a tumultuous ride.* (O'Donnell, 2021, in-text citations omitted)

Law, positioned as it is in this chapter as both the legal rules and the institutions that uphold these, has an important role in shaping social receptiveness and behavioural change and thereby enabling the necessary adaptation to compound, cascading and protracted disasters (Barnett et al., 2014; McDonald & Styles, 2014; O'Donnell & Gates, 2013; Raikes et al., 2021; Thom, 2019). However, for coastlines, attachments to place and to dominant, individuated ideas of private property protection continue to hinder more progressive adaptation responses (Lawrence et al., 2020; O'Donnell, 2020a; Siders & Ajibade, 2021). Legal and policy frameworks must provide the necessary economic and societal guard rails to guide human behavioural changes, but they, too, are problematic—such is the nature of cascading, compounding and protracted disasters.

Policy and regulatory frameworks must provide the necessary guidance to manage coastal property owners' expectations, while protecting—as far as is now possible—our unique and highly valued coastlines. Resorting to litigation (or the threat thereof) in and of itself is not a panacea for enabling long-lasting and effective climate change adaptation (O'Donnell, 2020b). Rather, climate change is an unlikeable protagonist in repositioning the importance of place within and through legal and social systems (O'Donnell, 2018). If that reorientation is taken seriously, people might be protected from themselves through government intervention that ought to be aimed at enabling the highest and best use of coastal land until the risks and climate change impacts are too great. Sack et al. (2020) discuss recent cases before the NSW Land and Environment Court where the state government successfully intervened to achieve such aims. Of course, resorting to litigation to safeguard our collective futures, including that of the coast herself, perhaps ought to be seen as a failure.

Conclusion

Australia with its federated government system, history of conflicting policies with respect to coastal management and governance, and propensity for coastal development is at high risk of a compounding, cascading and protracted coastal disaster event. The Australian experience of attention to prepare for this eventuality, by investing in coastal adaptation, is an impressive example of early-ish intervention. However, the complexity of the "melting pot", the unintegrated legal and policy systems, people's values and attachments, and political pressures all combine to undermine the effectiveness of this early intervention.

In addition, the demise of the globally notable and practitioner-supported NCCARF in 2018 as a result of no further federal funding left universities and other institutions without the structured support to maintain momentum in adaptation analysis, with flow-on effects to disasters-related research (see also Chapter 17 of this volume). As O'Donnell and Mummery put it in 2017, "*lives and money will be saved by strong climate adaptation measures*". Indeed, as climate and environmental changes continue to reach and surpass planetary boundaries as we reach tipping points (see also Chapter 4 of this volume), the need for strong adaptation to a changing climate has not gone away. As argued by Flannery (2020):

> *The climate challenges we now face are so great that we [Australia] must resurrect the coherent, holistic approach destroyed by the Abbott government. For no matter how successful we are at reducing greenhouse gas emissions, and indeed at drawing carbon down from the atmosphere, the task of adapting to inevitable climatic change remains enormous.* (p. 155)

The compounding, cascading and protracted disasters that contested coastlines face, including increasing development pressures combined with multiple erosion, storm and sea-level rise risks, is one of the ultimate adaptation challenges of our time. We now await the roll-out of NCRAS. Here lies a new opportunity for the Australian federal government to offer leadership in adaptation planning and, ideally, properly integrate public policy areas such as disaster preparedness and response with climate change adaptation and resilience. In the meantime, we can expect more incidents of social conflict and litigation that do little to resolve competing values of coastal land use in the new climate era.

References

Adger, N. W. (2016). Place, well-being, and fairness shape priorities for adaptation to climate change. *Global Environmental Change, 38*, A1–A3.

Barnett, J., Mortreux, C., Adger, N. W. (2013). Barriers and limits to adaptation: Cautionary notes. In S. Boulter, J. Palutikof, D. Karoly, & D. Guitart (Eds.), *Natural Disasters and Adaptation to Climate Change*. Cambridge University Press.

Barnett, J. et al. (2014). A local coastal adaptation pathway. 4 *Nature Climate Change,* 1103.

Barnett, J., Graham, S., Quinn, T., Adger, N. W., & Butler, C. (2021). Three ways social identity shapes climate change adaptation. *Environmental Research Letters, 16*, 124029.

Clennell, A., Houghton, J. & Gleeson, A. (2016, 8 June). Sydney storm: Collaroy residents blame local council and the greens for stopping a sea wall being built more than a decade ago. *The Daily Telegraph*. https://www.dailytelegraph.com.au/news/nsw/sydney-storm-collaroy-residents-blame-local-council-and-the-greens-for-stopping-a-sea-wall-being-built-more-than-a-decade-ago/news-story/5603af94ebe7e3bddaf76ce7acca0623

Cornwall, D. (2018, 1 September). Our fragile coast: Residents plead for beachfront homes. *The Australian*. https://www.theaustralian.com.au/nation/inquirer/our-fragile-coast-residents-plead-for-beachfront-homes/news-story/cf66187c4f8d61c4aa3245333d96f9c6

Coultan, M. (2016, 8 June). Collaroy-Narrabeen's stormy history of beach homes versus waves. *The Australian*. https://www.theaustralian.com.au/nation/nation/collaroynarrabeens-stormy-history-of-beach-homes-versus-waves/news-story/e4fe5240b30eb5a37bbebed9ee8d72db

DCC. (2009). *Climate change risks to Australia's coast—A first pass national assessment*. Department of Climate Change, Australian Government.

Dawson, A. (2017). *Extreme cities: The perils and promise of urban life in the age of climate change*. Verso.

Elrick-Barr, C., & Smith, T. (2022). Problem framing for Australian coastal management. *Environmental Science and Policy, 127*, 218–227.

Flannery, T. (2020). *The climate cure solving the climate emergency in the Era of COVID-19*. Text Publishing.

Frohlich, M. F., Smith, T. F., Jacobson, C. Fidelman, P. Carter, R. W., & Baldwin, C. (2019). Towards adaptive coastal management: Lessons from a "legal storm" in Byron Shire, Australia. *Ocean and Coastal Management, 179*, 104909.

Gibbs, M. T. (2016). Why is coastal retreat so hard to implement? Understanding the political risk of coastal adaptation pathways. *Ocean and Coastal Management, 130*, 107–114.

Hanslow, D., Morris, B., Foulsham, E., & Kinsela, M. (2018). A regional scale approach to assessing current and potential future exposure to tidal inundation in different types of estuaries. *Science Reports, 8*(1), 7065.

Harvey, N. (2016). The contribution-lock effect blocking integrated coastal zone management in Australia: The role of governance and politics. *Ocean Science Yearbook, 30*, 1–23.

Harvey, N. (2019). Protecting private properties from the sea: Australian policies and practice' *Marine Policy, 107*, 103566.

Harvey, N., & Clarke, B. (2019). 21st Century reform in Australian coastal policy and legislation. *Marine Policy, 103*, 27–32.

Harvey, N., & Smithers, S. (2018). How close to the coast? Incorporating coastal expertise into decision-making on residential development in Australia. *Ocean and Coastal Management, 157*, 237–247.

Houghton, J. (2016, 9 June). Sydney storm: Massive 2002 protest stopped Sea wall being built, but at what cost? *The Daily Telegraph*. https://www.dailytelegraph.com.au/news/sydney-storm-massive-2002-protest-stopped-sea-wall-being-built-but-at-what-cost/news-story/2bcbcfeda125681313380894e846ab71

Hurlimann, A., et al. (2014). Urban planning and sustainable adaptation to Sea-level rise. *Landscape and Urban Planning, 126*, 84–93.

Kay, K. (2016). Breaking the bundle of rights: Conservation easements and the legal geographies of individuating nature. *Environment and Planning A, 48*(3), 504–522.

Kinsela, M., Morris, D., Linklater, M., & Hanslow, D. (2017). Second-pass assessment of potential exposure to shoreline change in New South Wales. *Journal Marine Science and Engineering*, 5(4).

Kurmelov, R. (2021, 24 October). A 7m wall has gone up on a Sydney beach: Are we destroying public space to save private property? *The Guardian*. https://www.theguardian.com/environment/2021/oct/24/a-7m-wall-has-gone-up-on-a-sydney-beach-are-we-destroying-public-space-to-save-private-property

Lawrence, J., Boston, J., Bell, R., Olufsun, S., Kool, R., Hardcastle, M., & Stroombergen, A. (2020). Implementing pre-emptive managed retreat: Constraints and novel insights. *Current Climate Change Reports, 6*, 66–80.

Lukasiewicz, A., Dovers, S., Robin, L., McKay, J., Schilizzi, S. G. M., & Graham, S. (2017). Current status and future prospects for justice research in environmental management. In A. Lukasiewicz, S. Dovers, L. Robin, J. McKay, S.

Schilizzi, & S. Graham (Eds.), *Natural resources and environmental justice: Australian perspectives* (pp. 263–266). CSIRO Publishing.
McDonald, J., & Styles, M. (2014). Legal strategies for adaptive management under climate change. *Journal of Environmental Law, 26*(1), 25–53.
Northern Beaches Council. (2016, December). Coastal Zone Management Plan for Collaroy-Narrabeen Beach and Fisherman's Beach, Report, December 2016.
O'Donnell, T. (2016). Legal geography and coastal climate change adaptation: The vaughan litigation. *Geographical Research, 54*(3), 301–312.
O'Donnell, T. (2018). Climate adaptation on the Australian east coast. In M. H. Bruun, P. J. L. Cockburn, B. Risager, & M. Thorup (Eds.), *Contested property claims: What disagreement tells us about ownership*. Routledge.
O'Donnell, T. (2019a). Coastal management and the political-legal geographies of climate change adaptation in Australia. *Ocean and Coastal Management, 175*, 127–135.
O'Donnell, T. (2019b). Contrasting land use policies for climate change adaptation: A case study of political and geo-legal realities for Australian coastal locations. *Land Use Policy, 88*, 104145.
O'Donnell, T. (2020a). Don't get too attached: Property-place relations on contested coastlines. *Transactions of the Institute of British Geographers, 45*(3), 559–574.
O'Donnell, T. (2020b). Climate change adaptation litigation: A pathway to justice, but for whom? In A. Lukasiewicz, & C. Baldwin (Eds.), *Natural Hazards and Disaster Justice* (pp. 117–132). Palgrave Macmillan.
O'Donnell, T. (2021).Coastal lawscape: A framework for understanding the complexities of climate change adaptation. *Marine Policy*, 129, 104532.
O'Donnell, T. (2022). Managed retreat and planned retreat: A systematic literature review. *Philosophical Transactions of the Royal Society B: Biological Sciences, 377*(1854). https://doi.org/10.1098/rstb.2021.0129
O'Donnell, T., & Gates, L. (2013). Getting the balance right: A renewed need for the public interest test in addressing coastal climate change and sea level rise. *Environment and Planning Law Journal, 30*(3), 320–335.
O'Donnell, T. & Mummery, J. (2017, 11 May). The 2017 budget has axed research to help Australia adapt to climate change. *The Conversation*. https://theconversation.com/the-2017-budget-has-axed-research-to-help-australia-adapt-to-climate-change-77477
O'Donnell, E. L., & Talbot-Jones, J. (2018). Creating legal rights for rivers: Lessons from Australia, New Zealand, and India. *Ecology and Society, 23*(1), 7.
Pearman, G. (Ed.). (1988). *Greenhouse. Planning for climate change*. Brill.
Raikes, J., Smith, T. F., Baldwin, C. & Henstra, D. (2021). Linking disaster risk reduction and human development. *Climate Risk Management*, 32, 100291.

Rush, E. (2018). *Rising: Dispatches from the new American Shore*. Milkweed Editions.

Sack, B., Allen, T., & Thom, B. (2020). Coastal management and protecting the public interests: Recent NSW land and environment court decisions. *Environment Planning and Law Journal, 37*, 128–135.

Short, A. (2020). Australian coastal systems. Beaches, barriers and sediment compartments, Vol. 2, Springer.

Siders, A. R., & Ajibade, I. (2021). Introduction: Managed retreat and environmental justice in a changing climate. *Journal of Environmental Studies and Sciences, 11*, 287–293.

Stark, K. (1988). Designing for coastal structures in a greenhouse age. In G. Pearman (Ed.), *Greenhouse: Planning for climate change*. Brill.

Stocker, L. & Kennedy, D. (2009). Cultural models of the coast in Australia: Towards sustainability. *Coastal Management*, 37, 387–404.

Thom, B. (2012). Climate change, coastal hazards and the Public Trust Doctrine. *Macquarie Journal of International and Comparative Environmental Law, 8*(2), 21.

Thom, B. (2019). Future adaptive coastal management. In L. D. Wright, & C. Reid Nichols (Eds.), *Tomorrow's coasts: Complex and impermanent* (pp. 305–318). Springer.

Thom, B. (2020). Future challenges in beach management as contested spaces. In D. Jackson, & A. Short (Eds.), *Sandy beach morphodynamics* (pp. 711–732). Elsevier.

Thom, B. (2021). *Coastal climate change adaptation: Roles for the Australian Government*. Blog 201, 10 October 2012, www.australiancoastalsociety.org

Thom, B. G., Eliot, I., Eliot, M., Harvey, N., Rissik, D., Sharples, C., Short, A., & Woodroffe, C. D. (2018). National sediment compartment framework for Australian coastal management. *Ocean and Coastal Management, 154*, 103–120.

Thompson, R. (2007). Cultural models and shoreline social conflict. *Coastal Management, 35*, 211–237.

Woodroffe, C., and others. (2012). Approaches to risk assessment on Australian coasts: A model framework for assessing risk and adaptation to climate change on Australian coasts. Final Report, National Climate Change Adaptation Research Facility, Gold Coast.

Wright, L. D., & Reid Nichols, C. (Eds.) (2019). *Tomorrow's coasts: Complex and impermanent*. Springer.

Wright, L. D. & Thom, B. (2019). Promoting resilience of tomorrow's impermanent coasts. In L. D. Wright, & C. Reid Nichols (Eds.), *Tomorrow's coasts: Complex and impermanent* (pp. 341–354). Springer.

PART II

Compound Risks and Impacts

CHAPTER 6

Ten Years on: Brisbane's Compounding Flood Risk

Margaret Cook and Ella Harrison

Introduction

The Brisbane River floodplain is the "*most flood impacted*" catchment in Australia (BRSFMP, 2019, p. 8). Despite a long history of flooding in the region, many Southeast Queensland residents remain unaware, deny or ignore the flood hazard as they enjoy living near the river. Myths endure that Brisbane is flood-proof, protected by Wivenhoe Dam 80 kms (by road) upstream from the city, despite a substantial flood in 2011 demonstrating the dam can only do so much. This chapter addresses the flood legacy and the compounding hazard caused by urban growth and climate change, further exacerbated by public misunderstandings of the current

M. Cook (✉)
University of Queensland, Brisbane, QLD 4072, Australia
e-mail: mcookhistory@gmail.com

E. Harrison
WMA Water Pty Ltd, Tingalpa, QLD 4173, Australia
e-mail: ellamharrison@yahoo.co.uk

A. Lukasiewicz and T. O'Donnell (eds.), *Complex Disasters*, Disaster Risk, Resilience, Reconstruction and Recovery,
https://doi.org/10.1007/978-981-19-2428-6_6

flood risk. It identifies the government measures that have been implemented to reduce or better manage the risk since 2011 and assesses their potential efficacy, as well as raising the compounding factors of climate change and urban growth that could significantly increase future flood risk.

Recent academic historical scholarship has examined Brisbane's flood history (Cook & McKinnon, 2019), and there have been decades of flood analysis and modelling by flood practitioners on the region's present and future flood risk that considers the impact of climate change. We bring the two together to examine strategies or impediments to improving the region's flood resilience. Although, as Giddens reminds us, humans constantly live with risk (Giddens, 1991), with Beck (1992) coining the phrase "*risk society*" in his thinking on the environment and human relations in modern societies, the understanding of Southeast Queensland's flood risk by the wider community is varied. But as James Short (2001) argues climate change is the "*ultimate risk*", a compounding factor largely missing from the public discourse on future floods.

Beck (1992) is especially critical of the role of the media in disseminating concepts of risk, and others have highlighted its capacity to increase resilience through public education (McEwen et al., 2017; Miles & Morse, 2007). We agree that the media could play a greater role in Southeast Queensland in public education, but also acknowledge that the cycles of drought and flood, and population growth counteract public awareness. This erodes public memory, an acknowledged tool in hazard reduction (Colten & Sumpter, 2009; Garde-Hansen et al., 2017; McKinnon, 2020) but so does a deep-seated myth of flood immunity (Cook, 2018). The 2011 floods and its aftermath exposed the struggle to comprehend the existing risk. Given that, it seems unlikely that the Southeast Queensland community is ready to make the necessary changes to accommodate the risks associated with climate change.

Southeast Queensland's Flood Legacy

The Brisbane River system is 309 kms long with a 13,500 km^2 catchment, three major tributaries (the Stanley and Bremer Rivers and Lockyer Creek) and numerous creeks (Fig. 6.1). For centuries the Turrbal and Jagera people lived on the floodplain of these sub-tropical waterways and respected the life cycles of drought and flood. Their stories recorded history of flooding and in the 1820s British explorers John Oxley

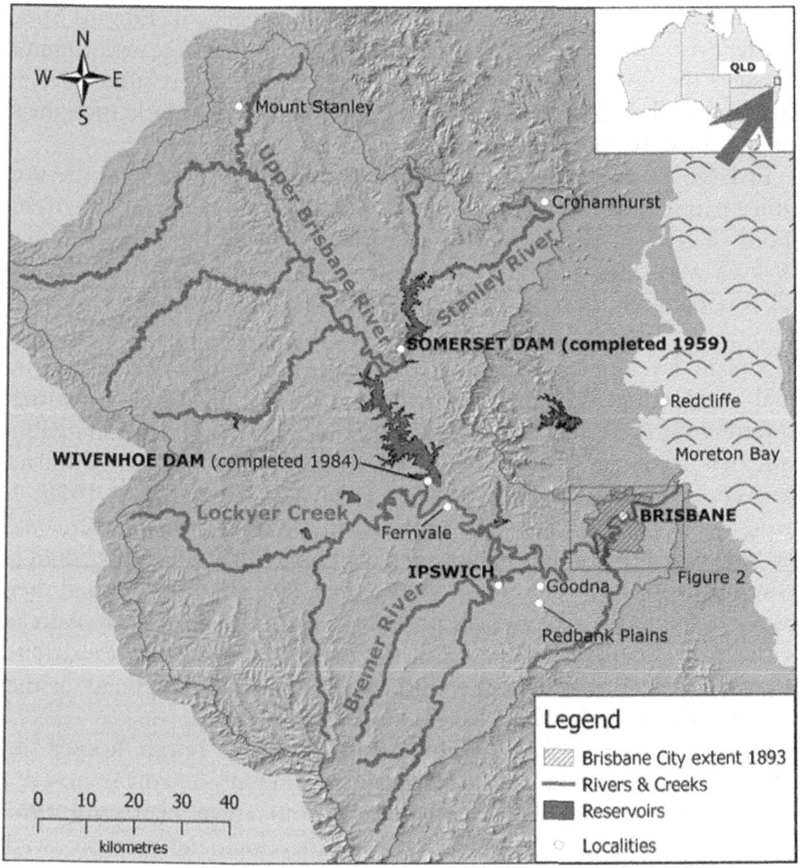

Fig. 6.1 Map of Brisbane River catchment drawn by Nick Cook (*Source* Cook, N. 2021)

and Alan Cunningham noted evidence of severe floods (Steele, 1984, pp. 160–174). Despite that knowledge, the colonists chose the floodplain for a penal colony in 1824. Infused with notions of human superiority over nature and Eurocentric understandings of the environment as a utilitarian resource (Bolton, 1981; Griffiths, 2001), the British established a new settlement on the riverbank and bestowed the name Brisbane on the town and its meandering river. The fledgling town was flooded in 1841 to a calculated height of 8.43 m at the Port Office gauge in Edward Street in the Brisbane CBD but with a small population, there was minimal damage, the flood went largely unheeded. The land was subdivided to the water's edge and sold for property development with little thought to the flood risk being created.

In 1890 Brisbane experienced a devastating flood (5.33 m at the Port Office gauge) but this did not temper growth; nor did the more extreme floods of 8.35 m and 8.09 m within a fortnight in 1893, even though 35 lives were lost, hundreds of homes washed away, and the damage bill was between £1 and £2 million (Henderson, 1897, p. 2). Floodplain occupants could no longer ignore the flood risk, but rather than halt floodplain development, the society looked to control the river with structural engineering, dredging and riverbend truncation. Attention turned towards construction of a dual-purpose water supply and flood mitigation dam that commenced in 1935 as a Depression employment stimulant. Delayed by World War II, Somerset Dam was completed in 1959. In favourable conditions, when the water fell in the Stanley River upstream of the Somerset Dam, engineers calculated a flood of 1893 heights could be reduced by up to four metres in Brisbane's CBD, although warned there would be minimal reduction in flood height if the rain fell downstream of the dam (Shepherd, 1972). But for many, the dam's mere existence offered sufficient reassurance to embrace floodplain development, as they were deluded into thinking that floods had been tamed.

During Brisbane's post World War II building boom houses and industry spread out on the floodplain with limited controls imposed as the city had weak building regulations and no Town Plan until 1965. Queensland developed without floodplain development controls, while adjacent states, New South Wales (NSW) and Victoria both introduced a state-wide flood policy in the 1970s. In Queensland a "*societal avarice for development*" (Cook, 2019, p. 76) increased Southeast Queensland's flood risk as the city grew. By 1971 Brisbane had a population of 867,784 and 251,036 occupied dwellings, many constructed in new subdivisions on the floodplain (BCC Annual Report, 1973/4, p. 31). Decades without a substantial flood made city planners reluctant to "lock up" land and development opportunities for an infrequent flood event.

Brisbane's 70 largely flood-free years ended in 1974, with a peak flood level of 5.45 m AHD recorded at the Port Office gauge. Fourteen people died and more than a third of the city was submerged as thousands of homes were destroyed. With the myth of flood immunity shattered, Brisbane town planners could have seized the opportunity to implement greater floodplain regulation. But there was seemingly no need as the government had already approved a second flood mitigation dam in 1971, Wivenhoe Dam, three times the capacity of Somerset. Engineer Trevor Grigg estimated that this dam might reduce a 1974 flood by 4 m at the Port Office Gauge (Grigg, 2011, p. 97), but only if the rain fell upstream, within the 50% of the catchment that was regulated; a caveat largely ignored by city dwellers.

In the 1970s the Australian Water Resources Council adopted the American practice of determining an annual exceedance probability (AEP) for floods, the probability of a flood of a similar size being equalled or exceeded in any year. A flood that is likely to occur once every 100 years when averaged over a very long time period has a 1 in 100 (1%) AEP; this was the equivalent of an 1893 flood. The 1974 flood was calculated as 1 in 60 AEP event. National flood guidelines recommended local authorities develop a modelled defined flood event (DFE) and a corresponding defined flood level (DFL). NSW's state flood policy restricted future development below the 1 in 100 AEP and actively encouraged removal of structures below this level to reduce the legacy of earlier development (NSW Statement of Policy, 1978). In 1978 the Brisbane City Council (BCC) adopted a DFL of 1 in 100 AEP which was lower than the most recent floods, relying on Wivenhoe Dam to reduce flood levels by three metres despite knowing that can only mitigate rainfall upstream of the dam (Smith & Handmer, 1984, p. 29). BCC rejected a higher DFL believing more severe restrictions would have "*high costs on the city's urban, social and economic fabric*" (Joint Flood Taskforce, 2011, p. 34), without recognising the costs devastating floods have on people, property, the economy and the environment. Without a policy of removing vulnerable structures below this DFL, 90% of Brisbane's houses built before the 1978 policy presented a huge legacy issue of flood vulnerability that was not addressed.

Government employees working in flood and disaster management drafted a State Flood Risk Management Policy Discussion paper in 2002 that recognised the state's piecemeal approach to flood management. However, when a state-wide policy was introduced in 2003 (SPP1/03) it was a much weaker policy than recommended in the discussion paper (Queensland Government, 2003). It did not deal with existing risk, only regulating new development, and allowed compensation to developers with an existing development application (Queensland Government, 2002, pp. 6–8). Under the 2009 *Sustainable Planning Act*, local governments were encouraged to undertake a flood study and implement appropriate flood management levels, but this was discretionary.

Between 1995 and 2006, Brisbane was Australia's fastest growing city, with a growth rate of 2.3% between 2004 and 2009 (ABS, 2007–2011). Between 2005 and 2011 the BCC approved 1811 development applications within the 1974 flood extent, as dwelling numbers grew from 712,500 in 1973 to 2,065,998 in 2011 (ABS, 2007–2011 and profile.id). The increasing flood vulnerability was exposed in 2011 when prolonged and extensive rainfall fell over the catchment, and a flood reached 4.46 m AHD at the Port Office Gauge. An estimated 14,100 Brisbane properties were affected (1203 houses inundated), and the total damage bill was estimated in the billions (QFCI Interim Report, 2011, p. 27). The community was shocked, describing the floods as unexpected, with many people completely unaware their homes were on one of the major floodplains in Australia.

The floods of 2011 highlight two key points—the almost complete lack of awareness and understanding of the Brisbane River flood behaviour, and the absence of risk-aligned development control in the floodplain. Although this flood height was less than the 1974 flood, Premier Anna Bligh conceded that the damage was so much worse than in previous floods because the city was bigger, more populated and many suburbs had been built since 1974 (Parnell, 2011, p. 1). The BCC calculated that 10% of the residential properties flooded in 2011 were built after the 1974 flood (QFCI Final Report, 2012, p. 146).

The floods became a major catalyst for change across all levels of government. Faced with devastation and growing community discontent, the government established the Queensland Floods Commission of Inquiry (QFCI) on 17 January 2011 to investigate the preparation, planning and response to the floods at government levels. Its final report in 2012 made 177 recommendations on state and local government flood management. From this, substantial bodies of work were undertaken, including the initiation of the Brisbane River Catchment Flood Studies. This four-phase project—represented in Fig. 6.2—considered the Brisbane River from a catchment perspective for the first time, rather than broken into arbitrary boundaries of the local government areas.

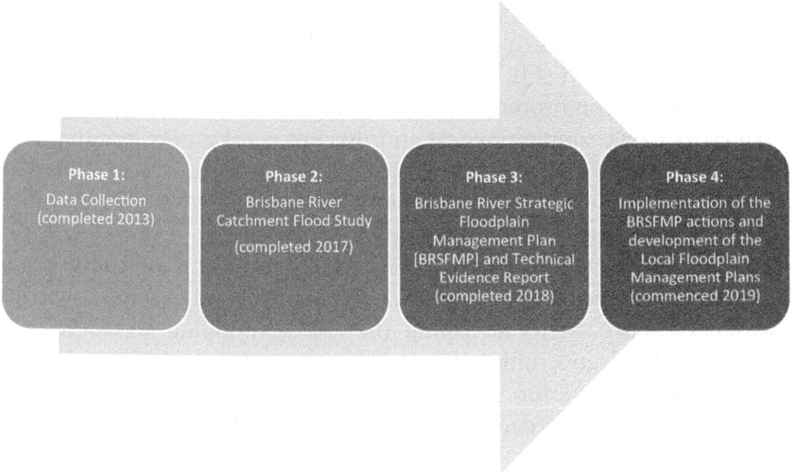

Fig. 6.2 The Brisbane River Catchment Flood Studies (*Source* Queensland Reconstruction Authority, 2019)

The project was conducted as a partnership between governmental staff and industry practitioners, requiring investment of more than $7 m from state and local governments and taking over seven years to complete. The Phase 2 Flood Study ultimately delivered a two-dimensional hydraulic model covering the Brisbane River floodplain downstream of Wivenhoe Dam from which the flood behaviour was defined. It modelled levels, flows, velocities, timings and physical extents of 11 design flood events, ranging from the small and frequent 1 in 2 (50%) AEP, to the extremely large and rare, 1 in 100,000 (0.001%) AEP. The subsequent Phase 3 Strategic Floodplain Management Plan then developed the regional framework to assess the flood risk and the full range of potential regional management options.

The findings confirmed observations made during the 2011 floods—extensive urban development throughout the Brisbane River floodplain over the past century had exacerbated the damage and made it the most densely developed floodplain in Australia (QRA, 2019, p. 25). As of 2017, an estimated 134,000 buildings (predominantly residential) occupied the Brisbane River floodplain, approximately 75% in Brisbane, 22% in Ipswich, 3% in Somerset and 0.5% in Lockyer Valley. A 1 in 10 AEP flood would inundate 51 buildings in Brisbane above habitable floor level, a number that increases to 12,000 in a 1 in 100 AEP flood and 32,000 in a 1 in 500 AEP flood (QRA, 2019, p. 25). These impacts have cascading affects as floods compromise transport and services and create intangible social costs (for example physical and psychological illness, trauma, relationship breakdown and poverty). Financial loss from a 1 in 100 AEP flood (in 2019 values) is estimated at $6760 million, costs that are borne by the wider society (QRA, 2019, pp. 27–28).

As a result of the Queensland Floods Commission of Inquiry and the findings of the Brisbane River Catchment Flood Studies, a multidisciplinary approach to managing the risk of flooding was developed at state and local government levels. The integrated catchment planning approach seeks to ensure appropriate linkages are made between disciplines and planning process to deliver multiple benefits (QRA, 2019, p. 16). It considers the "problem" and the "solution" through the lenses of a range of disciplines—including engineering, land use planning, disaster management, communications and engagement, environmental management, infrastructure management, economics and architecture—to provide a holistic approach to flood risk management. With limited opportunities for further physical structural mitigation options identified, it was collectively agreed that a stronger, more risk-aligned planning policy, enhanced emergency management capabilities, more resilient buildings and public engagement with at-risk communities to improve flood awareness were essential to avoid a repeat of the 2011 flood damage.

However, this work has mostly remained within the local and state government, taking years to develop and implement, with several more required before actual on the ground change commences. Meanwhile, community folklore and perpetuation of the "dam myth" prevails, stalling locally led solutions to living with floods.

The Dam Myth Prevails

In the popular mindset floods can be prevented. With the completion of Wivenhoe Dam in 1984, a deep-seated community myth flourished that "Wivenhoe will save us", the fallacy exposed by the 2011 floods. Rather than the calculated 3 m height reduction expected from Wivenhoe, the flood height in Brisbane's CBD was reduced by 2 m; all peak flood heights in the 2011 floods were higher than the DFL. In all, 14,100 Brisbane properties were affected across 94 suburbs between 11 and 12 January (QFCI Interim Report, 2011, 27).

As the Floods Commission of Inquiry noted (QFCI, 2012, p. 31) flood complacency "over many years" had contributed to the damage: "*Years of drought did not promote rigour in flood planning, whether in relation to disaster response, dam management or land use*". So too had faith in Wivenhoe Dam, described as the city's "*flood shield*" (*Australian*, 2011). As Marg Nixon from the suburb of Sherwood told the *Courier-Mail*, "*this was never going to happen again they told us. A lot of good Wivenhoe did*" (*Courier-Mail*, 2011, see Cook, 2018). Journalist Hedley Thomas in the *Australian* raised the spectre of dam mismanagement (*Australian*, 2011a) and the inquiry and media coverage fuelled a myth that the flood was largely the product of water released from the dam. The flood was framed as an "avoidable catastrophe" (*Weekend Australian*, 2011b), despite the dam reaching capacity as rain, the equivalent of two 1974 floods, flowed into the dam and demanded release, as well as the substantial rainfall downstream. The media largely ignored the reality that the dam had reduced the flood heights. The rhetoric of the "man-made flood" perpetuated a communal delusion that floods can be prevented by dams if operated correctly.

In 2014 the state government approved changes to the Wivenhoe Dam operation manual to change the water release strategy, a more gradual process rather than the mass release of water as occurred in the 2011 flood. Premier Campbell Newman announced that the change *could* mean a reduced flood peak by up to 10%. But by going further—"*It is designed to save businesses and homes from a repeat of the heart break suffered in 2011*"—Newman fuelled ideas that floods would be prevented (Queensland Cabinet, 2014). While the modifications could potentially reduce flood heights, the reduction may be minimal and only effective in specific hydrological events. But the action reaffirmed the faith in the Wivenhoe prevention myth and did nothing to increase communal understandings of flood risk.

With many convinced of dam mismanagement, in July 2014 a class action of 6800 claimants was lodged in the Supreme Court of NSW against Seqwater, SunWater and the State of Queensland for negligence in the operation of Wivenhoe and Somerset Dams in reducing the impact of dam releases downstream in the 2011 floods. On 29 November 2019 the court found in favour of the plaintiff and claimants, that the engineers had not complied with the flood manual and breached their duty of care (Beech-Jones, 2019) with a settlement of $440 million agreed to by the State of Queensland and SunWater on 26 February 2021. Seqwater appealed.

The media reported that the engineers had failed to "*correctly manage and follow operating procedures*" for Wivenhoe Dam or take into account rainfall forecasts and in doing so worsened the downstream inundation (Goodenough & Wuth, 2021). Supreme Court Justice Robert Beech-Jones did not find that flooding would not have occurred, yet after the judgement, principal lawyer Rebecca Gilsenan told the waiting media the focus was now "*on making sure Queenslanders were never again at the mercy of mismanaged dams meant to mitigate flooding*" (Australian Associated Press, 2019). Media accounts that suggested "*avoidable*" damage (Ritchie, 2021) became extrapolated to become an avoidable flood. Those involved in property sales were quick to seize on the decision reporting to potential buyers that "*it was eventually determined that the sudden flooding was a result of the Wivenhoe Dam gates being opened and water being released faster than they should have, which makes the tragedy preventable to some extent*" (Inspectmyhome, 2021). For those successful in the class action, now confident of compensation, the discussions of reduced flooding and qualifying phrases like "*to some extent*" were lost in the rhetoric that the floods could have been prevented. Media accounts of the judgement and real estate agents have further buttressed the myth of flood prevention as Brisbane's flood hazard compounds.

With the dam operators identified as villains, the floods became a bureaucratic error, successfully deflecting examination of government culpability, individual responsibility, planning laws and building codes, that had contributed to the flood damage. Although Premier Bligh, and others, reminded Queenslanders that "*dams do not stop floods*", they can only mitigate and minimise the impact, many refused to accept this (*Weekend Australian Inquirer*, 2011). In popular thinking Wivenhoe Dam remains the guarantee of protection if operated correctly.

Managing Development

On 20 January Premier Bligh declared that tough planning decisions would be made to reduce Brisbane's flood vulnerability. The "*last thing we want to do*", she informed the *Courier-Mail*, "*is rebuild in the same place*" (*Courier-Mail*, 2011, p. 2). But that is exactly what happened.

Development increased on the floodplain, leaving Southeast Queensland more vulnerable to unpredictable, but certain, future flooding.

Despite dire warnings that floods might cause a 30% drop in property values, any decline proved short term as owners rebuilt their properties and advertised their modern makeovers in sale brochures (Lutton, 2015). Only two years after the floods Josephine Johnston-Rowell from Johnston Dixon real estate specialists in riverfront properties declared in a special river report: "*While 2013 still hasn't seen a full resurgence in river values, confidence in the river continues to improve as memories of the devastating floods wane*" (Dixon, 2013, p. 1). As early as 1942, American geographer Gilbert White in his seminal work on floods warned that "*flood hazard tends to wax and wane in the public mind in direct relation to the occurrence of high water*" (White, 1942, p. 51). In Southeast Queensland as the 2011 floodwaters receded and houses were rebuilt, the memories faded for those who had not endured flooding. Successive dry years further allowed people to forget or underestimate floods, or to assume they will never happen again. The poor communal understanding of the 1 in 100 AEP terminology has created a misunderstanding that a flood of this height will only occur once in a century, an absolute rather than a 26% chance of repetition in thirty years or 75% chance in 100 years (Pielke, 1999, p. 416). Homeowners counted on a 99-year interval, while new Southeast Queensland residents arrived with little flood knowledge or followed a trend further noted by Gilbert White whereby they simply eliminated the hazard by ignoring it (White, 1942, p. 52).

Floods do not appear to have dampened people's desire to own riverside property. As real estate agent Patrick Dixon wrote in 2015 "*the general consensus from buyers is that* [the 2011 flood] *was a bureaucratic bungle that shouldn't have happened*" (Lutton, 2015). In 2021 real estate agent Alex Jordon reported that "*floods are simply not part of the conversation any more*", the "*most prestigious properties*" are in flood zone areas with homes snapped up at record prices (Webb, 2021). Property analysts noted in 2021 that five years after the floods, the "*worst-hit markets were back to new record heights*", citing the lure of the river and city centre location "enough to overcome any sort of issues related to the potential for flooding" (Pollard, 2021). Real estate agent Darren Piper had thought the flood would create a perception that homes in "*flood-prone suburbs*" were "*risky*", but, he wrote, "*anecdotally we've seen the complete opposite*". For many other buyers, the floods "*were now a distant memory*" (Tilly, 2018). Real estate agents framed the flood as a "*freak of nature*", an uncommon event, that may not happen again for 50 years or downplayed them as "*an inconvenience not a life-changing event*" (Queensland Property Investor, 2016). Referring to the "*mismanaged disaster*", homeowners Jonathan and Tracey Rubens expressed the thoughts of many: "*It shouldn't have happened but it did*". A risk manager by trade, Jonathon stated "*the chances of happening again are so small because the conditions were so rare*". With the market over the "*flood hurdle*" homeowners have added flood insurance to dam management as their flood mitigation strategy (Webb, 2021).

After the 2011 floods, the cost of insurance escalated in flooded postcodes, even in those not flooded. Some homeowners reported a five-fold premium increase, forcing some to abandon flood insurance and take the risk. John Payne, whose house flooded in 2011, cancelled his policy, telling the *Courier-Mail*: "*I'm betting the (Wivenhoe) dam will be managed more appropriately in future*" (Thompson, 2013). For those who could afford flood insurance, there was little incentive to adapt designs or building materials to mitigate future flood damage, including government-enforced building codes. Insurers are "*unimpressed*", left with a loss value of $1.53 billion after the 2011 floods, they are calling for better planning measures that emphasise mitigation and resilience rather than the constant cycle of "*clean-up and rebuild*". Currently only 3% of funding is spent on prevention, rather than recovery (Walker & Scott, 2021). Work in recent years to adapt houses to minimise flood damage, as showcased in the Flood Resilient Building Guidance for Queensland Homes (QRA, 2019) and implemented by the BCC through the Flood Resilient Homes Program (BCC, 2021) and others, has led the way in achieving reduced insurance premiums for resilient housing. This could potentially transform insurance from a "*band-aid*" solution to an incentive for behavioural change (Suncorp, 2018).

Shaping SEQ, the regional plan for Southeast Queensland indicates that population is expected to grow by an additional 1.8 million over the next 20 years (Queensland Government, 2017, p. 32). Some of this growth includes expansion within the floodplain with 60% forecast to be accommodated through consolidation of existing areas. Escalating urban density within the floodplain will increase population numbers exposed to floods. The Brisbane River Catchment Flood Studies found that unmanaged development could increase flood heights by metres in peak flood levels for future flood events, advocating collaboration between the catchment's planning authorities to understand where, when and how future growth could be accommodated in region.

Changing Climate

While many in the community struggle to understand the current hazard, there is the looming reality of climate change as a compounding risk factor. Australia is recognised as one of the most vulnerable to the effects of climate change amongst developed nations (Garnaut Review, 2008). There is increasing evidence that human-induced climate change is changing precipitation extremes, and that extreme flooding has increased globally over the twentieth century (Trenberth, 2011). There is confidence that global warming will result in changes in the frequency, intensity, spatial extent, duration and timing of extreme weather and climate events (IPCC, 2012). The impacts of this on flood behaviour will be dependent on each catchment's response to these changes.

The Brisbane River catchment is highly sensitive to changes in rainfall. As part of the Brisbane River Catchment Flood Studies, sensitivity testing of potential future climate change conditions was undertaken, using two scenarios determined by the Intergovernmental Panel on Climate Change (IPCC). Representative Concertation Pathways (RCPs) describe four different pathways of greenhouse gas emissions and atmospheric concentrations, air pollutant emissions and land use. RCP4.5 assumes emissions will peak by 2045 and then start declining, leading to lower increases in median temperatures than RCP8.5 which assumes emissions will continue to rise throughout the 21st century.

Using the regional hydraulic model of the Brisbane River, increases in rainfall intensity (that is, the amount of rain falling over a set time period, typically measured in millimetres per hour) as a result of increased median temperature under both RCP4.5 and RCP8.5 were modelled, for the 2050 and 2090 timeframes. The results show significant increases in flood levels in some locations. In the 1 in 100 AEP event, flood levels at the Port Office gauge in Brisbane increase between 1.2 and 2.5 m, depending on the scenario. A similar trend is observed at David Trumpy Bridge in Ipswich, with flood levels increasing by 0.9–2.4 m, while further upstream near Fernvale, levels increased by more than 4 m in some locations. Under the IPCC's RCP8.5 scenario, the 1 in 100 AEP flood level in 2050 is projected to be comparable to the current day 1 in 200 AEP level. Consequently, the tangible flood damages are forecast to more than double by 2090, from an estimated $187 million per year to $430 million per year, even without allowance for development growth (QRA, 2019, p. 33).

If Southeast Queenslanders cannot accept the present flood risk, how are they going to comprehend the increased risk of climate changes and the potential of floods 2.5 m higher than the last? Bohensky and Leitch analysed 12,000 newspaper articles published in the six months after the 2011 floods and found only 109 referred to climate change: linking (43), denying (44), confusing (15) or proving (7) the connection with floods and climate change (Bohensky & Leitch, 2014, p. 479). This supports our contention that public education on climate change and future risk is problematic. Writing in 2014, Bohensky and Leitch maintain that public consideration of climate change and adaptative strategies depends on government responses.

In 2019 the Queensland Reconstruction Authority (QRA) and partner agencies and councils undertook further assessment to better understand the cumulative impact of development on the Brisbane River flood behaviour, under both existing and future climates (WMA Water, 2021). They found that without further mitigation and/or management, the number of properties impacted could increase by 80% and flood damages by 360% under future climate conditions. The Global Facility for Disaster Reduction and Recovery (2018) recognises that "*The combination of urbanisation and climate change will be marked by an intensification in the frequency and severity of hydrometeorological hazards, with a dramatic increase in loss potential*". The Brisbane River catchment, already the floodplain with the highest flood damage potential in Australia, is likely to get much worse.

In 2017 the Queensland Government published its revised State Planning Policy, which included the state interest for "*Safety and resilience to hazards*". This approach implemented the recommendations of the Queensland Floods Commission of Inquiry, and requires planning authorities to consider the full range of design flood events, as well as climate change, and ensure land uses align with the potential risks from flooding. It articulates the desire to avoid areas of unacceptable risk wherever possible and requires mitigation of the risk to a tolerable level where avoidance cannot be achieved. A move to risk-aligned land use planning, where land uses are tailored to the potential risk of flooding and critical assets and the more vulnerable people and places are moved to less flood affected areas is ideal, but challenging to implement in already developed areas such as the Brisbane River floodplain. And unless located beyond the extent of the Probable Maximum Flood, a chance of flooding remains. This nuance, that risk-aligned does not mean risk-free, is difficult to communicate and the potential remains that it will again be ignored. Furthermore, as policies require years to develop and implement on the ground, it would seem that the flood hazard will increase with little popular understanding of the climate change-induced risk.

A Compounding Disaster

The Brisbane River floodplain is a living example of compounding risk. Legacy development and continued expansion in the floodplain are intensified by lack of awareness of the flood behaviour combined with a false optimism in our ability to tame the river through structural intervention. Further amplified by the effects of human-induced climate change, the most flood-impacted catchment appears set to continue with an ongoing focus on flood recovery, rather than prevention and preparedness.

The Global Facility for Disaster Reduction and Recovery (2018) argues that with urban expansion and climate change "*the resulting challenges will require the deployment of every tool in the disaster management tool box—from the technologies of the present and the future, to the illuminating and instructive lessons of the past*".

Despite a long history of flooding in the region many Southeast Queenslanders have little understanding of the present flood risk. Historically, developers, local and state governments, real estate agents and homeowners have denied or downplayed the hazard so as not to impede the property market (Cook, 2017). Disregarding the reality of the 2011 floods, there remains a propensity to support the Wivenhoe Dam myth of flood immunity (Jackman, 2011; Lutton, 2015; Thompson, 2013; Thompson & Robertson, 2011). Increasing development on the floodplain as occurring in Southeast Queensland is further exacerbated by property owners' tendency to disregard or insure against the existing risk, or believe it will be reduced by the changes to the operating manual for Wivenhoe Dam. The class action lawyers and successful claimants feel vindicated in their claim that dams can prevent floods, real estate agents profit from promoting the myth that riverside properties are safe, and governments can divert criticism of poor planning laws and building codes that do not mitigate the damage.

Resistance to adaptation is more pronounced in communities with a dependency on structural mitigation, as with Wivenhoe Dam (Colten & Sumpter, 2009) and the class action that found the 2011 dam operations had exacerbated the flood has done nothing to dissuade the myth that Wivenhoe Dam can save Brisbane from floods. Yet in reality it cannot, with its finite capacity and impotence if rainwater falls downstream, and, with the reality of climate change, its defences will be more limited in future. False optimism in the dam's potential is compounded by the rapid expansion of Southeast Queensland creating residents with no prior flood knowledge. With a climate of denial and a climatic cycle that moves from flood to drought (and increasingly bushfires), floods are soon forgotten which makes flood preparedness and knowledge acquisition seemingly irrelevant. Strategies of educating the public are essential. Messages need to clearly, and unashamedly, communicate the potential for flooding and its devastating consequences.

Ten years after the 2011 flood much has changed at a government organisational level as the whole catchment approach to flood management has created more collaboration within the region with regard to knowledge sharing and creation. Since the 1970s, flood managers have adopted a suite of hazard reduction strategies—public education, mapping, warnings and land use planning—to be used in conjunction with engineering solutions (Handmer & Dovers, 2007; O'Keefe et al., 1976; SCARM, 2000). This multifaceted approach underpins the Brisbane River Strategic Floodplain Management Plan (BRSFMP) and is the globally recognised approach. The BRSFMP acknowledges that climate change could significantly increase flood levels and risk across most of the floodplain and that the RCP8.5 climate change scenario should be adopted as a "*conservative*" approach where there are limited mitigation options. Building regulations, town planning and Defined Flood Events should incorporate an additional climate change factor allowance or use resilient building design. Community awareness of flood risk is a pillar of the strategy (QRA, 2019, p. 38), as scholarship has highlighted how historical knowledge creates social memory that helps foster resilience and adaptation (Colten & Giancarlo, 2011; Colten & Sumpter, 2009).

The BRSFMP cites public education in its nine desired outcomes but the challenge is to overcome dominant societal myths of flood immunity and climate change denial, as well as being able to communicate the technical disciplines of risk and risk management. The State government's emphasis on working with community to increase public awareness and its moves towards greater adaptive responses is commendable but has yet to face the test of implementation. The future of safe, resilient communities in the Brisbane River floodplain depends on overcoming these barriers.

References

Australian Associated Press. (2019, November 29). Queensland 2011 floods: Thousands of victims win class action over handling of dams. The Guardian, online edition. Retrieved from https://www.theguardian.com/australia-news/2019/nov/29/thousands-queenslanders-win-class-action-dams-2011-floods

Australian Bureau of Statistics (ABS). (2007–2011). Region Population Growth.

Beck, U. (1992). *Risk society: Towards a new modernity*. Sage Publications.

Beech-Jones, J. (2019). Rodriguez & Sons Pty Ltd v Queensland Bulk Water Supply Authority trading as Seqwater (No 22) [2019] NSWSC 1657. New South Wales: Judgment Summary, Supreme Court. Retrieved from https://www.mauriceblackburn.com.au/media/5105/19-11-29-rodriguez-judgment-summary.pdf

Bohensky, E. L., & Leitch, A. M. (2014). Framing the floods: A media analysis of themes of resilience in the 2011 Brisbane Flood. *Regional Environmental Change*, *14*, 475–488. https://doi.org/10.10007/s10113-013-0438-2

Bolton, G. (1981) *Spoils and spoilers*. Allen and Unwin.

Brisbane City Council Annual Report. (1973/4).

Brisbane City Council (BCC). (2021). *Flood Resilient Homes Program*. Retrieved from https://www.sustainablebrisbane.com.au/floodwise

Caldwell, A. (2011, January 20). Danger Zones may not be redeveloped. *Courier-Mail*, 2.

Colten, C. E., & Giancarlo, A. (2011). Losing Resilience on the Gulf Coast: Hurricanes and Social Memory. *Environment: Science for Policy for Sustainable Development*, *5*(4), 6–19.

Colten, C. E., & Sumpter, A. R. (2009). Social memory and resilience in New Orleans. *Natural Hazards*, *48*, 355–364. https://doi.org/10.1007/s11069-008-9267-x

Cook, M. (2017). Vacating the floodplain: Urban property, engineering, and floods in Brisbane (1974–2011). *Conservation and Society*, *15*(3), 344–354. https://doi.org/10.4103/cs.cs_16_95"\t"_blank

Cook, M. (2018). "It will Never happen Again": The Myth of flood immunity in Brisbane. *Journal of Australian Studies*, *42*(3), 328–342. https://doi.org/10.1080/14443058.2018.1487871

Cook, M. (2019). *A river with a city problem: A history of Brisbane Floods*. University of Queensland Press.

Dixon, J. (2013). *Special Floods Class Action Supplement River Report*.

Garde-Hansen, J., McEwen, L., Holmes, A., & Jones, O. (2017). Sustainable flood memory: Remembering as resilience. *Memory Studies*, *10*(4), 384–405. doi/https://doi.org/10.1177/1750698016667453

Garnaut Review. (2008). *Targets and trajectories: Supplemental draft report*. Garnaut Climate Change Review.

Giddens, A. (1991). *Modernity and self-identity; self and society in the late modern age*. Stanford University Press.

Global Facility for Disaster Reduction and Recovery. (2018). *After Shocks: Remodelling the Past for a Resilient Future*. Retrieved from https://geowb.maps.arcgis.com/apps/MapJournal/index.html?appid=96a7377dbc6d49c1b57a5fe3c10f7b4f

Goodenough, C., & Wuth, R. (2021, May 7). QLD flood victims battle for compensation. *The Young Witness*. Retrieved from https://www.youngwitness.com.au/story/7242678/qld-flood-victims-battle-for-compensation/

Griffiths, T. (2001). *Forests of ash: An environmental history*. Cambridge University Press.

Grigg, T. cited in Cousins, S. (2011, February 16). Wivenhoe Releases Could've Prevented Floods? Nonsense Says Experts. *Crikey*.

Handmer, J., & Dovers, S. (2007). *Handbook of disaster policies and institutions: Improving emergency management and climate change adaptation*. Routledge.

Henderson, J. B. (1897). Floods in the Brisbane River, and Schemes for Abatement of their Disastrous Effects. Annual report of the Hydraulic Engineer for Water Supply, 1896. *Queensland Votes and Proceedings*, 4, 2.

Inspect my Home. (2011, March 2). What to consider before buying property in a Brisbane flood zone. Retrieved from https://www.inspectmyhome.com.au/blog/buying-property-in-a-brisbane-flood-zone

IPCC (Intergovernmental Panel on Climate Change). (2012). Workshop Report of the Intergovernmental Panel on Climate Change Workshop on Socio-Economic Scenarios. *IPCC Working Group III Technical Support Unity*. Potsdam Institute for Climate Impact Research.

Jackman, C. (2011, January 13). Children of the Flood Watch their Certainty Washed Away. *Australian*, 3.

Joint Flood Taskforce. (2011). Joint Flood Taskforce Report. Prepared for the Brisbane City Council by the Joint Flood Taskforce.

Lunn, S. (2011, January 15–16). The untameable torrent. *The Weekend Australian, Inquirer*, 1.

Lutton, E. (2015, November 10). Floods and property: Brisbane Five Years On. *Domain Media*. Retrieved from https://www.domain.com.au/news/floods-and-property-brisbane-five-years-on-20151110-gksi0s/

McCarthy, J. (2011, January 15–16). Damage worse than 1974. *Courier Mail*, 5.

McEwen, L., Garde-Hansen, J., Holmes, A., Jones, O., & Krause, F. (2017). Sustainable flood memories, lay knowledges and the development of community resilience to future flood risk. *Transactions of the Institute of British Geographers*, *42*, 14–28.

McKinnon, S. (2019). Remembering and forgetting 1974: The 2011 Brisbane floods and memories of an earlier disaster. *Geographical Research*, *57*(2), 204–214.

McKinnon, S. (2020). Placing memories of unforgettable fires: Official commemoration and community recovery after the 2003 ACT firestorm. In S. McKinnon, & M. Cook (Eds.), *Disasters in Australia and New Zealand: Historical Approaches to Understanding Catastrophes*. Palgrave Macmillan.

Miles, B, & Morse, S. (2007). The role of news media in natural disaster risk and recovery. *Ecological Economics*, *63*, 365–373.

NSW Planning and Environment Commission. (1978). Statement of Policy. Circular 15, 16/8/798.

O'Keefe, P., Westgate, K., & Wisner, B. (1976). Taking the naturalness out of natural disasters. *Nature*, *260*, 566–567. https://doi.org/10.1038/260566a0

Parnell, S. (2011, January 13). The River that Turned the City into a Mudflat. *The Australian*, 1.

Pielke, R. A. Jr. (1999). Nine Fallacies of Floods. *Climate Change*, *42*, 416. Retrieved from https://pdfs.semanticscholar.org/8fdb/ab58563a76f87eff84c3ddac0951a3203f86.pdf

Pollard, E. (2021, January 12). Brisbane residents who stayed put after 2011 flood claw back house values. *ABC News*, online edition. Retrieved from https://www.abc.net.au/news/2021-01-12/queensland-2011-brisbane-flood-residents-claw-back-house-values/13043086

Profile.id. (2011). City of Brisbane: Brisbane LGA 2011 Census Results. Retrieved from https://profile.id.com.au/assets/profile/brisbane/PDF/Brisbane%20LGA.pdf

Queensland Cabinet and Ministerial Directory. (2014, April 1). Changes to Wivenhoe Dam operations. Retrieved from https://statements.qld.gov.au/statements/74405

Queensland Floods Commission of Inquiry Interim Report. (2011).

Queensland Floods Commission of Inquiry Final Report. (2012).

Queensland Government. (2002). State flood risk management policy: summary discussion paper. Brisbane: Natural Resources and Mines.

Queensland Government. (2003). State Planning Policy SPP1/03 Mitigating the Adverse Impacts of Flood, Bushfire and Landslide.

Queensland Government. (2017). *Shaping SEQ; South East Queensland Regional Plan 2017*. Retrieved from https://apo.org.au/sites/default/files/resource-files/2017-08/apo-nid102626.pdf

Queensland Property Investor. (2016). The flood prone suburbs buyers are flocking to. Retrieved from https://qldpropertyinvestor.com.au/the-flood-prone-suburbs-buyers-are-flocking-to/

Queensland Reconstruction Authority. (2019). Brisbane River Strategic Floodplain Management Plan (BRSFMP). Retrieved from https://www.qra.qld.gov.au/sites/default/files/2019a-04/plan_-_brisbane_river_strategic_floodplain_management_plan_april_2019a.pdf

Queensland Reconstruction Authority. (2019). Flood Resilient Building Guidance for Queensland Homes. Retrieved from https://www.qra.qld.gov.au/resilient-homes/flood-resilient-building-guidance-queensland-homes

Ritchie, E. (2021, August 30). Do-nothing dam plan led to Brisbane flood disaster. Court told. *The Australian*. Retrieved from https://www.theaustralian.com.au/nation/politics/donothing-dam-plan-led-to-brisbane-flood-disaster-court-told/news-story/b454f9e5da0aee1b1c189934344e73f9

SCARM Report 73. (2000). Floodplain Management in Australia: Best practice principles guidelines. CSIRO Publishing.

Scott, L. (2011, January 22–23). QWeekend, The Flood 2011. *Courier-Mail*, special issue, 24.

Shepherd, E. M., Co-Ordinator General's Department. 1972, February 24. Flood Conditions in the St Lucia University Campus. Queensland State Archives ID 569498.

Short, F. Jr. (2001). Foreword in Jaeger. In C. Carlo et al. (Eds.), *Risk, uncertainty and rational action*. Earthscan Publications.

Smith, D. I., & Handmer, J. (1984). Urban flooding in Australia: Policy development and implementation. *Disasters*, *8*(2), 105–117. https://doi.org/10.1111/j.1467-7717.1984.tb00859.x

Steele, J. (1984). *Aboriginal pathways in Southeast Queensland and the Richmond River*. University of Queensland Press.

Suncorp. (2018, October 23). Designing homes that accept water on the floodplains of Brisbane River. Retrieved from https://www.suncorpgroup.com.au/news/features/creating-homes-that-accept-water-on-the-floodplains-of-brisbane-river

Thomas, H. (2011a, January 15–16). Damned if they Do, Damned if they Don't. *The Australian*, 13.

Thomas, H. (2011b, January 22–23). The Great Avoidable Catastrophe. *The Weekend Australian*, 10.

Thompson, T. (2013, January 7). Skyrocketing insurance premiums hit Queensland flood victims two years on. Courier-Mail, online edition. Retrieved from https://www.couriermail.com.au/news/queensland/skyrocketing-insurance-premiums-hit-queensland-flood-victims-and-others/news-story/d3fafa9eb9bc8b4b739c5c5603005229

Thompson, T., & Robertson, J. (2011, April 14). Bad planning decisions blamed. *Courier-Mail*, 6.

Thomas, H., & Uren, D. (2011, January 18). Dam in Sights of Flood Inquiry', *The Australian*, 1.

Tilly, E. (2018, October 20). Buyers rush to buy riverfront property as prices float to the top. Courier-Mail, online edition. Retrieved from https://www.realestate.com.au/news/buyers-rush-to-buy-riverfront-property-as-prices-float-to-the-top/

Trenberth, K. E. (2011) Changes in precipitation with climate change. *Climate Research. 47*, 123–138. https://doi.org/10.3354/cr00953

Walker, J., & Scott, M. (2021, August 30). Queensland floods: It'll happen again. The Australian, online edition. Retrieved from https://www.theaustralian.com.au/inquirer/queensland-floods-its-going-to-happen-again/news-story/5be22bfe99b859ac0c32d5b15a024e89

Webb, S. (2021, January 11). The Brisbane floods and the property market, 10 years on. *Domain*. Retrieved from https://www.domain.com.au/news/the-brisbane-floods-10-years-on-the-property-market-a-decade-on-1013575/

White, G. F. (1942). Human adjustment to floods. PhD Thesis. Chicago: University of Chicago.

WMA Water. (2021). Brisbane River Cumulative Impact Assessment—Final Project Report. Queensland Reconstruction Authority.

CHAPTER 7

Accounting for the Compounding Effects of Climate Change on Coastal Residents

Sonia Graham

Introduction

Sea-level rise is one of the most certain impacts of climate change. By 2050 it is estimated that at least 340 million people globally will be affected by coastal flooding (Kulp & Strauss, 2019). Yet the analyses underpinning such predictions focus on people's usual place of residence (e.g., Agrawala et al., 2003; Crowell et al., 2010). Where population movements are acknowledged, the scale of analysis is usually at regional or larger spatial scales and over long time periods (McGranahan et al., 2007). While some studies explore the effects of sea-level rise on temporary residents, such as tourists (e.g., Uyarra et al., 2005) and seasonal residents (e.g., Novaczek et al., 2011), the analyses focus on small scales and short

S. Graham (✉)
School of Geography and Sustainable Communities, The University of Wollongong, Northfields Ave, Wollongong, NSW, Australia
e-mail: sgraham@uow.edu.au

A. Lukasiewicz and T. O'Donnell (eds.), *Complex Disasters*, Disaster Risk, Resilience, Reconstruction and Recovery,
https://doi.org/10.1007/978-981-19-2428-6_7

time periods. There are few, if any, holistic analyses that focus on the individual and the compound impacts that people are likely to synchronously experience in their home and other places where they live, work, and play. A new approach is needed that accounts for the impacts of sea-level rise, and other climate change impacts, on the multiple places to which individuals are attached. Such a multi-place approach can help answer the question: Which individuals are most vulnerable to the compound risks of climate change?

The extensive geography and environmental psychology literature on place attachment is concerned with the bonds that exist between people and places (Davenport & Anderson, 2005; Devine-Wright et al., 2015; Esteban-Guitart et al., 2013), where places are defined by both their physical and social characteristics (see Lewicka, 2011 for a comprehensive review of the place attachment literature). Overwhelmingly, this literature focuses on the attachments that people feel to their primary place of residence (Esteban-Guitart et al., 2013). A much smaller body of research explores the attachments that people feel to other places, such as where they work (e.g., Le Roy & Rioux, 2012), shop (e.g., Ujang, 2012), and recreate (e.g., Ednie et al., 2010). There are only a handful of studies that consider multi-place bonds. To date, multi-place research has focused on attachments to multiple residences (e.g., McIntyre et al., 2006) or attachments at multiple scales—neighbourhood, town, regional, country, and/or global (e.g., Bonnes et al., 1990; Devine-Wright et al., 2015; Esteban-Guitart et al., 2013). There has been limited research that explores attachments to multiple non-residential places at a particular scale (Di Masso et al., 2019).

In the context of climate change, there is a body of geography research that has investigated the impacts of sea-level rise and adaptation policies on people's place attachments (e.g., Adger et al., 2013; Agyeman et al., 2009; Köpsel et al., 2017; O'Neill & Graham, 2016). Such research has highlighted that sea-level rise will not only impact the physical and built environment, but also the non-material things people value, such as their social interactions, social status, identity, and recreational opportunities (Graham et al., 2013). Like the place attachment literature more broadly, these applied studies are spatially constrained; there is an implicit assumption that place attachment is concentrated where people reside, "*with the presumption that these are the only places that people value and form relations of belonging with*" (Devine-Wright et al., 2015, p. 69).

When research on climate change and multiple place attachments has coincided, the focus has been on evaluating the effects of multi-place attachments on individual CO_2 emissions (Hiltunen, 2007) or the effects of multi-scalar place attachments on attitudes towards climate change (Devine-Wright et al., 2015), rather than the impacts of climate change on multi-place attachments. The aims of this chapter are to: (1) use a qualitative dataset to reveal the multi-places to which individuals are attached; (2) evaluate the compound impacts of climate change on residents' multi-place attachments; and (3) consider how such compound impacts can be accounted for in future climate change policy and research.

Coastal Residents Vulnerable to Compound Climate Change Impacts in Victoria, Australia

This chapter is part of a broader research project that sought to understand the impacts of sea-level rise and adaptation policy on five coastal communities in south-east Australia (see Barnett et al., 2014). The five communities are located along the length of the Gippsland East coast of Victoria and vary in size, with permanent populations of 24, 104, 298, 322, and 4810 people for Manns Beach, McLoughlins Beach, Port Albert, Seaspray and Lakes Entrance, respectively (ABS, 2016).

All five communities have similar demographics. The populations are 14 to 27 years older than the Australian median age of 38 years (ABS, 2016) because of out-migration of youth and in-migration of retirees. The median household incomes for all five communities are significantly lower (21.1–54.1%) than the Australian median (ABS, 2016). The communities all have high proportions (26.7–62.7%) of unoccupied private dwellings, reflecting the large number of seasonal residents who use these locations for recreation and leisure.

The stretch of coastline upon which the communities are located is one of the most vulnerable to sea-level rise in Australia (Department of Climate Change, 2009). Almost all the infrastructure in the four smaller communities is located within 1 m of mean sea level. Thus, according to traditional notions of vulnerability, the residents' of these communities are physically and socially vulnerable to coastal flooding. In addition to the risk of sea-level rise, climate change is also likely to exacerbate the risk of bushfires and increase the frequency of extreme events in the region, such as intense precipitation and associated riverine flooding (Brooke &

Hennessey, 2005). The next section describes how the study sought to understand residents' multiple place attachments and how this altered their individual vulnerability to sea-level rise.

Describing Multiple Place Attachments

Thirty-seven semi-structured interviews were conducted with 42 people, i.e., five interviews involved two participants, who live in the communities on a permanent or seasonal basis. A semi-structured approach was taken to allow participants to explain, in their own words and in their own way, their attachment to the place that they permanently or seasonally live, other places that they regularly spend time in, places they travel to, places where they have previously lived and any other places that they perceive to be special (see Barnett et al., 2014 for the interview questions). These questions are consistent with questions usually asked in place attachment studies, which ask people about their special places (Eisenhauer et al., 2000), better places to be (Stedman, 2003) and places they would like to show a visitor (Schiavo, 1988). The interview questions did not require that participants only name local places, thus participants were able to raise and discuss attachments at regional, national, and global scales (Devine-Wright et al., 2015).

Approval to conduct the interviews was obtained from the University of Melbourne Human Research Ethics committee prior to conducting the interviews. Interviews were conducted between April and May 2012. While the data was collected almost a decade ago, the extreme weather events that have occurred since data collection help to demonstrate the compound effects of climate change that residents have experienced and point to a need for further research and data collection on this topic. A purposive snowball sampling strategy was used to gauge the views of as many different types of residents as possible, using characteristics that have previously been found to be important in place attachment studies, including their length of residence (Lewicka, 2011), age, level of education (Bonaiuto et al., 2004) and income (Bonnes et al., 1990). Interviewees were recruited by calling publicly listed community groups, approaching business owners, visiting community halls, and by chain-referral.

The interview transcripts were initially analysed by identifying all the places listed by interviewees and undertaking a thematic analysis of the reasons why interviewees mentioned these places. Data about the places residents valued was then entered into a spreadsheet and cross-checked against the Climate Risk Australia website (coastalrisk.com.au) to evaluate whether the special places in Australia are also at risk of sea-level rise (0.74 m in 2100, consistent with adaptation planning benchmarks across Australia). These qualitative and quantitative analyses form the basis of the results presented next.

Multiple Place Attachments

When interviewees were asked about the places that are special to them, most (79%) spoke of their home. The special places beyond the home that interviewees mentioned most often included the beach (68%), workplace (37%), and garden (26%). The reasons why these places were special were because of their natural beauty, the peace and quiet, and the strong social bonds that they enabled (Graham et al., 2014, 2018). Interviewees also appreciated the ease with which they could move around their localities and get to other places, such as the mountains and snow. For residents who liked to travel further afield, home was important because it helped them feel "*settled*" and gave them a "*base to come back to*" (McLoughlins Beach interviewee).

Cumulatively, there were 139 special places that interviewees mentioned as significant beyond the towns in which they resided. These were places where interviewees' families and friends lived, where they had previously lived, where they spent time accessing essential services—shops, eateries, and medical facilities, where they worked, went on day trips, and where they recreated and holidayed.

On average, interviewees mentioned seven special places beyond their towns, ranging from 1 to 22 other places. About one-third of interviewees mentioned five or fewer places, half mentioned six to eight places, and one-sixth mentioned ten or more places. Approximately three-quarters of the places mentioned were in the state of Victoria, one-sixth elsewhere in Australia, and one-tenth outside Australia.

The places mentioned spanned a range of spatial scales, from specific manmade ($n = 3$) or environmental features ($n = 18$)—beaches, bays, rivers, lakes, inlets, waterfalls, coastlines, islands, forests and mountains—to towns and suburbs ($n = 90$), cities ($n = 6$), regions ($n = 3$), states and territories ($n = 9$), countries ($n = 8$) and continents ($n = 2$). The rest of the results focus only on special places at the local scale (towns and suburbs or smaller scales) because the effects of sea-level rise, and other climate change impacts, are acutely experienced at local scales (Hess et al., 2008).

Compound Risks of Sea-Level Rise to Individuals with Multiple Place Attachments

When residents' multiple local place attachments were categorised according to whether they are at risk of sea-level rise, it was possible to identify four levels of vulnerability among interviewees.

Most vulnerable: There were three interviewees for whom all the places they were currently attached—Ninety Mile Beach, Seaspray, and/or Sale—were at risk of sea-level rise. All three interviewees only named one or two places as being special to them. One person at most risk had a long-term attachment to the area and family networks nearby:

> *We used to camp down here and the camp got blown away so we bought a house instead... it's close to my family... I love the area, that's why I like walking around and going to all these little places just around here, walking the town and checking it out, that's all.* (Seaspray interviewee)

The other two most vulnerable interviewees indicated that they may not continue living in the area in the future. One was a young person who explained that "*I don't want to be here. I don't know it's just - it's a nice place it's just not really that many opportunities here*" and the other person was renting. Thus, if these two interviewees form attachments to other places, they may become less vulnerable.

Moderately vulnerable: The second group of interviewees (n = 17) was those for whom at least half of the places they were attached to were at risk of sea-level rise. Almost all (27/32) of the special places at risk of sea-level rise mentioned by this group of interviewees were located along the east coast of Victoria. A further eight places that are special to this group are located near the coast but are not at risk of sea-level rise by 2100. Members of this group often talked about their recreational activities, such as boating, which were central to their relationships with coastal places. For example, one interviewee explained, "*It's ideal. We've got a jet ski as well. It's just an ideal place because you can go right round to Port Albert, to Port Welshpool, round to McLoughlins [Beach] on the jet ski without ever having to go out into the Bass Strait because it's all inlets and all that*". Another explained:

> *Special in that they're familiar, I suppose. You just feel comfortable when you're there. Not especially in a sense it's something magical has happened there—well, I suppose things do happen all over the town, but there's not one particular place that I would say it is particularly special, although there is a place called The Bluff which is out at Lake Tyers, and if I had to pick a spot, if I went away, I reckon that's the spot that I'd always think about. The surf's good there, the beach is beautiful. So that's probably one spot is cool. The other spot I really enjoy to go to, and I don't get there as often as I should, is just the Entrance, just to do that walk and then stand up on the rock wall and just look at that, just the water coming and going. It's just phenomenal, it's an awesome body of water. So that's probably another spot that I can stand in awe and watch for hours. So probably those two spots.* (Lakes Entrance interviewee)

Somewhat vulnerable: The third group ($n = 14$) were those for whom at least one, but less than half of their special places, were at risk of sea-level rise. For these interviewees, almost all (11/14) of the special places at risk of sea-level rise were near—in the adjoining towns or suburbs—where they lived. Their other 40 place attachments stretched across Victoria, with some in NSW and South Australia. Many of these other place attachments were to peri-urban, rural or mountainous regions.

There were similarities between the people in the moderate and somewhat vulnerable groups. People in the moderately vulnerable group valued multiple coastal places "*most of the trips we do are to other little coastal spots. So, it's usually between here and Paynesville when we're going that way. Now and then we take drives to Buchan*" (Lakes Entrance interviewee). People in the somewhat vulnerable group were more likely to identify other places where they could have their needs, such as recreational activities or social interactions, met. For example, one interviewee explained, "*all the places I go to, they're all... fishing places. That's the same. But here—I'm living here. Look, I like—Robe is probably—it's bigger. It's got more possibilities*" (McLoughlins Beach interviewee).

Least vulnerable: The last group ($n = 4$) were at minimal risk of sea-level rise because none of the places they were attached to beyond the home were at risk of flooding. More than half of the special places identified by these interviewees were in, or adjacent to, forested areas. While these places are not at risk of sea-level rise, they are highly vulnerable to bushfires (Symons et al., 2016). Indeed, of six—Benambra, Deptford, Nowa Now, Omeo, Orbost, and Youngs Creek—of the fourteen places that this group identified as special were threatened or burned during the 2019–2020 bushfires (Forest Fire Management Victoria, 2020). The places that the least vulnerable group mentioned as being special to them are also at risk of riverine flooding. For example, Morwell, a special place for the lowest risk group, was at risk of flooding in July 2021 (VIC Emergency, 2021).

While people in the least vulnerable group identified that "*no, there's nowhere really special*" (Lakes Entrance interviewee) beyond the places where they live, they did emphasise their deep connections to where they call home, "*I have connections spiritually, culturally and just everyday*" (Lakes Entrance interviewee), or "*this is probably the best place I've lived in*" (McLoughlins Beach interviewee).

The four groups were similar with respect to the gender, age, level of education, household size and composition, or length of residence. However, four of the five interviewees who had multi-generational ties to the case study communities belonged to the moderately vulnerable group, i.e., those for whom at least half of the places they are attached to are at risk of sea-level rise.

Discussion

The emphases of the place attachment and climate change literatures on people's usual place of residence are important; almost four-fifths of residents mentioned their home when discussing special places. Yet, the results reveal that focusing on home at the exclusion of other special places, results in a partial understanding of what people value about where they live and how they will be affected by sea-level rise and climate change more generally. In the discussion that follows, consideration is given to the extent to which results are consistent with, and go beyond, other research on multi-place bonds and how the results could inform the design of future climate change policy and research that takes into account compound risks.

Past multi-place research has revealed that people can be categorised based on their place attachment preferences. One of the key distinctions made is between people whose place attachment is centred on the home versus those whose focus is on the neighbourhood, city centre or periphery (Bonaiuto & Bonnes, 1996; Bonaiuto et al., 2004; Bonnes et al., 1990). This has parallels with the local/cosmopolitan distinction that was developed in the 1950s (Gustafson, 2006). Other, more recreation-oriented research has found that residents and visitors can be categorised based on the types of physical and cultural features that they prefer (Ednie et al., 2010). In this study, there was evidence of both types of place attachment preferences. There were some individuals for whom the home was central. These tended to be individuals in the most and least vulnerable categories. Those in the moderately vulnerable category held strong place attachments to mostly coastal places, and were therefore vulnerable to compound risks of sea-level rise. Those in the somewhat vulnerable category were more likely to hold place attachments to forested and/or mountainous places, and are therefore vulnerable to compound risks arising from other climate change impacts, such as bushfires.

Another important distinction found between the groups was the role that length of association had on the likelihood that residents held multiple coastal place attachments. Past research has repeatedly found that the length of residence affects place attachment (Lewicka, 2011). In this study it was the length of association that was important. Past research has found the duration of visits (lengths of stays) to other places affects place attachment (Hailu et al., 2005) and that higher levels of association result in more frequent visits (Moore & Graefe, 1994). While it is not surprising that people whose families have been associated with East Gippsland for multiple generations (e.g., through the ownership of second homes) have stronger attachments to diverse coastal locations, it is important to consider how strong multi-place attachments exacerbates their vulnerability to sea-level rise.

Policy Implications of Recognising Compound Risks

The distinction between people's preferences for home, coastal or forested/mountainous regions as well as the length and strength of their attachments is important in preparing for sea-level rise and climate change more generally. One of the main policy tools implemented in Australia, and other countries internationally, are planning regulations that target the robustness of existing infrastructure (Tol et al., 2008), such as requiring new buildings and renovations to take into account future sea-level rise predictions (Hurlimann et al., 2014). For people in the most vulnerable category with strong attachments to coastal homes, and few other attachments, such policies have the potential to considerably reduce their vulnerability to sea-level rise. However, for those people who have multiple non-residential attachments to coastal places, such policies are likely to have a limited impact on their compound vulnerability to sea-level rise.

Two other common sea-level rise policy options are protection or retreat (Tol et al., 2008). There has been limited development of protection or retreat adaptation policies in Australia, particularly for small coastal towns. People with strong place attachments to mountainous or forested regions may be more amenable to sea-level rise retreat policies, which not only reduce their compound vulnerability to sea-level rise but enable them to move closer to the environmental features that they value. Yet, retreat policies also need to account for the compound risks that such areas are exposed to from other climate change impacts, such as bushfires. Therefore, retreat policies that account for compound risks not only need to consider the compound risks of one climate change hazard, such as sea-level rise, but the compound risks of multiple climate change hazards. Further research is needed that investigates if and how multi-place attachments can inform the co-design of, as well as the viability and acceptability of retreat policies.

For residents who feel strong emotional bonds to multiple coastal places, protection policies could reduce their compound vulnerability but is unlikely to be implemented for small and remote coastal places, and retreat is likely to result in disconnection, grief, loss, and anxiety arising from disruption to strong place attachments (Agyeman et al., 2009; Hess et al., 2008). This is particularly likely for the moderately vulnerable group of residents, as many have multi-generational ties to places along the East Gippsland coast. Such residents are resistant to evidence that sea-level rise is occurring and/or problematic due to their familiarity with, and acceptance of, current coastal flooding (Fincher et al., 2014). In other similar coastal communities, where adaptation policy options have been presented as a fait accompli, residents with multi-generational place attachments have been found to disengage with, challenge and/or prevent the development of adaptation options (Williams, 2016). It is with these residents that co-produced research and policy, that critically reflects on perceptions and expectations of the coast (McFadden, 2008), may have the greatest impact.

Conclusions

There is an emerging body of geography research that explores the impacts that sea-level rise is likely to have on the things that people value about where they live. Yet, there is a need to understand the compounding effects that sea-level rise, and other impacts of climate change, are likely to have on individuals beyond their homes and neighbourhoods. This is particularly true for residents who live in regional towns and small cities, who regularly spread their activities across multiple places.

This project provides preliminary evidence that understanding the multiple places that people are attached to can reveal unique patterns of compound vulnerability that are not only based on people's sociodemographic characteristics, but also their preference for spending time in the home, at coastal or inland places and the strengths of their associations to such places. While the focus of the data collection and analysis involved recognising the compound risks that sea-level rise poses to people's multi-place attachments, the discussion has highlighted that the compound risks of other climate change impacts also need to be taken into account. Recognising the significance of multiple place attachments to individuals can inform policy responses that are sensitive to the needs of diverse groups who reside in and visit different places. While past research has tended to focus on the role of nested (vertical) place attachments, from the local to the global, here we find that networked (horizontal) place attachments also need to be taken into consideration.

New research is needed that expands existing climate change and place attachment research, to the networks of place attachments. This will require multi-place attachment studies that integrate the currently disparate research on either homes and secondary homes, multiple recreational places, or nested neighbourhood/city/periphery/national/global analyses. Similarly, climate change research would benefit from greater and new engagement with multi-place attachment research. This will require novel spatially diverse methods of data collection and analysis as well as new theories that explain how multi-place bonds emerge and are sustained over time, especially in the face of compounding environmental changes.

Acknowledgements This research was funded through Australian Research Council Linkage Grant LP100100586, with support from the East Gippsland

Shire Council, the Gippsland Coastal Board, the Victorian Department of Planning and Community Development, the Victorian Department of Sustainability and Environment, and Wellington Shire Council. I acknowledge the work of colleagues Professor Jon Barnett, Professor Ruth Fincher, Associate Professor Anna Hurlimann and Dr Colette Mortreux on the broader project of which this paper is part. I would like to thank Chandra Jayasuriya for creating the map. Finally, thank you to all the people who took the time to participate in the interviews.

References

ABS. (2016). *Census of population and housing*. Australian Bureau of Statistics.

Adger, N. W., Barnett, J., Brown, K., Marshall, N., & O'Brien, K. (2013). Cultural dimensions of climate change impacts and adaptation. *Nature Climate Change, 3*(2), 112–117.

Agrawala, S., Ota, T., Ahmed, A. U., Smith, J. A., & van Alast, M. (2003). Development and climate change in Bangladesh: Focus on coastal flooding and the Sundarbans. OECD.

Agyeman, J., Devine-Wright, P., & Prange, J. (2009). Close to the edge, down by the river? Joining up managed retreat and place attachment in a climate changed world. *Environment and Planning A, 41*, 509–513.

Barnett, J., Fincher, R., Hurlimann, A., Graham, S., & Mortreux, C. (2014). *Equitable local outcomes in adaptation to sea-level rise: final project report*. University of Melbourne.

Bonaiuto, M., & Bonnes, M. (1996). Multiplace analysis of the urban environment: A comparison between a large and a small Italian city. *Environment and Behavior, 28*(6), 699–747.

Bonaiuto, M., Bonnes, M., & Continisio, M. (2004). Neighbourhood evaluation within a multiplace perspective on urban activities. *Environment and Behavior, 36*(1), 41–69.

Bonnes, M., Mannetti, L., Secchiaroli, G., & Tanucci, G. (1990). The city as a multi-place system: An analysis of people-urban environment transactions. *Journal of Environmental Psychology, 10*, 37–65.

Brooke, C., & Hennessey, K. (2005). *Climate change impacts in Gippsland*. CSIRO.

Crowell, M., Coulton, K., Johnson, C., Westcott, J., Bellomo, D., Edelman, S., & Hirsch, E. (2010). An estimate of the U.S. population living in 100-year coastal flood hazard areas. *Journal of Coastal Research, 26*(2), 201–211.

Davenport, M. A., & Anderson, D. H. (2005). Getting from sense of place to place-based management: An interpretive investigation of place meanings and perceptions of landscape change. *Society and Natural Resources, 18*, 625–641.

Department of Climate Change. (2009). *Climate change risks to Australia's coast*. Department of Climate Change.

Devine-Wright, P., Price, J., & Leviston, Z. (2015). My country or my planet? Exploring the influence of multiple place attachments and ideological beliefs upon climate change attitudes and opinions. *Global Environmental Change, 30*, 68–79.

Di Masso, A., Williams, D. R., Raymond, C. M., Buchecker, M., Degenhardt, B., Devine-Wright, P., Hertzog, A., Lewicka, M., Manzo, L., Shahrad, A., Stedman, R., Verbrugge, L., & Von Wirth, T. (2019). Between fixities and flows: Navigating place attachments in an increasingly mobile world. *Journal of Environmental Psychology, 61*, 125–133.

Ednie, A. J., Daigle, J. J., & Leahy, J. E. (2010). The development of recreation place attachment on the Maine coast: User characteristics and reasons for visiting. *Journal of Park and Recreation Administration, 28*(1), 36–51.

Eisenhauer, B. W., Krannich, R. S., & Blahna, D. J. (2000). Attachments to special places on public lands: An analysis of activities, reason for attachments, and community connections. *Society and Natural Resources, 13*(5), 421–441.

Esteban-Guitart, M., Monreal-Bosch, P., & Vila, I. (2013). A qualitative study on transnational attachment among eight families of foreign origin. *Psyecology, 4*(3), 245–266.

Fincher, R., Barnett, J., Graham, S., & Hurlimann, A. (2014). Time stories: Making sense of futures in anticipation of sea-level rise. *Geoforum, 56*, 201–210.

Forest Fire Management Victoria. (2020). Gippsland Fires—Fire Extent 2019/2020. Department of Sustainability and Environment.

Graham, S., Barnett, J., Fincher, R., Hurlimann, A., Mortreux, C., & Waters, E. (2013). The social values at risk from sea-level rise. *Environmental Impact Assessment Review, 41*, 45–52.

Graham, S., Barnett, J., Fincher, R., Hurlimann, A., & Mortreux, C. (2014). Local values for fairer adaptation to sea-level rise: A typology of residents and their lived values in Lakes Entrance, Australia. *Global Environmental Change, 29*, 41–52.

Graham, S., Barnett, J., Mortreux, C., Hurlimann, A., & Fincher, R. (2018). Local values and fairness in climate change adaptation: Insights from marginal rural Australian communities. *World Development*.

Gustafson, P. (2006). Place attachment and mobility. In N. McIntyre, D. R. Williams, & K. E. McHugh (Eds.), *Multiple dwelling and tourism: Negotiating place, home and identity* (pp. 17–31). CAB International.

Hailu, G., Boxall, P. C., & McFarlane, B. L. (2005). The influence of place attachment on recreation demand. *Journal of Economic Psychology, 26*, 581–598.

Hess, J. J., Malilay, J. N., & Parkinson, A. J. (2008). Climate change. The importance of place. *American Journal of Preventive Medicine, 35*, 468–478.

Hiltunen, M. J. (2007). Environmental impacts of rural second home tourism—case lake district in Finland. *Scandinavian Journal of Hospitality and Tourism, 7*(3), 243–265.

Hurlimann, A., Barnett, J., Fincher, R., Osbaldiston, N., Mortreux, C., & Graham, S. (2014). Urban planning and sustainable adaptation to sea-level rise. *Landscape and Urban Planning, 126*, 84–93.

Köpsel, V., Walsh, C., & Leyshon, C. (2017). Landscape narratives in practice: Implications for climate change adaptation. *The Geographical Journal, 183*(2), 175–186.

Kulp, S. A., & Strauss, B. H. (2019). New elevation data triple estimates of global vulnerability to sea-level rise and coastal flooding. *Nature Communications, 10*, 4844.

Le Roy, J., & Rioux, L. (2012). The mediating role of workplace attachment in the relationship between organizational commitment and organizational citizenship behavior. *International Review of Social Psychology, 25*(3), 211–233.

Lewicka, M. (2011). Place attachment: How far have we come in the last 40 years? *Journal of Environmental Psychology, 31*, 207–230.

McFadden, L. (2008). Exploring the challenges of integrated coastal zone management and reflecting on contributions to "integration" from geographical thought. *Geographical Journal*, *174*(4), 299–314.

McGranahan, G., Balk, D., & Anderson, B. (2007). The rising tide: Assessing the risks of climate change and human settlements in low elevation coastal zones. *Environment and Urbanization, 19*(1), 17–37.

McIntyre, N., Williams, D. R., & McHugh, K. E. (2006). *Multiple dwelling and tourism: Negotiating place, home and identity*. CAB International.

Moore, R. L., & Graefe, A. R. (1994). Attachments to recreation settings: The case of rail users. *Leisure Sciences, 16*(1), 17–31.

Novaczek, I., MacFadyen, J., Bardati, D., & MacEachern, K. (2011). *Social and cultural values mapping as a decision-support tool for climate change adaptation*. University of Prince Edward Island.

O'Neill, S. J., & Graham, S. (2016). (En)visioning place-based adaptation to sea-level rise. *Geo: Geography and Environment*, *3*(2).

Schiavo, R. S. (1988). Age differences in assessment and use of a suburban neighborhood among children and adolescents. *Children's Environments Quarterly, 5*(2), 4–9.

Stedman, R. C. (2003). Is it really just a social construction? The contribution of the physical environment to sense of place. *Society and Natural Resources, 16*(8), 671–685.

Symons, J., Jones, R., Milne, R., & Macleod, A. (2016). *Economic geography of bushfire and flood vulnerability in Victoria at the statistical local area scale*. Victoria University.

Tol, R. S. J., Klein, R. J. T., & Nicholls, R. J. (2008). Towards successful adaptation to sea-level rise along Europe's coasts. *Journal of Coastal Research, 242*, 432–442.

Ujang, N. (2012). Place attachment and continuity of urban place identity. *Procedia - Social and Behavioral Sciences, 49*, 156–167.

Uyarra, M. C., Côté, I. M., Gill, J. A., Tinch, R. R. T., Viner, D., & Watkinson, A. R. (2005). Island-specific preferences of tourists for environmental features: Implications of climate change for tourism-dependent states. *Environmental Conservation, 32*(1), 11–19.

VIC Emergency. (2021). *Incidents and warnings*. https://emergency.vic.gov.au/respond. Accessed 12 July 21.

Williams, R. D. (2016). *Community resilience and adaptation to coastal change: A case study from Middle Beach, South Australia*. CSIRO.

CHAPTER 8

A Health Geography of the Compound Effects of Contaminated Sites and Extreme Weather Events on Mental Health

Rupert Legg

Introduction

A prominent outcome of climate change is the expected increase in the frequency of natural disasters, such as extreme weather events. Extreme weather events "*are meteorological, hydrological, climatological, or related incidents that cause widespread damage, either in terms of human lives, property damage or both*", such as floods and droughts (Simpson, Weissbecker, & Sephton, 2011, p. 58). The Intergovernmental Panel on Climate Change (IPCC) has predicted that a global temperature rise of 1.5 °C would accompany an increase in frequency, intensity and amount of heavy precipitation events globally, whilst drought events are likely to

R. Legg (✉)
Institute for Sustainable Futures, University of Technology Sydney, Ultimo, NSW 2007, Australia
e-mail: rupert.legg@uts.edu.au

A. Lukasiewicz and T. O'Donnell (eds.), *Complex Disasters*, Disaster Risk, Resilience, Reconstruction and Recovery,
https://doi.org/10.1007/978-981-19-2428-6_8

be longer in the most vulnerable regions (IPCC, 2018). For the purposes of this chapter, the impacts of extreme weather events are considered a kind of disaster. Disaster researchers have long proposed that the impacts of such disasters will not be dispersed equally, with disadvantaged communities often located in areas where extreme weather events are expected to increase most dramatically (Tuitjer, 2021; Watts et al., 2018). For health geographers, particularly, this has manifested in an interest with how vulnerability and resilience in relation to such extreme weather events affects health inequality (Curtis & Oven, 2012; Cutchin, 2007).

At the same time, environmental contamination is fast becoming "*a ubiquitous, if uneven, condition*" (Liboiron et al., 2018, p. 331). Unregulated and unmitigated chemical emissions over the past several decades have resulted in many hundreds of thousands of sites around the world being contaminated by heavy metals alone (Landrigan et al., 2018). Environmental contamination has been referred to as a "*creeping*" (Bleicher & Groß, 2013) or "*slow-onset*" disaster (Cline et al., 2010), and a kind of "*slow violence*" (Davies, 2022; Nixon, 2011) in reference to the gradual and persistent effects it has on the environment and humans who live nearby. Such conceptualisations also incorporate the possibility that the contamination can suddenly "*burst*" into public consciousness at times throughout its duration (Bleicher & Groß, 2013). As Edelstein (2018, p. 132) notes, this occurs during "*periods of crisis involving new discoveries, challenges, announcements, conflicts, and decisions*". Much environmental justice research has documented that contamination tends to occur in vulnerable and disadvantaged areas, often those placed near industrial activity (Downey & Van Willigen, 2005; Gee & Payne-Sturges, 2004; Pastor et al., 2001; Zhang, 2010). As with extreme weather events, then, it is unlikely that the health effects of environmental contamination will occur evenly across society.

Given these global trends, it is unsurprising that the co-occurrence of environmental contamination with extreme weather events is starting to increase (Göransson, 2016; Sandifer & Walker, 2018). Broadly, such compound disasters, defined by Gissing et al. (2022, p. 2) as "*more than one disaster event, occurring contemporaneously or within a short timeframe*" (p. 2), have also been observed to be rising in frequency (Gissing et al., 2022). There is every indication that these trends will continue into the future, particularly for contamination and extreme weather events. For instance, as industrial activity is predominantly concentrated in areas near waterbodies and on floodplains, the likelihood of contaminated sites being flooded increases (Göransson, 2016). Likewise, as contamination is a slow-onset disaster, which often takes many years to remediate, the chance of another catastrophe occurring during this period is only increasing as climate change makes disasters more frequent. Adding to this, emerging contaminants, such as per- and poly-fluoroalkyl substances (PFAS), do not break down easily, are stubborn to remediation and thus remain in the environment for a long time. As Sandifer and Walker (2018) emphasise, disasters are changing and this requires preparing for new and complicated impacts.

Despite this, risk is often assessed on the assumption that disasters happen individually (Gissing et al., 2022) and there has been relatively little investigation of compound disasters in the literature. This chapter aims to address this by asking what the impacts of compound disasters, in this case environmental contamination and extreme weather events, are on mental health. Further, it also questions whether these impacts are purely additive, or whether the interaction of each individual disaster compounds to produce its own effect. First, literature examining extreme weather events, environmental contamination and their impact on mental health is examined. Second, a framework for evaluating how compound disasters affect mental health and wellbeing is proposed. After this, the research background and method for this chapter are outlined, followed by a presentation of the results. The chapter finishes with a discussion of how compound disasters affect mental health and with some conclusions for disaster management.

Extreme Weather Events, Environmental Contamination and Mental Health

There has been much recent research exploring the connection between various disasters and mental health (see, for instance, North and Pfefferbaum [2013] for a review). In regards to extreme weather events, numerous studies have found that mental health is adversely affected by floods, with psychological distress, depression, anxiety and trauma all more common amongst survivors (Cruz et al., 2020; Shukla, 2013; Simpson et al., 2011). A review of such literature reported that long-term mental ill-health was most strongly associated with an individual's extent of exposure to the flood, and the amount of damage to physical infrastructure and the home (Zhong et al., 2018). Similarly, droughts have been observed to correlate with depression and poor wellbeing for farmers and rural residents (Edwards et al., 2015; O'Brien et al., 2014). The strongest associated factors include the duration of the drought and its economic and migratory effects (Vins et al., 2015). Depending on the severity of the extreme weather event, then, the mental health of those exposed is likely to suffer.

There has been less investigation of the mental health and wellbeing effects of living on or near contaminated sites, although such research has increased over the past few years. A recent systematic review found consistent evidence that the experience of living on or near environmental contamination has a small-to-moderate effect on anxiety, stress and depression (Schmitt et al., 2021). Building on this review, Sullivan et al. (2021) reported that psychological distress is exacerbated by residents' perception of the risk the contamination poses to their physical health and the extent to which they feel abandoned or ignored by government. Whilst there are similarities to the mental health effects of extreme weather events, then, some differences are also worth emphasising.

At a surface level, Picou (2009) suggests that a difference lies in the environment that each type of disaster affects and the duration: natural disasters tend to damage the built environment temporarily, whilst some technological disasters affect the biophysical environment long term. Edelstein (2018) proposes that these long-lasting effects of technological disasters, such as environmental contamination, make them more likely to lead to adverse psychological effects. Further, there is usually political and social consensus around what causes natural disasters, and how victims should be supported. On the other hand, technological disasters often result in disputes as there are many parties responsible for the regulation, use, containment and control of hazardous substances. This can lead to blame and obfuscation, which exacerbate the psychological distress of victims (Edelstein, 2018). Schmitt et al. (2021), considering this particular element of technological disasters, proposed that contamination will likely have a larger effect on depression than natural disasters as a result. Considering these different characteristics of each disaster, it is unclear how their compounding together may shape mental health.

Some research on natural and technological disasters occurring together, or "*natech*" disasters (Picou, 2009), can contribute to answering this question. Natech disasters are technological disasters that are initiated or triggered by a natural disaster. Matsumoto et al. (2014), in an examination of the co-occurring tsunami and nuclear disasters at Fukushima, observed that the disasters combined "*the direct psychological shock of a natural disaster with the stress of persistent anxiety of sustained exposure to radioactive contamination*" (p. 186). Likewise, Picou (2009) reported that survivors of Hurricane Katrina who perceived they had been exposed to toxic chemicals as a result of flooding spreading contamination across the environment displayed higher levels of worry than survivors who did not perceive they had been exposed. Contrastingly, a study by Solomon et al. (1989) found that those who were exposed to dioxin contamination as a result of flooding in St Louis in 1982, exhibited lower anxiety levels than those purely exposed to the flood. It was theorised that this was a result of blame: flood victims blamed themselves, whereas those exposed to the contamination blamed government and industry, which seemed to result in less adverse psychological symptoms. This is contrasting to the evidence examined previously on environmental contamination, suggesting there is a need for closer study of how compound disasters may affect mental health.

Whilst there has been some research examining natech disasters involving environmental contamination, then, there has been less research examining a disaster that co-occurs with contamination but has a separate cause. In light of this, it is worth distinguishing between natech disasters and compound disasters. Compared to natech disasters, which are brought about by a technological disaster, compound disasters can involve disasters of different causes and they may occur within a short timeframe, not necessarily simultaneously. This distinction is important, as it is possible that a compound disaster can occur when a community is trying to recover from a previous disaster and this can have distinct effects as a result, especially for mental health. The co-occurring disasters explored in this chapter, contamination and extreme weather events, can be considered compound disasters because of their separate causes. The contamination was not caused by the extreme weather, instead existing beforehand. Examining compound disasters such as these could also allow greater unravelling of the separate effects of the different disasters and determining of whether or not their interaction acts to produce effects of their own. As Pain (2021) suggests in her analysis of the spatial and temporal aspects of trauma, "*different traumas often work not in isolation but layer up and accumulate, not simply adding to the intensity of trauma but enabling interaction between traumas over time*" (p. 7). Unravelling such layering and accumulation requires a multiscalar and spatial perspective of disasters and mental health. It is proposed in this chapter that a health geography framework is capable of doing this.

Conceptualising a Geography of Compound Disasters and Mental Health

Health geography is a useful lens to approach compound disasters. As Curtis and Oven (2012) suggest, health geographers pay particular attention to the scalar and spatial processes of a disaster and how they interact with society and human activities to produce variable effects. That is, they emphasise that people's relationships with the places around them, at various scales, constantly shape whether they have an enabling or disabling effect on their health (Kearns & Moon, 2002). In relation to environmental contamination, then, health geographers have considered how the event disrupts people's relationship with their surroundings. For instance, Prior et al. (2019), in building off Edelstein's (2002) theoretical framework of the "*lifescape*", propose that contamination disrupts people's relationships with their health, self, home, environment and society. These lifescape disruptions are inherently stressful and they adversely affect mental health as long as they persist (Edelstein, 2018).

Expanding on this conceptualisation, health geographers have focused on individual scales within the lifescape. This involves demonstrating how a changing relationship to the environment or home, for instance, may affect mental health. Smith, Luginaah, and Lockridge (2010) determined that natural environments that were held sacred to First Nations people in Ontario, Canada, transitioned from having therapeutic benefits to being actively harmful when contaminated. Likewise, Prior et al. (2019) discussed residents' descriptions of their homes changing from focusing on notions of safety and relaxation to fear and danger after the presence of contamination in the surrounding environment became known. In both these cases, the authors emphasise that mental health and wellbeing are not only being adversely affected by the contamination, but also through residents' deteriorating relationships with the places that once supported their mental health. The lifescape as a framework, then, allows health geographers to examine how place and spatial processes, across different scales, interact to produce psychological outcomes.

Beyond environmental contamination, some disaster research has also considered the lifescape in examining effects on mental health. Straub et al. (2020), in reference to areas prone to floods, earthquakes and wildfires, state that *"the ever-present threat of so many diverse and potentially devastating natural hazards alters the 'lifescape' of residents and those charged with managing natural hazard events when they occur"* (p. 107).

Again, it is the disruption to residents' relationships to the world around them that produces psychological effects. Whilst Straub et al. (2020) describe this in regard to natural disasters, it is important to emphasise that some disaster researchers suggest that lifescape disruptions are unique to technological disasters. Edelstein (2018) indicates that all disasters may disrupt the lifescape to some extent, but that technological disasters have the capacity to violate people's everyday assumptions about life. Gill et al. (2016) expand on this suggesting that loss of trust, prolonged dread and uncertainty are elements that create the greatest lifescape change and are most commonly present in technological disasters. It is worth examining whether compound disasters follow a similar pattern. As such, this health geography framework focusing on the lifescape is capable of analysing how contaminated land and extreme weather events occurring together may shape residents' mental health.

Method

This chapter applies this framework to two compound disasters: the long-term environmental contamination of two sites—Williamtown and Richmond—and the occurrence of floods and droughts during the contaminated period. The contamination resulted from the decades-long use of firefighting foams by the Australian Defence Force (Defence) on their bases at Williamtown and Richmond—both semi-rural suburbs are located adjacent to these bases (Department of Defence, 2015). Contained within the firefighting foams were PFAS, the chemicals that have recently gathered global attention for their likely carcinogenic properties and high durability in the environment (Bavas & Davoren, 2019). Through Defence practising putting out fires since the 1970s, the PFAS has seeped into groundwater and travelled into the residential areas (Department of Defence, 2015). There are a number of routes through which nearby residents could be exposed to the chemicals, such as through dust, air, food and water. At Williamtown, residents were notified about the presence of the contamination in 2015 (Taylor & Cosenza, 2016), whilst the Richmond residents were notified in 2018 (Calderwood, 2018). Remediation at both sites is ongoing, and due to the high persistence of PFAS, no dates have been provided for when the environment will be contaminant free.

Richmond was in the middle of experiencing a drought when the PFAS contamination was announced (Duffin, 2018). As Richmond is semi-rural, almost all interviewees had daily physical contact with their local environment, either as farmers or hobby gardeners. Likewise, both Williamtown and Richmond are located on floodplains. In February 2020, heavy rainfall in Richmond led to the water level of the Hawkesbury River rising over its bank, causing the surrounding floodplains to be submerged in water (Falson, 2020). A similar rainfall event in the Hunter Region in July 2020 caused flash floods in Williamtown, resulting in its floodplains being similarly inundated (Bielby, 2020). Both floods caused evacuations, road closures and property damage.

To explore how these compound disasters relate to mental health, 12 interviews with residents across the two sites, two with officials from local councils and one with a regulator from the NSW State Government, conducted between August 2020 and July 2021, are drawn upon. Residents were aware of the PFAS contamination and were knowledgeable of the potential health effects the chemicals could have if they had been exposed, whilst the government officials were involved in shaping the overall management response to the contamination through participation in expert panel meetings. The interviews with residents covered a range of topics, including their mental health, how they had been affected by living near environmental contamination, what events were perceived to affect their mental health, such as droughts or floods, and what actions they implemented to minimise exposure. Interviews with regulators and government officials, on the other hand, focused on what was being done or what could be done to mitigate the potential adverse mental health effects experienced by residents and the overall regulatory approach to the contaminated sites. Ethical approval was obtained from the University of Technology Sydney's Ethics Department. Newspaper articles and documents from the *Inquiry into the management of PFAS contamination in and around Defence bases* by the Joint Standing Committee on Foreign Affairs, Defence and Trade (2018) are also drawn upon as supplementary evidence at various stages.

To unravel how residents' mental health was affected by their experience of the compound disasters, the self-reported effects reported by residents were analysed, particularly with reference to the emotive or affective terms they used to describe their mental state. Qualitative methods such as this are capable of unravelling residents' experiences and views (Cutchin, 1999), which is particularly relevant when evaluating mental health impacts (Crowe et al., 2015). As Ladd et al. (2007,

p. 53) state, "*employing narrative accounts, testimonies, and oral histories in research can contribute immensely to a greater multidimensional understanding of how disasters can not only disrupt survivors' lifescapes, but also diminish trust, generate feelings of recreancy, create resource loss, and produce psychological stress and trauma*". The next section explores these experiences and narratives through analysis of the interviews.

Results

The interview results are reported here in regard to how the compound disasters were reported by residents to affect their mental health, and also acted to shift the way they perceived their home, local environment and society. First, how the contamination solely affected mental health is examined. Afterwards, the psychological outcomes of the compound disasters are explored.

Contamination and Mental Health

Throughout the interviews, residents described experiencing poor mental health because of the contamination. This began the moment they found out about the contamination, with interviewees describing the sudden and immediate psychological impact they felt at the onset. Residents used words like "*shock*" and "*alarm*" to describe their reaction to finding out about the contamination, and many referred to a physical manifestation of this, such as a sinking of the stomach. After this initial period, residents' descriptions of their mental health were more focused on their cognitions and the unceasing and ever-present nature of the stress they were experiencing. For instance, several residents suggested that their current state was constant and that these feelings did not go away:

> *I just get really anxious . . . and angry all the time and I can't cope with it.*

These psychological outcomes accompanied changes to the way residents related to the various dimensions of their lifescape. For instance, it was common for residents to report how the contamination changed their relationship to their home. Particularly, residents described their homes before they found out about the contamination with words like "*peace*" and "*comfort*", whereas afterwards their language shifted. Words like "*prison*" and "*trapped*" were common, indicating the home was now an uncomfortable place for residents to spend time:

> *I do not want to be at home . . . my daughter doesn't spend a lot of time here; she'll stay at a friend's place. My husband, well, he works away a lot . . . There's no interest in this place at all.*

Similarly, residents described the environment as "*beautiful*" and "*quiet*" prior to the contamination. Afterwards, it became a source of worry, with interviewees concerned about how nature would be impacted and if the chemicals could be remediated.

Finally, there was also a considerable sense of disillusionment, predominantly towards the government, but also some other community members. For instance, there was incredulity at what the intentions of the government were:

> *The whole purpose of what they [the government] were doing was to limit their liability. That's all it was. It was to limit liability. It had nothing to do with fixing the problem or helping the people.*

Others mentioned getting into disagreements with other residents and being told to stop drawing attention to the contamination. Residents described these occurrences as angering and saddening them, respectively. In summary, the contamination shifted residents' lifescape considerably, adversely affecting their mental health and wellbeing.

Contamination, Extreme Weather Events and Mental Health

Whilst these mental health and wellbeing effects seemed to remain relatively constant whilst residents lived in the contaminated region, many made references to times when their stress was elevated. Particularly, a drought in Richmond, and the floods in both Richmond and Williamtown, were focal points in a number of the interviews conducted.

The drought at Richmond prior to the announcement of the contamination was predominantly a concern for farmers. As one farmer described:

> *Well, it was actually a really difficult time already because we were in the drought. And things were really, really quite desperate. So, things were really bad already, and then to read about the contamination was just devastating.*

However, the drought also contributed to residents' concerns about the contamination more generally. As the topsoil was left dry and vulnerable, residents perceived there was an increased likelihood of them being exposed to the contaminated dirt and dust:

> *When it's windy, it [the PFAS] is in the dust, and we're inhaling it and we know this. And Defence is going ahead with the expansion of their base and that just stirs up even more dust that we get covered in. It's an absolute living hell, right here, right now.*

Beyond these concerns relating to the body and exposure, the drought and its associated dust was also perceived to penetrate the home. Interviewees described being concerned about dust in their houses, despite the government advising that exposure from dust was unlikely to be a significant problem.

In this sense, the environmental effects of the drought not only brought the contamination front of mind, it also worsened residents' relationships with their homes. The drought and contamination, then, compounded together, exacerbating residents' already negative relationships with their homes.

Adding to this, the drought also disrupted the financial situation of farmers and those dependent on the land, indirectly affecting mental health. For instance, farmers described how the financial effects of the contamination and drought combined contributing to their poor mental health:

> *We had nothing. We were broke because we had already been through the drought, and then this happened . . . we had to pay thousands of dollars to put the cattle on town water . . . In the end, we had to sell them because we had no money and our business had been destroyed . . . Because it was a drought, the cattle were sold off very cheaply. We lost everything for very little money.*

These financial effects stemmed not only from the steps taken to avoid cattle and livestock being exposed to contaminated water, but also from the relatively low remuneration farmers received when selling their product. Interviewees speculated that the combination of drought and exposure to PFAS resulted in their cattle and produce being unhealthy. Consequently, this combining of the different disasters rendered residents more vulnerable to the individual impacts of each disaster and created compounded psychological effects.

The floods at Richmond and Williamtown also led to elevated periods of poor mental health. In a similar manner to the drought, the environmental effects of the floods brought the contamination front of mind, often exacerbating mental health outcomes. For instance, one resident described how when it floods, "*I feel sick because I can see the water coming out of the ground, and know that it's contaminated*". Leading on from this, a focus of concern around the floods was on how it shaped the environment and whether it would make it more likely that residents could be exposed to the contamination:

> *[The flood] worried me about how much bigger the plume may have been. I don't understand the plume or how it spreads but I do know there were several major floods before that were several metres high. I want them to clean it up so that it doesn't spread if there is another flood. I want them to do a remediation and clean up wherever they can.*

However, the floods also shaped residents' lifescape in ways different to that of the drought. There were no mentions of how the inside of residents' homes might be affected by floods in the interviews, perhaps because the floodwaters did not actually enter homes. However, in the *Inquiry into the management of PFAS contamination in and around Defence bases* residents described flood waters accumulating on people's properties for months as a result of the poor drainage leading off the Defence base at Williamtown (Joint Standing Committee on Foreign Affairs, Defence and Trade, 2018). This had further ramifications. Particularly, residents in the interviews blamed the government for contributing to the spreading of the contamination, despite the floods being natural in origin. There was often a sense that more could be done by the government to prevent the flooding:

> *I don't think Defence* [the federal government] *even knew that this area used to flood. There are seasonal floods. Just because you put the dam walls up and it hasn't flooded since 1992 doesn't mean it won't do it again. Then, this year it finally flooded. Defence didn't even know there was a risk of flood, they just assumed it doesn't flood anymore. After the flood I was worried that the contamination could spread.*

The regulators interviewed were aware that this was a considerable source of concern for residents and recognised that the nature of the residential areas, in floodplains, meant they were more likely to flood:

> *When there's a flood it's drawn back up into the surface through the pressure and that spreads even more. There's a couple of major drains that flow off the base towards the waterways and towards or over people's properties. This all meant there was a lot of frustration from the community.*

However, there was a sense amongst the various levels of the government that this flooding was inevitable and that nothing could be done to prevent it (Joint Standing Committee on Foreign Affairs, Defence and Trade, 2018). These mismatched perspectives only lead to a further sense of abandonment, exacerbating residents' anger at the government. As one resident described:

> *Living near the contamination was crap, but nowhere near as bad as what it's been like living with the disgusting way that we've been treated.*

As with the drought, then, the floods made people more aware of the contamination and caused them to worry more about it, adversely affecting their mental health. However, these mental health effects were further compounded by the blame and abandonment that resulted.

Discussion

These results indicate that the adverse effects of living near contamination on residents' self-reported mental health and wellbeing are compounded by extreme weather. Particularly, the extreme weather events appeared to elevate residents' concerns about their bodies, homes, the environment and government, further disrupting the way they think about and relate to each other. These changes to the lifescape of residents undoubtedly contributed to the severity of the mental health and wellbeing impacts. Such findings align with other disaster research examining the co-occurrence of disasters that finds the impacts of both combine to produce stronger effects (Matsumoto et al., 2014; Picou, 2009; Solomon et al., 1989).

Of particular interest from these results was the difference between how the drought and floods interacted with the contamination. Concerns about the drought were more related to financial difficulties and exposure to the PFAS chemicals via dust, whereas the floods created greater acrimony between residents and government, whilst there was still concern about exposure occurring through interaction with floodwaters. Given the nature of the qualitative data, it is difficult to determine which form of extreme weather had a greater effect on mental health; however, the level of disillusionment all residents interviewed felt towards the government suggests that anything that contributes to this is likely to have considerable psychological impact. In light of this, as the floods exacerbated residents' anger at the government about the contamination, it is possible they had a greater influence. This aligns with other research that finds that when a sense of blame is present, mental health outcomes tend to be worse (Mason et al., 2010; Schmitt et al., 2021; Sullivan et al., 2021).

Edelstein's (2018) concept of the lifescape can further illuminate the ways mental health effects, as self-reported by residents, differed between the compound disasters. The drought seemingly exacerbated the effects of the contamination across the scales of the body and home. For instance, residents tended to describe how the perceived increased presence of dust in the home added to their psychological distress, as it increased the likelihood of them being exposed to the PFAS chemicals. The floods, on the other hand, intensified the impacts of the contamination at other scales of residents' lifescapes. For instance, residents' concerns about possible exposure to the PFAS chemicals were largely concentrated in the local environment, as opposed to the home, as this was where the contaminated water pooled and posed a threat of exposure. Further, residents described how their perceptions of broader society, particularly the role the state plays in protecting its citizens, were damaged by the floods, leading to feelings of abandonment. In this sense, the extreme weather events altered the way residents experienced the contamination, causing different components of the accompanying lifescape disruptions to be exacerbated.

The benefits of this health geography approach, then, are that the effects of the compound disasters can be unravelled, revealing that the psychological outcomes, as reported by residents, were altered and transformed by the disasters compounding. That is, the compounding of the disasters likely influenced mental health in and of itself because of the way the disasters become entangled. For instance, other research has found that blame and anger are very uncommon responses to floods (Zhong et al., 2018). Yet, the presence of blame here indicates an entangling of the compound disasters. This entangling altered the mental health effects reported by residents, seemingly generating new and additional stress beyond the sole impacts of a single disaster. Reflecting further on the disconnect between regulators' and residents' views on the risks of extreme weather events, it is apparent that more could be done to alleviate residents' concerns. Whilst a small increase in the risk of floods may seem like just that to government officials, to residents it represents a whole lot more. As was clear from residents' descriptions of the impacts of the floods, even a small increase in the chance of a flood could carry a considerable mental health impact. Aligning with other recent research on compound disasters, then, the need for those involved in disaster risk reduction to consider the possibility other disasters may change the nature of an initial disaster is necessary (Cutter, 2018).

Conclusion

Given the likelihood that compound disasters will only increase in occurrence in the future, it is essential that their impacts are better understood. This research has contributed to this, determining that compound disasters can become entangled and produce their own effects on residents' mental health, disrupting their lifescape in unique ways. It is currently not common in discussions of how to manage contaminated sites to include measures to address the elevated psychological distress that emerges from this extreme weather. In fact, it is rare for them to consider that this distress will likely vary over time and have periods of elevation. It is critical that those managing environmental hazards consider the possibility that they become compounded by different events and have plans in place to protect residents from these more detrimental mental health effects.

A limitation of this chapter was its focus on individual cases, indicating that care should be taken when generalising these results. A further limitation was that the interviews were conducted some months after the occurrence of the extreme weather, whereas it would have been ideal to conduct interviews both before and after their impacts were felt. Future research could adopt such a methodology and examine the impacts of other kinds of natural disaster compounding with environmental contamination, such as earthquakes or cyclones.

References

Bavas, J., & Davoren, H. (2019, October 22). What is being done about the risk of PFAS contamination? *ABC News*. Retrieved from https://www.abc.net.au/news/2019-10-22/pfas-explainer/11623772

Bielby, N. (2020, July 28). SES gets hundreds of calls for help. *The Newcastle Herald*. Retrieved from https://www.newcastleherald.com.au/story/6851715/dramatic-rescues-and-hundreds-of-calls-for-help-amid-flash-floods/

Bleicher, A., & Groß, M. (2013). Response and recovery in the remediation of contaminated land in Eastern Germany. In B. Allen, & R. A. Dowty Beech (Eds.), *Dynamics of Disaster* (pp. 207–222). Routledge.

Calderwood, K. (2018, November 7). PFAS contamination near Richmond Air Base prompts egg and meat consumption warning. *ABC News*. Retrieved from https://www.abc.net.au/news/2018-11-07/pfas-contamination-warning-near-richmond-raaf-base/10473326

Cline, R. J., Orom, H., Berry-Bobovski, L., Hernandez, T., Black, C. B., Schwartz, A. G., & Ruckdeschel, J. C. (2010). Community-level social support responses in a slow-motion technological disaster: The case of Libby, Montana. *American Journal of Community Psychology, 46*(1), 1–18.

Crowe, M., Inder, M., & Porter, R. (2015). Conducting qualitative research in mental health: Thematic and content analyses. *Australian & New Zealand Journal of Psychiatry, 49*(7), 616–623.

Cruz, J., White, P. C., Bell, A., & Coventry, P. A. (2020). Effect of extreme weather events on mental health: A narrative synthesis and meta-analysis for the UK. *International Journal of Environmental Research and Public Health, 17*(22), 1–17.

Curtis, S. E., & Oven, K. J. (2012). Geographies of health and climate change. *Progress in Human Geography, 36*(5), 654–666.

Cutchin, M. P. (1999). Qualitative explorations in health geography: Using pragmatism and related concepts as guides. *The Professional Geographer, 51*(2), 265–274.

Cutchin, M. P. (2007). The need for the "new health geography" in epidemiologic studies of environment and health. *Health & Place, 13*(3), 725–742.

Cutter, S. L. (2018). Compound, cascading, or complex disasters: What's in a Name? *Environment: Science and Policy for Sustainable Development, 60*(6), 16–25.

Davies, T. (2022). Slow violence and toxic geographies: 'Out of sight' to whom? *Environment and Planning C: Politics and Space, 40*(2), 409–427.

Department of Defence. (2015). *RAAF Base Williamtown: A history of AFFF*. Retrieved from https://www.defence.gov.au/Environment/PFAS/docs/Williamtown/Factsheets/RAAFWilliamtownStage2FactSheet14Sep15.pdf

Downey, L., & Van Willigen, M. (2005). Environmental stressors: The mental health impacts of living near industrial activity. *Journal of Health and Social Behavior, 46*(3), 289–305.

Duffin, P. (2018, August 8). Drought bites 100 per cent of NSW. *The Hawkesbury Gazette.* Retrieved from https://www.hawkesburygazette.com.au/story/5574051/drought-bites-100-per-cent-of-nsw/

Edelstein, M. R. (2002). Contamination: The invisible built environment. In R. B. Bechtel & A. Churchman (Eds.), *Handbook of environmental psychology* (pp. 559–588). Wiley, J.

Edelstein, M. R. (2018). *Contaminated communities: Coping with residential toxic exposure.* Routledge.

Edwards, B., Gray, M., & Hunter, B. (2015). The impact of drought on mental health in rural and regional Australia. *Social Indicators Research, 121*(1), 177–194.

Falson, S. (2020, February 10). Richmond, Windsor bridges flooded for first time in 28 years as Hawkesbury lashed with rain and wind. *The Hawkesbury Gazette.* Retrieved from https://www.hawkesburygazette.com.au/story/6622342/richmond-windsor-bridges-submerged-for-first-time-in-28-years/

Gee, G. C., & Payne-Sturges, D. C. (2004). Environmental health disparities: A framework integrating psychosocial and environmental concepts. *Environmental Health Perspectives, 112*(17), 1645–1653.

Gill, D. A., Ritchie, L. A., & Picou, J. S. (2016). Sociocultural and psychosocial impacts of the Exxon Valdez oil spill: Twenty-four years of research in Cordova, Alaska. *The Extractive Industries and Society, 3*(4), 1105–1116.

Gissing, A., Timms, M., Browning, S., Crompton, R., & McAneney, J. (2022). Compound natural disasters in Australia: a historical analysis. *Environmental Hazards, 21*(2) 159–173.

Göransson, G. (2016). *Multi-hazard: Contaminated land vulnerable to natural hazards and effects of climate change.* Paper presented at the 17th Nordic Geotechnical Meeting, NGM 2016, Reykjavik, Iceland, 25–28 May, 2016.

IPCC. (2018). *Global warming of 1.5°C: An IPCC special report on the impacts of global warming of 1.5°C above pre-industrial levels and related global greenhouse gas emission pathways, in the context of strengthening the global response to the threat of climate change, sustainable development, and efforts to eradicate poverty*: Intergovernmental Panel on Climate Change.

Joint Standing Committee on Foreign Affairs, Defence and Trade. (2018). *Inquiry into the management of PFAS contamination in and around Defence bases.* Retrieved from Canberra, ACT: https://parlinfo.aph.gov.au/parlInfo/download/committees/reportjnt/024207/toc_pdf/InquiryintothemanagementofPFAScontaminationinandaroundDefencebases.pdf;fileType=application%2Fpdf

Kearns, R., & Moon, G. (2002). From medical to health geography: Novelty, place and theory after a decade of change. *Progress in Human Geography, 26*(5), 605–625.
Ladd, A. E., Gill, D. A., & Marszalek, J. (2007). Riders from the storm: Disaster narratives of relocated New Orleans college students in the aftermath of Hurricane Katrina. *Journal of Public Management and Social Policy, 13*(2), 51–80.
Landrigan, P. J., Fuller, R., Acosta, N. J. R., Adeyi, O., Arnold, R., Basu, N., … Zhong, M. (2018). The Lancet Commission on pollution and health. *The Lancet, 391*(10119), 462–512.
Liboiron, M., Tironi, M., & Calvillo, N. (2018). Toxic politics: Acting in a permanently polluted world. *Social Studies of Science, 48*(3), 331–349.
Mason, V., Andrews, H., & Upton, D. (2010). The psychological impact of exposure to floods. *Psychology, Health & Medicine, 15*(1), 61–73.
Matsumoto, J., Kunii, Y., Wada, A., Mashiko, H., Yabe, H., & Niwa, S.-I. (2014). Mental disorders that exacerbated due to the Fukushima disaster, a complex radioactive contamination disaster. *Psychiatry and Clinical Neurosciences, 68*(3), 182–187.
Nixon, R. (2011). *Slow violence and the environmentalism of the poor*. Harvard University Press.
North, C. S., & Pfefferbaum, B. (2013). Mental health response to community disasters: A systematic review. *JAMA, 310*(5), 507–518.
O'Brien, L. V., Berry, H. L., Coleman, C., & Hanigan, I. C. (2014). Drought as a mental health exposure. *Environmental Research, 131*, 181–187.
Pain, R. (2021). Geotrauma: Violence, place and repossession. *Progress in Human Geography, 45*(15), 972–989.
Pastor, M., Sadd, J., & Hipp, J. (2001). Which came first? Toxic facilities, minority move-in, and environmental justice. *Journal of Urban Affairs, 23*(1), 1–21.
Picou, J. S. (2009). Katrina as a Natech disaster: Toxic contamination and long term risks for residents of New Orleans. *Journal of Applied Social Science, 3*(2), 39–55.
Prior, J., Gorman-Murray, A., Connon, I., McIntyre, E., Adams, J., & Madden, B. (2019). A geography of residents' worry about the disruptive effects of contaminated land. *Geographical Research, 57*(1), 52–76.
Sandifer, P. A., & Walker, A. H. (2018). Enhancing disaster resilience by reducing stress-associated health impacts. *Frontiers in Public Health, 6*(373), 1–20.
Schmitt, H. J., Calloway, E. E., Sullivan, D., Clausen, W. H., Tucker, P. G., Rayman, J., & Gerhardstein, B. (2021). Chronic environmental contamination: A systematic review of psychological health consequences. *Science of the Total Environment, 772*, 145025.

Shukla, J. 2013. Extreme weather events and mental health: Tackling the psychosocial challenge. *International Scholarly Research Notices, 2013*, 1–7.
Simpson, D. M., Weissbecker, I., & Sephton, S. E. (2011). Extreme weather-related events: Implications for mental health and well-being. *Climate Change and Human Well-Being*, 57–78.
Smith, K., Luginaah, I., & Lockridge, A. (2010). *'Contaminated' therapeutic landscape: the case of the Aamjiwnaang First Nation in Ontario, Canada.* Paper presented at the Geography Research Forum.
Solomon, S. D., Regier, D. A., & Burke, J. D. (1989). Role of perceived control in coping with disaster. *Journal of Social and Clinical Psychology, 8*(4), 376–392.
Straub, A. M., Gray, B. J., Ritchie, L. A., & Gill, D. A. (2020). Cultivating disaster resilience in rural Oklahoma: Community disenfranchisement and relational aspects of social capital. *Journal of Rural Studies, 73*, 105–113.
Sullivan, D., Schmitt, H. J., Calloway, E. E., Clausen, W., Tucker, P., Rayman, J., & Gerhardstein, B. (2021). Chronic environmental contamination: A narrative review of psychosocial health consequences, risk factors, and pathways to community resilience. *Social Science & Medicine, 276*, 113877.
Taylor, M. P., & Cosenza, I. (2016). *Review of the New South Wales Environment Protection Authority's Management of Contaminated Sites*. Retrieved from NSW, Australia: https://www.epa.nsw.gov.au/~/media/EPA/Corporate%20Site/resources/epa/Contaminated-Sites-Review-2016.ashx
Tuitjer, L. (2021). Exploring urban verticality during the 2011 flood in Bangkok, Thailand. *Area, 53*(2), 336–344.
Vins, H., Bell, J., Saha, S., & Hess, J. J. (2015). The mental health outcomes of drought: A systematic review and causal process diagram. *International Journal of Environmental Research and Public Health, 12*(10), 13251–13275.
Watts, N., Amann, M., Arnell, N., Ayeb-Karlsson, S., Belesova, K., Berry, H., ... Costello, A. (2018). The 2018 report of the Lancet Countdown on health and climate change: shaping the health of nations for centuries to come. *The Lancet, 392*(10163), 2479–2514.
Zhang, Y. (2010). Residential housing choice in a multihazard environment: Implications for natural hazards mitigation and community environmental justice. *Journal of Planning Education and Research, 30*(2), 117–131.
Zhong, S., Yang, L., Toloo, S., Wang, Z., Tong, S., Sun, X., ... Huang, C. (2018). The long-term physical and psychological health impacts of flooding: A systematic mapping. *Science of The Total Environment, 626*, 165–194.

PART III

Cascading Disasters and Risks

CHAPTER 9

Living with Floods in Informal Settlements: Compounding and Cascading Risks in Makassar, Indonesia

Erich Wolff and Diego Ramírez-Lovering

Introduction

As more complex understandings of disasters gain currency, the literature shows that more studies are required to better understand how hazards can lead to cascading and compounding disasters in rapidly urbanising areas (Colvin et al., 2020). This is particularly important in quickly urbanising contexts in which residents of low-lying settlements have historically

E. Wolff (✉)
Monash Art, Design and Architecture, Monash University, Caulfield East, VIC, Australia
e-mail: wolff.erich@gmail.com

D. Ramírez-Lovering
Director of the Informal Cities Lab (ICL), Monash University, Caulfield East, VIC, Australia
e-mail: diego.ramirez@monash.edu

A. Lukasiewicz and T. O'Donnell (eds.), *Complex Disasters*, Disaster Risk, Resilience, Reconstruction and Recovery,
https://doi.org/10.1007/978-981-19-2428-6_9

coexisted with floods such as in the peripheries of many cities in Southeast Asia.

While locally developed adaptation measures such as constructions on stilts and canals have been well documented (Acharya & Prakash, 2019; Chatterjee, 2010; Dangol & Day, 2017; Marfai et al., 2015), urbanisation is continuously changing the nature of disasters in these areas. Importantly, while significantly improving living conditions, the expansion of networked infrastructure has the potential of amplifying cascading effects of floods in these areas. This process means that flash flooding can cause significant secondary effects, for instance, triggering the failure of telecommunications and sanitation systems. As such, the proliferation of infrastructure networks (such as electricity, telecommunications, transport and sanitation) requires a deeper consideration of the compounding and cascading nature of floods that affect the residents of quickly urbanising areas.

Cascading disasters are often exemplified by situations in which the impacts of one hazard unfold and trigger secondary effects of varying magnitudes or even other hazards, such as when the energy dissipated by an earthquake initiates a landslide (Kappes et al., 2012). Also used to classify disasters whose effects can be felt across connected systems, this concept is particularly relevant to understanding how the consequences of local disruptions can be amplified by technological failures in infrastructure networks. The concept of compounding disasters, in turn, acknowledges that the processes through which disasters unfold can be combined as different hazards can take place simultaneously having compounded effects (Pescaroli & Alexander, 2018).

This chapter contributes to this body of knowledge by presenting a firsthand case study on how compounding and cascading effects of floods are understood and addressed at the neighbourhood scale by communities living in the city of Makassar, Indonesia. This case study aims to provide insights into how complex disasters are experienced by communities used to flooding in rapidly urbanising contexts.

Understanding Cascading and Compounding Disasters at the Neighbourhood Scale

Following a tradition of authors that argued for a more reflexive (Beck, 1992) and complex approach to risk science (Perrow, 1999), the notion of cascading and compound disasters has often been linked to the literature on infrastructure risk management (Pescaroli & Alexander, 2018). More recently, however, the concept of cascading risks has been increasingly used to encompass the economic, social and cultural dimensions of disasters. As these more complex framings of risk gain currency (Lukasiewicz & Baldwin, 2020), the literature on the nuances of how cascading disasters unfold at the neighbourhood scale is still incipient, particularly in historically disadvantaged contexts. Due to political and economic biases (Sandoval & Sarmiento, 2020), these sites are often less studied, despite concerns with growing vulnerabilities to the impacts of climate change.

Case studies exploring the multiple interacting processes triggered in disasters at the neighbourhood scale are still rare. Kappes (2012) argues that the understanding of cascading effects is still very limited, undermining multi-hazard risk studies that seek to investigate interactions between different hazards. The need for additional studies on the neighbourhood-scale effects of complex disasters is illustrated by how most case studies on cascading risks focus on critical infrastructure systems such as bridges or power plants and their vulnerabilities (Pescaroli & Alexander, 2018). Focusing mainly on technical aspects, these studies only partially explore how social and cultural forces, such as unequal access to resources, can lead to cascading impacts following disruptions (Chandler, 2019; Tierney, 2015).

A growing body of knowledge, however, highlights the importance of accounting for the experiences of communities with cascading disasters in policy and management. The case study of the effects of Hurricane Katrina by Little (2002), for example, exemplifies the challenge of addressing cascading effects that are economic, technical and social in nature. Other authors also acknowledged that the direct damage caused by the hurricane in itself was limited when compared to the secondary and tertiary effects of infrastructure collapse and its systemic consequences (Hartman & Squires, 2006).

Similarly, other studies have used the framework of cascading and compounding processes to refer to the multiple and cross-scalar effects of disasters whose impacts range from disruptions to existing livelihoods to increased complexities in the deployment of disaster relief efforts (Berariu et al., 2015). Exploring these concepts, this chapter discusses how residents of informal settlements make sense of the cascading and compounding nature of major floods in their neighbourhoods, as illustrated in Fig. 9.1.

The Context of Makassar and Indonesia

The city of Makassar is the capital of the province of South Sulawesi in Indonesia, a fast-growing country whose largest cities are located in low-lying areas along the coast (Anon, 2016). The rapid growth of Indonesia is exemplified by the metropolitan area of Jakarta, which has boomed over the last decades becoming the third biggest global metropolis in 2015 (OECD/European Commission, 2020). Following similar urbanisation rates, the metropolitan areas of the main Indonesian cities, including Makassar, have significantly expanded over the last few decades aggravating historic vulnerabilities (Marfai et al., 2015; Padawangi, 2012; van Voorst, 2014).

The South Eastern suburbs of Makassar have been gradually urbanised over the last forty years, a period in which the city's peripheries sprawled over previously rural areas (BPS-Statistics of Makassar Municipality, 2020). While the number of residents has grown significantly throughout the years, the drainage network can be traced back to rice plantations and early settlements established among a network of canals, typical of infrastructure systems developed during the Dutch colonial era (Ravesteijn & Kop, 2008). In the 1990s, the construction of the Bili-Bili dam, upstream from Makassar, aimed to regulate water flows and alleviate flooding risks in the city.

Accounts from community members revealed that several residents in the area moved to the surroundings of the Batua canal after being evicted from other locations in the 1980s. Having uncertain land tenure and receiving little compensation for their houses, these families settled on affordable land in flood-prone areas considered inappropriate for habitation by the local authorities (Jusman, 2010). Some residents have shared during interviews that they have dealt with major floods in the settlements annually, sometimes for over 30 years. While these accounts might seem, representative of extreme cases of urbanisation over flood plains, these situations are not unusual and have been documented in many other countries in the Global South (Choudhury & Haque, 2016; Molle, 2009).

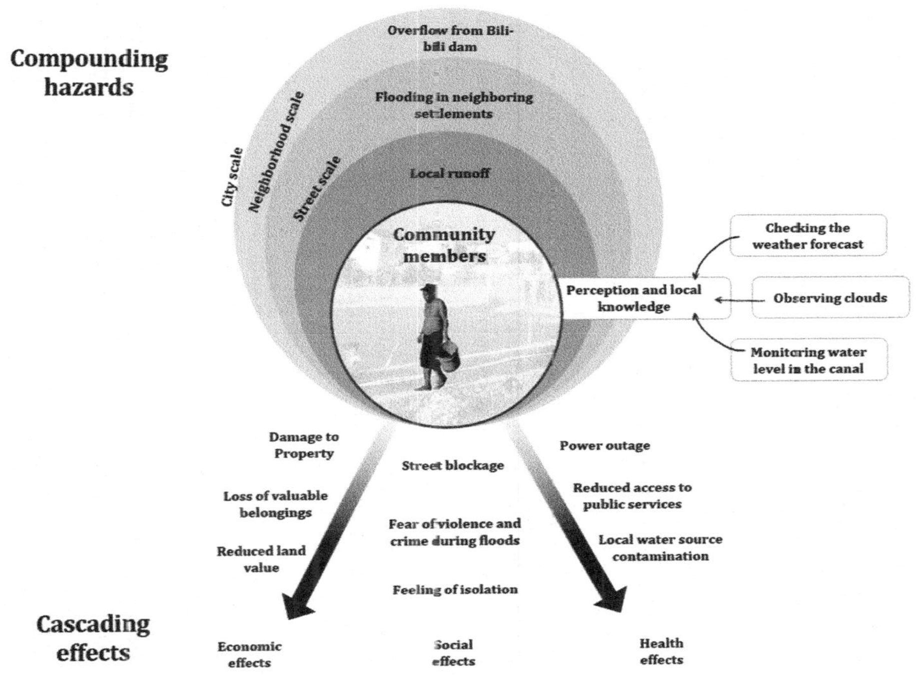

Fig. 9.1 Compounding hazards capable of aggravating floods in Makassar and their cascading effects according to the accounts of residents living in the Tallo catchment (*Source* Author construct)

While the challenges of informal urbanisation in flood-prone sites are well documented, much is still unknown about how communities understand the multiple dimensions of disasters and their cascading effects (Hazarika et al., 2018; Kelman et al., 2015; Langill & Abizaid, 2020; Šakić Trogrlić et al., 2019). Further studies on how communities conceptualise complex and multidimensional disasters are particularly important in the context of the Asia–Pacific, a region frequently affected by flooding and storms (UNDRR/CRED, 2020). The next section introduces the method used to understand cascading and compounding effects of floods in Makassar.

Method

The memories and descriptions of flood impacts that underpin this chapter were collected through semi-structured interviews conducted in April 2019 with residents of settlements participating in the Revitalising Informal Settlements and their Environments (RISE) Program. RISE is a transdisciplinary program implementing and examining the benefits of nature-based infrastructure in the context of informal settlements in Indonesia and Fiji (Leder et al., 2021; Ramirez-Lovering et al., 2020). Understanding floods in the participating settlements is an important aspect of RISE as water level variations have been identified as a source of contamination and a threat to the maintenance of the infrastructure system being tested (French et al., 2021; Wolff et al., 2019).

Drawing from interviews with 54 residents of the settlements, this chapter is part of a larger investigation on flooding across eight different neighbourhoods of Makassar participating in the RISE Program. The selected quotes presented in this chapter were all provided by members of the communities living in the Tallo catchment, the main urbanised watershed in the metropolitan area of Makassar (Sudirman et al., 2018). Starting with community leaders and members who have been engaged with RISE, the selection of participants used a "snowballing" approach in which one interviewee recommended others. This process aimed to provide not only qualitative data but also a nuanced and context-specific framing of how floods are conceptualised and responded to in these communities.

The interview guide was designed to work as an open-ended questionnaire (Gilbert, 2008) as proposed by other qualitative studies that investigated floods and droughts from a community-based perspective

(Hazarika et al., 2018; Marchezini et al., 2017; McEwen & Holmes, 2017; Rickards, 2012). The questions offered residents the opportunity to talk openly about environmental hazards in their settlements and to describe floods (intensity, frequency and duration) and their effects (such as impacts on health and damage to furniture or house).

The interviewees were also encouraged to share additional experiences beyond the scope of the initially proposed questions. In most of the interviews, for instance, we had the opportunity to have conversations about the community and their strategies to prepare for floods, mitigate damages and recover after them. The notes derived from the interviews were transcribed, coded and analysed qualitatively to facilitate the identification of common connections between ideas and groups of concepts. The method followed an inductive approach (Hodkinson, 2008) in the sense that the categories of grouping and analysis were not predetermined, but emerged from the experiences shared by the residents. The findings were then synthesised and exemplified through selected quotes to reveal the compounding and cascading processes that characterise floods in the settlements.

While this kind of immersive and reflexive investigation of floods is not usual in traditional studies on risk assessment, we considered it essential to have a clearer understanding of the lived experiences of floods and their impacts in the settlements. Understanding this kind of nuance from a grounded perspective was essential to help us make sense of floods in Makassar as multidimensional and cascading phenomena. All of the sensitive information in this chapter, including names and the location of the settlements, is presented in an anonymised format to protect the identities of the participants.

The Description of Compounded Floods in Makassar

The interviews showed that the flood-prone nature of the settlements and the complex and interconnected effects of water level variations are well-known to the residents living along the Batua canal. Notably, several interviewees demonstrated a sophisticated knowledge of the compounding effects of local and regional floods that require both an understanding of the city-scale water flows and a grounded experience of local floods. The findings, illustrated in Fig. 9.1, show how compounding hazards and cascading effects are understood by the residents of Makassar.

The description of floods as complex and connected processes is one the most important insights shared by residents during the interviews. As such, the process that residents describe as "banjir" (floods in Bahasa Indonesia), might not necessarily align with what risk specialists would conventionally consider riverine flooding in flood plains (Olesen et al., 2017).

Aware of the varying degrees of flooding experienced in Makassar over the last few decades, residents describe floods not as singular, but as rather complex phenomena influenced by multiple different types of hazards. For example, when asked to describe floods in his settlement, one resident living near the Batua canal shared that:

> *Floods here are of two types: we have the ordinary flood and the extreme one. The ordinary* [banjir biasa] *happens every year, sometimes once every month of the rainy season… this means maybe 3 or 4 times a year. The extreme one happens usually not more than once a year.* […] *The ordinary one is just annoying, makes the street muddy. The extreme one is when we need to protect the house and sometimes go somewhere else.*

The categorisation of floods is common among the residents and indicates that they possess a clear understanding of the differences between nuisance and extreme floods. Furthermore, it reveals a sophisticated ability to identify the multiple hazards that are often grouped within the umbrella concept of "flooding". The interviews show that this differentiation demands a capacity to describe floods not only as water level variations affecting their houses but also at the street, neighbourhood and city scales.

Several residents demonstrated having developed a neighbourhood scale understanding of floods based on the observation of the water flows beyond their immediate surroundings. While many residents were able to broadly describe the hydrological characteristics of the settlements, at least three interviewees described the properties of their neighbourhood specifically as a "*big bowl*" that receives water from neighbouring terrains and then releases it back into the Tallo River. In the words of one of the residents:

> This area floods a lot… You know, the water comes from all neighbours and stays here until it drains through the dirt to the canal.

This description reveals that the residents understand how the neighbourhood-scale topography influences the natural drainage patterns in the area. On another scale, this resident also shared during the interview that this process is tightly linked to the flow of water on the streets and, consequently, in the private areas. Similarly, several interviewees described in great detail the drains and culverts where water flows and accumulates before reaching their individual properties during flash floods. One resident stated:

> *The man next door expanded the house a few years ago and did some landfilling. Since then, when there is strong rain the water flows from his backyard straight into my living room. In the floods this year, I had to put some sandbags in front of the door and then the flow went into the other neighbour's plot!*

This kind of account shows a particular way of conceptualising water that is based on the understanding of the small-scale interventions derived from different households' attempts to prevent inundation. This quote also indicates that residents are aware of the risk of flash floods at the street scale and have developed adaptations to address them. While not unique to Makassar, this way of understanding floods is deeply grounded in observation and experimentation as a daily practice as described by other authors. For instance, Šakić Trogrlić et al. (2019) discussed a similar situation when explaining how communities conceptualise floods in Malawi. They argue that communities "*demonstrated a high level of understanding of flood dynamics, by detailing the ways in which waters reach and spread throughout their villages*" (Šakić Trogrlić et al., 2019: 8).

Several residents also discussed the city-wide processes that influence the water level variations in their settlements. In particular, many interviewees mentioned the relationship between the operation of the Bili-Bili dam upstream, at the outskirts of Makassar, and the severe floods observed in the settlements. While residents often mention the dam as a successful intervention capable of reducing floods, many also indicated that the overflow of water from the dam can be compounded by local flash floods causing disastrous results. One of the interviewees shared that:

> *Before the dam, we had floods all the time during the rainy season. After Billi-billi we have fewer floods, but when they happen, they are much quicker,*

> *so we cannot prepare ourselves. In 2019 I could not move my things upstairs because the water level changed too quickly* [...].

The identification of different processes through which floods take place indicates that, to some extent, residents recognise floods as phenomena caused by combinations of multiple factors. The description of more frequent and less severe floods in the past has been linked to a higher number of flash flooding events. The recent, less frequent and more intense floods, conversely, have been described by residents as a compounded effect derived from the combination of local flash floods with significant amounts of water released from the dam upstream. These descriptions are consistent with the finding that floods in the region are a result of runoff, tidal variations and river overflow that involve seasonal cycles as well as meteorological and neighbourhood-scale processes.

Mentioning concerns with the operation of the dam, residents have reported that they don't often feel threatened by street-scale water flows but the combined effects of these flows with the water from upstream can have devastating consequences. In order to prepare for these situations, community members living along the Batua canal shared that sharing flood knowledge about processes taking place at different scales was critical to estimating flood intensity. One resident, for instance, described the local mosque as a platform for sharing local knowledge and preparing for the compounded effects of floods:

> *We didn't know if there would be a* [compounded] *flood* [in 2019], *but some of my neighbours had shared with me that Bili-bili* [dam] *was full and there was a good chance that they would need to open the gates.* [...] *I coordinated a meeting with the ladies' group at the mosque to discuss who needed help if the* [compounded] *flood came.*

This passage suggests that local groups and institutions play a central role in how residents prepare for floods at the street and city scales. Another interviewee mentioned that the messaging application shared by the members of the mosque was particularly important for her as it had become a tool to share information about the dam operation during stormy days. Similar to the account shared by this resident, several interviewees shared that the communities rely on local knowledge to anticipate floods. When preparing for incoming floods, residents mentioned using not only informal communication tools, such as mobile messaging applications but also other information sources such as weather forecasts and official government warnings to predict flood intensity. Some interviewees also mentioned relying on their own observations of clouds, wind and other indications of incoming storms as a strategy to prepare for floods.

These examples provide insights into the different scales and processes through which communities conceptualise compounding floods in the area surrounding the Batua canal. This section has shown that residents understand floods through the neighbourhood's topography, through the street-scale knowledge of water flows and even through the operation of the dam located upstream. In the next section, we elaborate on how these emerging understandings relate to the cascading effects of floods in Makassar.

The Cascading Effects of Floods in Makassar

The interviews describing the combined effects of local runoff and the overflow of the dam suggest that compounded floods can lead to several cascading effects beyond the damage to properties in the affected areas. The interviews show that these effects can be felt in different time-frames and vary significantly in nature, ranging from economic, social and psychological consequences. Residents are concerned both with impacts during the floods, such as the fear of losing valuable belongings and after floods; such as health consequences including rashes from contact with floodwater and respiratory diseases from inhabiting a damp house.

Many residents also mentioned being afraid of being isolated at home due to street blockages caused by compounded floods. These situations have been linked to higher rates of domestic violence and other kinds of crime during prolonged floods. Other cascading effects of the isolation were described by one of the residents:

> *When there is a big flood, my biggest fear is to be hurt and not be able to go to the hospital because the streets get blocked. One of my neighbours once had a baby in the middle of a flood and it was very hard. It all went well, but since then I am worried about not being able to go out of the settlement during floods.*

This excerpt shows that residents often feel cut-off from the city and basic public services during floods. While residents can often leave their houses following minor floods, compounded floods are particularly concerning due to the blockage of streets that, in turn, prevents the provision of urgent medical care. In order to leave their houses during compounded floods that can last for up to a week, some residents mentioned using rafts and even building temporary bamboo bridges to other houses. These examples illustrate, therefore, not only the cascading physical consequences that result from compounded floods but also foreshadow the psychological and social consequences of these processes.

The notion that floods can cause effects beyond immediate infrastructure failures has been voiced by interviewees that discussed how historic water levels can influence property values in the area. One of the residents said that most of the people living in the area had settled there because they could not afford to pay for plots in safer areas of the city. Acknowledging recent compounded floods, he shared that the area might become even cheaper. In the interview, he said that:

> *I am safe from floods here, but down the* road [referring to houses a few metres away from his] *the people are in trouble when floods come. My house is higher than the others because I paid for a few additional meters of protection.*

This description shows that the effects of floods are not limited to physical damage to properties, but also include political and social aspects such as implications to land value. This suggests that residents are also aware of the effects that floods can have in the long term, further aggravating the vulnerabilities of historically disadvantaged communities. The awareness of flood risks and their consequences were also described as a factor that shapes the behaviour and the daily lives of residents during the rainy season.

For example, several interviewees mentioned that in the period between November and March they needed to constantly monitor the rain. Particularly in the case of communities living in low-lying areas, the vigilant observation of the weather was often described as a strategy to compensate for higher exposure to flood risks. These descriptions are consistent with other studies that have demonstrated that communities living in flood-prone areas often develop ways for assessing risks derived from their daily needs and perceptions (Acharya & Prakash, 2019; Marfai et al., 2015). These needs are often linked to the disruption of water and electricity supply, the most common cascading effect of floods mentioned by the residents.

Most of the residents of the area along the Batua canal who stated that they depended on local wells to have access to water supply demonstrated a concern with the well water quality after floods. Under such circumstances, floods represent an additional threat to health when they contaminate local water sources. One of the residents, for instance, explained:

> *In general, the water from the well is good, but I cannot drink it after the flood. The flood leaks into the well and the water gets smelly and dirty. Sometimes I have to use this water to shower or to wash dishes if it is not smelly, but I never use it to drink after the flood.*

Interviewees that demonstrated concern with water insecurity after floods shared that they depended on official information and local knowledge to decide on preventive measures in advance of compounding floods. Residents mentioned, for instance, that they checked official weather forecasts, observed the clouds and monitored the water level in the canal nearby as strategies to anticipate incoming floods. They mentioned that they would move valuable belongings to elevated floors and stockpile bottled water and food whenever they noticed signs of major floods. This is an important strategy for the communities, especially when residents were aware that the Bili-Bili reservoir was full and that the gates could be opened at any time.

The examples indicate that communities are aware of cascading impacts as a consequence of compounded floods in Makassar. The concerns expressed with the overflow from the Bili-Bili dam illustrate the well-known importance of effective and clear communication for the remediation of the effects of disasters. This discussion, further explored in the concluding section of this chapter, is particularly relevant in contexts in which residents are vulnerable to the compounding effects of multiple hazards operating in different scales and timeframes, such as the communities living along the Batua canal.

Conclusions

The accounts provided by the residents of Makassar indicate that floods are understood at the neighbourhood scale through personal and shared experiences as well as through their cascading effects. The details of how floods in Batua are aggravated by the overflow of excess water from the Bili-Bili dam serves as a rich example of the challenges of compounded and cascading disasters in informal settlements. In this case, the flash flooding in the settlements coincided with extreme rainfall at the Bili-Bili reservoir, therefore, generating a compound flood to the communities living downstream from the dam.

The accounts provided by residents suggest that the combined consequences of local flash floods and the release of the excess water at the Bili-Bili dam cascade downstream affecting transport networks, medical services and electricity supply and water availability in several neighbourhoods in the Tallo River catchment. These accounts are commensurate with other works that have documented how other processes triggered compounding phenomena causing major cascading effects. Pescaroli and Alexander (2018: 2253) describe such situations as cases in which the "disruption of the infrastructure and its cascading effects on society were subject to nonlinear escalation and became the primary source of crisis that needed to be managed". In the case of Makassar, while the communities have historically dealt with minor floods, the effects of flash floods are significantly aggravated when local runoff is compounded by the overflow from the Bili-Bili reservoir upstream.

One of the central findings of this research is that residents living along the Batua canal are concerned with the lack of reliable information regarding the operation of the Bili-Bili dam. According to the interviews, conflicting information about the operation of the dam can severely undermine the chances of immediate, responsive actions to control cascading effects. Similar experiences have been documented by other authors in the context of Indonesia showing that lack of communication is a common aspect affecting communities' resilience to floods. For example, a study by Van Voorst in Jakarta exemplified how informal communication between people in different parts of a catchment through the use of walkie-talkies was valued by residents when preparing for incoming floods (van Voorst, 2019). The examples discussed in this chapter show that residents in Makassar rely on similar information sources to prepare for and mitigate the cascading impacts of compounded floods in their settlements. These insights into the lived experiences of cascading disasters can be valuable to advance community-based strategies that are able to operate in dialogue with existing local knowledge.

The interviewees that discussed local knowledge revealed that communities conceptualise floods as phenomena influenced not only by processes operating at the street, neighbourhood or city scales but also by changes through time. For example, many of the residents discussed floods based on patterns of urbanisation and how the settlements have been occupied. Interviewees discussed that due to urban development along its banks, alterations in its shape, and the construction of gates, the Batua canal is now more susceptible to local flooding than it was twenty years ago. According to the interviews, the ongoing urbanisation process demands careful observation of signs that can help residents prepare for the potential cascading effects of floods. As a response, this chapter discussed how residents have developed ways of estimating flood intensity based on local knowledge relying on weather forecasts and observation of clouds and water levels. Similar conclusions on the importance of local knowledge have also been documented in the context of other communities that have historically coexisted with flooding (Acharya & Prakash, 2019; Šakić Trogrlić et al., 2018).

While local knowledge and access to information can improve preparedness and response to floods, it is essential to acknowledge that communication should not be seen as a definitive solution to the growing vulnerabilities of quickly urbanising areas. The examples presented in this chapter provided insights into the ways in which communities conceptualise and respond to the cascading effects of floods through community action and locally developed mitigation measures such as the construction of houses on stilts and the use of sandbags. While these strategies play an important role in how members of these communities have historically addressed the effects of floods, the interviews showed that more work is necessary to ensure that existing vulnerabilities are not aggravated by compounded hazards in the context of climate change. More importantly, these examples should not be used to justify actions that transfer the responsibility of identifying, mapping and monitoring risks to already disadvantaged communities.

These concerns are also discussed within the Sendai Framework for Disaster Risk Reduction (UNISDR, 2015), which highlights the importance of addressing growing vulnerabilities to disasters in island countries. These concerns are particularly relevant when examined from the perspective of complex disasters as the effects of climate change are expected to disproportionately impact historically marginalised communities in the Asia–Pacific. As such, it is important to acknowledge that the understanding of compounding and cascading disasters demands a transition in the methods utilised to plan for contingencies and to test scenarios (Wolff et al., 2021; Pescaroli & Alexander, 2018).

The complex nature of flooding in Makassar also indicates that conventional single-hazard analyses are not fully able to capture the nuances and context-specific aspects of cascading disasters. This chapter showed that a grounded exploration of the implications of floods in the context of informal settlements is key to allow for a better understanding of the cascading effects that can derive from compounded floods. This kind of study is paramount as governments gradually provide the basic services necessary to support the residents of quickly growing neighbourhoods and, in doing so, develop increasingly interconnected infrastructure systems.

In this context, it is expected that studies on cascading and compound disasters will be increasingly needed to address the new challenges posed by the vulnerabilities of interconnected infrastructure systems. For instance, the study of the failure of a power plant through its cascading effects sheds light not only on its direct effects on the local community but also on how it can subsequently impact substations, hospitals and other critical facilities. Through the accounts of residents, this chapter explored how flooding is understood on the ground as a multilayered phenomenon that has significant cascading effects ranging from loss of access to services to the reduction of land value.

The experiences discussed in this chapter contribute to a broader and more "people-centred" approach to disasters, one that considers communities as central stakeholders whose lived experiences must be considered in the elaboration of early warning systems and contingency plans. These experiences are expected to influence future disaster studies that champion community engagement in the characterisation, mapping and monitoring of risks through methods such as participatory mapping and citizen science (Porto de Albuquerque & Albino de Almeida, 2020; See, 2019; Wolff, 2021).

Acknowledgements The authors acknowledge the contributions of the RISE Consortium, particularly the colleagues who mediated and translated the interviews. We would also like to thank Associate Professor Shanti Sumartojo for the guidance and mentoring through this doctoral research.

The RISE program is funded by the Wellcome Trust [OPOH grant 205222/Z/16/Z], the New Zealand Ministry of Foreign Affairs and Trade, the Australian Department of Foreign Affairs and Trade, the Asian Development Bank, the Government of Fiji, the City of Makassar and Monash University, and involves partnerships and in-kind contributions from the Cooperative Research

Centre for Water Sensitive Cities, Fiji National University, Hasanuddin University, Southeast Water, Melbourne Water, Live and Learn Environmental Education, UN-Habitat, UNU-IIGH, WaterAid International and Oxfam.

Ethics review and approval was secured by the Monash University Human Research Ethics Committee (Melbourne, Australia), and the Ministry of Research, Technology and Higher Education Ethics Committee of Medical Research at Universitas Hasanuddin (Makassar, South Sulawesi, Indonesia); and Fiji National University Ethics Committee. The RISE trial is registered on the Australian New Zealand Clinical Trials Registry (ANZCTR) (Trial ID: ACTRN12618000633280). All study settlements, households, and caregivers/respondents provided informed consent.

References

Acharya, A., & Prakash, A. (2019). When the river talks to its people: Local knowledge-based flood forecasting in Gandak River Basin, India. *Environmental Development, 31*, 55–67. https://doi.org/10.1016/j.envdev.2018.12.003

Anon. (2016). *Indonesia's Urban Story*.

Beck, U. (1992). *Risk society: Towards a new modernity*. Sage Publications.

Berariu, R., Fikar, C., Gronalt, M., & Hirsch, P. (2015). Understanding the impact of cascade effects of natural disasters on disaster relief operations. *International Journal of Disaster Risk Reduction, 12*, 350–356. https://doi.org/10.1016/j.ijdrr.2015.03.005

BPS-Statistics of Makassar Municipality. (2020). *Kota Makassar Dalam Angka 2020 (Makassar Municipality in Figures)*. Section of Integration Processing and Statistics Dissemination.

Chandler, D. (2019). Resilience and the end(s) of the politics of adaptation. *Resilience, 7*(3), 304–313. https://doi.org/10.1080/21693293.2019.1605660

Chatterjee, M. (2010). Slum dwellers response to flooding events in the megacities of India. *Mitigation and Adaptation Strategies for Global Change, 15*(4), 337–353. https://doi.org/10.1007/s11027-010-9221-6

Choudhury, M.-U.-I., & Emdad Haque, C. (2016). "We are more scared of the power elites than the floods": Adaptive capacity and resilience of wetland community to flash flood disasters in Bangladesh. *International Journal of Disaster Risk Reduction, 19*, 145–158. https://doi.org/10.1016/j.ijdrr.2016.08.004

Colvin, R., Lewis, S., & Howden, M. (2020). Implications of Climate Change for Future Disasters. In A. Lukasiewicz, & C. Baldwin (Eds.), *Natural*

Hazards and Disaster Justice: Challenges for Australia and Its Neighbours. Springer Singapore.
Dangol, N., & Day, J. (2017). Flood adaptation by informal settlers in Kathmandu and their fear of eviction. *International Journal of Safety and Security Engineering, 7*(2), 147–156. https://doi.org/10.2495/SAFE-V7-N2-147-156
French, M., Fiona Barker, S., Taruc, R. R., et al. (2021). A planetary health model for reducing exposure to faecal contamination in urban informal settlements: Baseline findings from Makassar, Indonesia. *Environment International, 155*, 106679. https://doi.org/10.1016/j.envint.2021.106679
Gilbert, G. N. (2008). *Researching social life*. 3rd ed. Sage Publications.
Hartman, C., & Squires, G. D. (Eds.). (2006). *There is no such thing as a natural disaster: Race, class, and hurricane Katrina*. Routledge.
Hazarika, N., Barman, D., Das, A. K., Sarma, A. K., & Borah, S. B. (2018). Assessing and mapping flood hazard, vulnerability and risk in the upper Brahmaputra river valley using stakeholders' knowledge and multicriteria evaluation (MCE): Assessing and mapping flood hazard. *Journal of Flood Risk Management, 11*, S700–S716. https://doi.org/10.1111/jfr3.12237
Hodkinson, P. (2008). Grounded theory and inductive research. In *Researching social life*. Sage Publications.
Jusman, S. K. M. Si. (2010). *Flood and inundation-prone areas in Makassar (Rawan Genangan/Banjir Kota Makassar)*.
Kappes, M. S., Keiler, M., von Elverfeldt, K., & Glade, T. (2012). Challenges of analyzing multi-hazard risk: A review. *Natural Hazards, 64*(2), 1925–1958. https://doi.org/10.1007/s11069-012-0294-2
Kelman, I., Gaillard, J. C., Mercer, J., Crowley, K., Marsh, S. & Morin, J. (2015). Culture's role in disaster risk reduction: Combining knowledge systems on Small Island Developing States (SIDS). In F. Krüger, G. Bankoff, T. Cannon, B. Orlowski, & E. L. F. Schipper (Eds.), *Cultures and disasters: Understanding cultural framings in disaster risk reduction*. Taylor & Francis Group.
Langill, J. C., & Abizaid, C. (2020). What is a bad flood? Local perspectives of extreme floods in the Peruvian amazon. *Ambio, 49*(8), 1423–1436. https://doi.org/10.1007/s13280-019-01278-8
Leder, K., Openshaw, J. J., Allotey, P. et al. (2021). study design, rationale and methods of the Revitalising Informal Settlements and their environments (RISE) study: A cluster randomised controlled trial to evaluate environmental and human health impacts of a water-sensitive intervention in informal settlements in Indonesia and Fiji. *BMJ Open, 11*(1) e042850. https://doi.org/10.1136/bmjopen-2020-042850
Little, R. G. (2002). Controlling cascading failure: understanding the vulnerabilities of interconnected infrastructures. *Journal of Urban Technology, 9*(1), 109–123. https://doi.org/10.1080/106307302317379855

Lukasiewicz, A., & Baldwin, C. (Eds.). (2020). *Natural hazards and disaster justice: Challenges for Australia and its neighbours*. Springer Singapore.

Marchezini, V., Trajber, R., Olivato, D., Muñoz, V.A., Pereira, F. de O., & Luz, E. A. O. (2017). Participatory early warning systems: Youth, citizen science, and intergenerational dialogues on disaster risk reduction in Brazil. *International Journal of Disaster Risk Science, 8*(4), 390–401. https://doi.org/10.1007/s13753-017-0150-9

Marfai, M. A., Sekaranom, A. B., & Ward, P. (2015). Community responses and adaptation strategies toward flood hazard in Jakarta, Indonesia. *Natural Hazards, 75*(2), 1127–1144. https://doi.org/10.1007/s11069-014-1365-3

McEwen, L., & Holmes, A. (2017). *Sustainable flood memories: Developing concept, process and practice in flood risk management*. In F. Vinet (Ed.), (pp. 141–55). Elsevier.

Molle, F. (2009). *Contested waterscapes in the Mekong region: Hydropower livelihoods and governance*. Earthscan.

OECD/European Commission. (2020). The growth of metropolitan areas. In *Cities of the world, a new perspective on urbanisation*. OECD Publishing.

Olesen, L., Lowe, R., & Arnbjerg-Nielsen, K. (2017). Flood damage assessment: Literature review and recommended procedure. Socio-technical flood resilience in water sensitive cities—quantitative spatio-temporal flood risk modelling in an urban context (Project B4.1). Australian Government—Department of Industry and Science.

Padawangi, R. (2012). The right to flood-free homes: urban floods, spatial justice and social movements in Jakarta, Indonesia. Global Visions: Risks and Opportunities for the Urban Planet 199–211.

Perrow, C. (1999). *Normal accidents: Living with high-risk technologies*. Princeton University Press.

Pescaroli, G., & Alexander, D. (2018). Understanding compound, interconnected, interacting, and cascading risks: A holistic framework: A holistic framework for understanding complex risks. *Risk Analysis, 38*(11), 2245–2257. https://doi.org/10.1111/risa.13128

Porto de Albuquerque, J., & Albino de Almeida, A. (2020). Modes of engagement: Reframing "sensing" and data generation in citizen science for empowering relationships. In T. Davies, & A. Mah (Eds.), *Manchester toxic truths*. University Press.

Ramírez-Lovering, D., Prescott, M. F., Josey, B., Mesgar, M., Spasojevic, D., & Wolff, E. (2020). Operationalising research: Embedded Ph.D.'s in transdisciplinary, action research projects. In R. Barnacle, & D. Cuthbert (Eds.), *The Ph.D. at the end of the world: Provocations for the Doctorate and a future contested* (pp. 45–64). Springer Nature.

Ravesteijn, W., & Kop, J. (2008). *For profit and prosperity: The contribution made by Dutch engineers to public works in Indonesia, 1800–2000*. KITLV Press.

Rickards, L. (2012). *Critical breaking point: The effects of climate variability, change and other pressures on farm households*. Report. Birchip Cropping Group and the Sustainable Agriculture Initiative Platform Australia.

Šakić Trogrlić, R., Wright, G. B., Adebayo, A. J., Duncan, M. J. & Mwale, F. (2018). Taking stock of community-based flood risk management in Malawi: Different stakeholders, different perspectives, *Environmental Hazards*, *17*(2), 107–27. https://doi.org/10.1080/17477891.2017.1381582

Šakić Trogrlić, R., Wright, G., Duncan, M., van den Homberg, M., Adeloye, A., Mwale, J., & Mwafulirwa, J. (2019). Characterising Local Knowledge across the Flood Risk Management Cycle: A Case Study of Southern Malawi. *Sustainability*, *11*(6), 1681. https://doi.org/10.3390/su11061681

Sandoval, V., & Sarmiento, J. P. (2020). A Neglected issue: informal settlements, urban development, and disaster risk reduction in Latin America and the Caribbean. *Disaster Prevention and Management: An International Journal*, *29*(5), 731–745. https://doi.org/10.1108/DPM-04-2020-0115

See, L. (2019). A review of citizen science and crowdsourcing in applications of Pluvial flooding. *Frontiers in Earth Science*, *7*, 44. https://doi.org/10.3389/feart.2019.00044

Sudirman, S. T., Barkey, R. A., & Ali, M. (2018). Watershed identification and its effect toward flood (case study: Makassar city). *International Journal of Advanced Research*, *6*(5), 513–519. https://doi.org/10.21474/IJAR01/7059

Tierney, K. (2015). Resilience and the Neoliberal Project: Discourses, Critiques, Practices—and Katrina. In K. Frailing, & D. W. Harper (Eds.), *American Behavioral Scientist*, *59*(10), 1327–1342. https://doi.org/10.1177/0002764215591187

UNDRR/CRED. (2020). Human Cost of Disasters 2000–2019 Report–UN Office for Disaster Risk Reduction. United Nations Office for Disaster Risk Reduction.

UNISDR. (2015). Sendai Framework for Disaster Risk Reduction 2015–2030. UNISDR.

van Voorst, R. (2014). The right to aid: Perceptions and practices of justice in a flood-hazard context in Jakarta, Indonesia. *The Asia Pacific Journal of Anthropology*, *15*(4), 339–356. https://doi.org/10.1080/14442213.2014.916340

van Voorst, R. (2019). 'Wake up! Flood!' Inside Indonesia, May 22.

Wolff, E. (2021). The promise of a "people-centred" approach to Floods: types of participation in the global literature of citizen science and community-based flood risk reduction in the context of the Sendai framework. *Progress in Disaster Science*, *10*, 100171. https://doi.org/10.1016/j.pdisas.2021.100171

Wolff, E., Prescott, Michaela F., & Ramírez-Lovering, D. (2019). Design for ecological resilience: A case study of green infrastructures in informal settlements. In 2019 World Congress Resilience, Reliability and Asset Management Conference Proceedings (pp. 128–131). Singapore.

Wolff, E., French, M., Ilhamsyah, N., Sawailau, M. J., & Ramírez-Lovering, D. (2021). Collaborating with communities: Citizen science flood monitoring in urban informal settlements. *Urban Planning*, 6(4), 351–364. https://doi.org/10.17645/up.v6i4.4648

CHAPTER 10

Chokepoints: The Challenges of Improving Surveillance of Emerging Biological Hazards Across the Indo-Pacific Region

Jack Thomson, Paul Barnes, Nicholas Thomson, and Paul Arbon

Introduction

To better understand emerging hazard regimes post-COVID-19 and mitigate against cascading impacts and cumulative effects, future threat surveillance, containment and indeed vaccine development will need to

J. Thomson (✉) · P. Arbon
Torrens Resilience Institute, Adelaide, SA 5001, Australia
e-mail: jack.thomson@flinders.edu.au

P. Arbon
e-mail: paul.arbon@flinders.edu.au

P. Barnes
Faculty of Arts, Design and Architecture, UNSW, Sydney 2052, Australia
e-mail: paul.barnes@unsw.edu.au

A. Lukasiewicz and T. O'Donnell (eds.), *Complex Disasters*, Disaster Risk, Resilience, Reconstruction and Recovery,
https://doi.org/10.1007/978-981-19-2428-6_10

adopt a more integrated view of risk to expand strategic early warning across distant but connected locations. With a focus on the IPR, this chapter reviews a number of immediate considerations that are central to any proposals for future investment in improving surveillance, risk management, and containment planning:

1. Emergent Infectious Diseases (EID) vs Emergent Biological Hazards (EBH)
2. Networked Risk Landscapes (NRLs) and the challenges to anticipating hazards within Complex Adaptive Systems (CAS)
3. Current capabilities of Early Warning Systems (EWS) in the Indo-Pacific Region (IPR)
4. Chokepoints as a focus for improved Early Warning and Mitigation
5. Formal and informal governance and structures of authority across borderlands
6. Foresight and scenario planning as a means of anticipating risk and assessing appropriate technologies.

Emergent Infectious Disease (EID) vs Emergent Biological Hazards (EBH)

Emergent Infectious Diseases (EIDs) are diseases where the incidence could or has already increased in the previous twenty years. A minority of EIDs can lead to large-scale epidemics or global pandemics and their impacts have long challenged and destabilised societal resilience. In the IPR alone between 2003 and 2019, there were eight known localised EID outbreaks, including SARS, H5N1, H1N1, and ZIKA. These outbreaks occurred in parallel with ongoing pandemics of infectious disease such as TB, HIV, and Malaria and the persistent risk of antibiotic resistance (Montgomery et al., 2019; Morens et al., 2020). Contrasted against this history, the emergence and rapid global spread of COVID-19 cannot be considered a black swan event.

N. Thomson
Coral Bell School of Asia Pacific Affairs, Acton, ACT 2601, Australia
e-mail: nicholas.thomson@anu.edu.au

Epidemics and pandemics affect humans wherever they live, but the impacts have not always been distributed equitably. A key factor in the historic global expansion of economic empires was the devastation caused by introduced disease that spread along expanding maritime trade routes. This diffusion of disease continues to disproportionately affect the most vulnerable in society. Even before COVID-19 over two million people died annually from zoonotic diseases alone, most of whom were in poorer countries (UNEP-ILRI, 2020).

Zoonotic disease is the result of several factors, but the term is often used to describe the transmission of infection from animal to human, and the transition to disease pathogen within humans themselves. While many infectious diseases harmful to humans are zoonotic in origin, they are of greatest concern when they have been previously unseen, or their management and treatment are underdeveloped or impacts difficult to scale. These 'jumps' from animal to human do not take place in isolation. Crossover events are influenced not just by the nature of contact between a host and a causal agent, but a whole range of biological, physical, ecological, socio-political, and economic factors that also contribute to the increased risk of Emergent Biological Hazards (EBH) (Allen et al., 2017; Coker et al., 2011).

An understandable focus on zoonotic diseases such as Ebola has long distracted attention from what should be more broadly understood as Emergent Biological Hazards (EBH). EBH refers not just to EIDs (i.e., zoonotic), but also includes cyanobacteria, plant and animal pests, and diseases. The threat to population and/or animal and environmental health posed by EBH has been long known (Bottrell & Schoenly, 2012). A recent but now largely forgotten example from 2020 was the outbreak of desert locust (*Schistocerca gragaria*) that had the potential cascading effect of undermining the food, and therefore the political security of the Horn of Africa through to the Middle East and Indian Subcontinent (UNFAO, 2020).

Broader EBHs will continue to pose the threat of significant cascading disruption to global health and human security through both direct and indirect impacts. India, Thailand, and Vietnam for example, are responsible for most of the world's staple rice exports and are central to international food security, but rice production remains subject to shifting climate regimes, and existing and novel pests and diseases such as Brown Plant Hopper, Fall Army Worm, and fungal rust pathogens (Bottrell & Schoenly, 2012; de Groote et al., 2020; Keesing et al., 2010).

Networked Risk Landscapes (NRL) and the Challenges of Anticipating Hazards Risk in Complex Adaptive Systems (CAS)

Emergent Biological Hazard potential is amplified by landscape destruction, climate change, trade, environmental crimes, agricultural intensification, and human mobility (Coker et al., 2011). This trend is increasing as global connectivity facilitates what can be termed networked risk landscapes, wherein economic integration enables the rapid propagation of both direct and indirect hazards along transboundary transit and supply-chain pathways linking diverse and distant human-ecological systems (HES).

Framing diverse but interconnected human-ecological systems as networked risk landscapes provides insight into the challenges and potential options for improved surveillance and containment of potential EBH origins and transmissions. Specifically, the overlapping boundaries of linked HES frequently interconnect what the UNISDR (2009) identifies as two broad categories of disaster risk in development contexts: extensive and intensive. These are location specific social constructs.

Extensive risk is associated with rural areas and refers to '*the exposure of dispersed populations to repeated or persistent hazard conditions of low or moderate intensity*' (UNISDR, 2009). Conversely, intensive risk characterises cities and urban peripheries and is defined as '*the exposure of large concentrations of people and economic activities*' (UNISDR, 2009, p. 18). Extensive risk in rural areas is associated with current and rapidly escalated habitat fragmentation, and associated edge exposure which leads to varied EBH potential (Albers et al., 2020; Keesing et al., 2010). But conditions of intensive risk also pose significant Emergent Biological Hazards (EBH) potential; species that thrive in human-dominated landscapes such as rats, swine, mosquitoes, are established disease threats, particularly under conditions of urban, semi-urban, or rural poverty which is not uncommon across many geographies of the IPR (Han et al., 2016).

Both intensive and extensive risk presents significant, differentiated, but interconnected Emergent Biological Hazards potential. Within the IPR as elsewhere therefore, any future efforts to improve pandemic surveillance and early mitigation action will need to anticipate a multi-layered risk environment. These environments will be characterised by endemic poverty and locations where disasters and efforts to reduce disaster risk, take place against the background of daily hazards faced by communities with entrenched disadvantage (Sen, 1999).

The field work of Thomson (2019) across high-risk Emergent Biological Hazards (EBH) origin locations of the IPR provides an example of a complex feedback mechanisms and indirect/direct hazard transmission within and across the boundaries of diverse human-ecological systems. This work was conducted across the East–West Economic Corridor linking Vietnam-Lao PDR—Thailand and extending through Myanmar to the Bay of Bengal. Such transboundary corridors cut across all nations of mainland IPR and a mosaic of NDLs representing extensive and intensive risk.

In one case study, new roads built under recent transboundary cooperation agreements were found to allow far greater plantation establishment by foreign multinationals across multiple sites within the study region. This in turn led to a process akin to a tragedy of the commons; with no secure land tenure, communities rapidly moved in to harvest high-value timber and wildlife from lands otherwise about to be cleared. Money gained from such products sold or trafficked across national borders was used as seeding for outmigration (ex situ), predominately for young adults and children. Up to seventy nine per cent of this migration was illegal and in and of itself resulted in significant likelihood of harm to individuals and families, both in transit (human trafficking) and ex situ locations (labour exploitation and unsafe working conditions). In situ, such income allowed for some increase in agricultural productivity on existing paddy land through increased mechanisation and other inputs such as chemical fertiliser. But remittance and improved transboundary transport links in turn facilitated further migration, mechanisation, land clearance, and wildlife exploitation (Thomson, 2019; Thomson et al., 2018).

The ways in which households continually manage diversity of landscape and livelihood components and linkages are representative of highly complex adaptive systems (CAS) (Coetzee et al., 2016). The potential of CAS approaches to improving options for decision-making and strategy have been posited but not yet fully articulated or operationalised in Disaster Risk Reduction (DRR) praxis (Burger et al., 2021; Coetzee & Van Niekerk, 2018). Until this issue is addressed, a significant underlying challenge for improving EBH surveillance and early mitigation action will remain the very poor understanding of how diverse but interlinked HES differ in their ability to persist, adapt, or transform in response to different types of complex change (Cumming & Peterson, 2017). This challenge is particularly acute in the context of notable longstanding restrictions of existing Early Warning Systems (EWS) capability across the numerous chokepoints and remoter regions of the IPR.

Current Capabilities of Early Warning Systems (EWS) in the Indo-Pacific Region (IPR)

As far back as 2015, UNISDR's Global Assessment Report estimated that an annual global investment of US$ 6 billion in Early Warning Early Action (EWEA) would result in US$ 360 billion in disaster risk reduction benefits (UNISDR, 2015). Similar estimates for returns on investment have been reported for improved surveillance in One Health, with the World Bank estimating that 'an annual investment of USD3.4 billion in animal health systems worldwide would avert losses incurred through delayed or inadequate responses to zoonoses at almost double the preventative investment' (World Bank, 2021).

Multi-hazard early warning systems (MH-EWS) in particular have been heavily promoted by DRR agencies, which see them as contributing to Sustainable Development Goals (SDGs) and Priority 4 of the SFDRR 2015–2030. As defined by the UNISDR (2015, p. 13), MH-EWS can *'address several hazards and/or impacts of similar or different type in contexts where hazardous events may occur alone, simultaneously or cascading over time, and taking into account the potential interrelated effects.'* But there has been criticism that the term 'all-hazards' itself lacks a clear conceptual definition, and there has been very little empirical research into how such approaches can be implemented (Gall et al., 2015).

There are further several longstanding challenges that may undermine any promotion of MH-EWS for EBH surveillance and early mitigation action. A significant and persistent challenge is that a focus on known, large-scale, and usually singular rapid-onset hazards such as cyclones, has meant that most existing EWS have not sufficiently considered slow-onset hazards nor information that could improve livelihood security or avert environmental degradation, i.e., the very conditions that can lead to greater EBH potential (Birkmann et al., 2013; Thomson, 2019).

A further surveillance constraint is forecasting capability in and of itself, especially in Low-Mid Income Countries (LMIC). This applies to both the current capacity of mid to long-range hydro-meteorological forecasts globally, but also a lack of historical disaster records and a lack of monitoring infrastructure and trained personnel in LMICs (Baudoin et al., 2016; Zommers et al., 2014). Status reviews of Early Warning Systems (EWS) in the period leading up to SARS-COV-2 across the Association of Southeast Asian Nations (ASEAN) for example (Bisri, 2019), reported that the geographical coverage of singular hazards such as tsunamis, cyclones, and other hydro-meteorological threats had improved at the national level, but more work was needed at the local level. Importantly, this also included coverage of more remote rural communities and at regional levels to address slower onset and extensive risks.

In many LMIC settings, monitoring and detection are frequently restricted by structural and non-structural aspects of the technology involved. Just because a weather station is in place, for example, does not mean that it is functioning, or that local staff have been trained to, or collect and report, information regularly or accurately (CSIS, 2019; Thomson et al., 2018). Specific to EBH, emerging or novel zoonotic or crop disease may not be detected and assessed as a risk until it is too late because public health or customs staff may have neither the equipment nor training to know what to look for (Ager et al., 2014; Piot, 2014; Thomson, 2019).

In any attempt to improve strategic Early Warning we find ourselves faced with a significant foundational challenge. Where it is most needed, at the local level of high potential EBH origin and transmission sites, there is frequently no single authoritative source of hazard warning, forecasting, or information due to the local absence of formal and functioning DRR institutions or trained and capacitated personnel (CSIS, 2019; Thomson et al., 2018). Even for single onset hazard impacts such as floods, these issues are persistent and widespread, (Bisri, 2019; Thomson, 2019).

Chokepoints as a Focus for Improved Early Warning and Mitigation

A significant unresolved challenge for both Early Warning research and praxis remains the means of anticipation and warning of the onset of disasters in complex, interconnected, and dynamic systems. This is particularly so in the contexts of LMICs where there are frequently limited institutional, social, and financial capacities to both identify and prioritise future investment (Bisri, 2019; CSIS, 2019; Thomson, 2019). When we consider the needs for improved early warning and preparedness for the next EBH therefore, where would we start and who would we start with?

Within the IPR as elsewhere, the phenomenon of networked risk landscapes greatly accelerates EBH potential through recent, rapid, and ongoing transboundary development and infrastructure investment. The most recent example of this is the Regional Comprehensive Economic Partnership (RCEP) which, along with longstanding initiatives such as the Belt-and-Road and East–West/North–South Economic Corridors, link economies within the IPR and across the globe. These corridors have greatly accelerated habitat fragmentation, connectivity, and mobility through expanded road, rail, and marine transport routes, hydropower and irrigation development, commercial plantations, and agricultural extensification (Ishida, 2008).

Some transboundary corridor locations bring together multiple factors that can increase the risk of EBH origin and transmission. Within the IPR as elsewhere, a number of corridors meet at physical migration, trade, and transport chokepoints. Such physical chokepoints can include Special Economic Zones (SEZ) marine ports, border crossings, road-rail hubs, etc., overlap with known global EID hotspots characterised by tropical regions experiencing rapid and significant land-use changes (Allen et al., 2017).

The movement of people, commodities, and hazard transmission within or across many linked HES will be accelerated predominately, but not exclusively, along and through transboundary corridors and chokepoints. Increasing development, climatic variation, conflict, and/or deterioration of local livelihood systems, is already an established global phenomenon and under such drivers as climate change, populations of developing nations are predicted to become both increasingly susceptible and increasingly mobile in response to evolving disaster risk (Black et al., 2013; IPCC, 2014).

We suggest that these chokepoints, whether they be intranational such as peri-urban Special Economic Zones (SEZ), or international borders, hubs or ports, offer significant opportunity on a systemic and practical basis to improve strategic early warning and containment in otherwise restricted resource contexts capability. Sentinel sites at key locations are not necessarily a new proposition, there are a number of long-established locations for climate change, biodiversity, and to an extent public health monitoring (Glantz, 2003; IPCHS, 2019; TERN, 2016). But biological threat surveillance and containment capability at chokepoints remains underdeveloped (Hort et al., 2017; Montgomery et al., 2019; Okello et al., 2015). Yet before an inevitably rushed adoption of any 'technological fix' post-SARS-COV-2, we need to actually understand the interplay between the geographies, landscapes, and people that entangle at these chokepoints.

Formal and Informal Governance and Structures of Authority Across Borderlands

Chokepoints bring together multiple factors that can increase EBH origin and transmission potential. They therefore at first glance, offer an ideal strategic focus for improved EBH surveillance and containment capability as well as vaccination coverage in restricted capability contexts of LMICS. But in these locations and in these contexts, resilience in the face of disaster is traditionally determined by a variety of biophysical and socio-economic processes that are both complex and opaque. Complex because disaster risk varies within and between locations over time under the influence of both internal and external processes (Béné et al., 2014; Walker et al., 2012); and opaque to outsiders because they frequently occur in diverse cultural, and frequently data-poor contexts (Grol-Prokopczyk, 2013; Rigaud et al., 2018).

There can be many different communities within a geographically defined 'place', and hence there are different perceptions, as well as different levels of capacity, that could lead to disparities in overall disaster risk (Cutter et al., 2008). Chokepoints are located on the borderlands, whether they be national, or intranational such as per-urban SEZ that draw in mass rural migration (Thomson et al., 2014; Kanta et al., 2018). Borderland locations are often beyond the limits of central government authority. For hazard surveillance, risk communication, and mitigation in particular, the notion that such places function on normal systems and protocols of governance is far from the truth. Such borderlands operate in a very different manner, and a notable limitation to improving strategic early warning, is that authority is questioned and questionable and frequently not under the control of the nation state.

It is equally important not to generalise borderlands as 'wild frontiers'. Rather, while they are frequently contested spaces, in many instances they are also extremely sophisticated, networked, and economically vibrant. In these conditions, the notion that preparing and responding to EBH is sole remit of a local level field epidemiologist is not accurate. Rather, it requires an understanding that the dynamic of what are conventionally seen as an epidemiological or health issues, are more likely to be issues of security and livelihood. It is the informal and the different roles that people play on a daily basis that rule. For example, an individual may be a law enforcement officer, a tea shop owner on the side, and have a close relative who is part of a trafficking or smuggling operation (Thomson et al., 2018).

Improved understanding of EBH potential requires greater understanding of these networked risk landscapes and the interaction between people, not only in the way that they live and their place within any one location, but also who they are to each other and how they communicate across borders. This will pose immediate practical and ethical dilemmas with regard to local expectations and the use of sensitive information (Nordhagen et al., 2012), and overriding, higher-level themes that can include hierarchy and patron–client relationships, patriarchy, and the role of women and minorities. This is on top of very practical day-to-day issues such as appropriate means of data collection, site access, and the availability and openness of key respondents and those in positions of authority. In this context, we propose that culturally adaptive and iterative foresight approaches are a necessary first step in the assessment of any proposals for improved EBH surveillance.

Foresight and Scenario Planning as a Means of Anticipating Risk and Assessing Appropriate Warning Technologies

A stronger evidence base is needed to understand emerging hazard regimes post-SARS-CoV-2 and the increasing scale and complexity of future humanitarian crises. A key question in the context of this evolving disaster risk is how we do go about designing capability and agility into existing disaster risk management approaches and improving EBH surveillance and containment capability in complex governance systems?

There will be a rush of One Health and pandemic preparedness initiatives post-SARS-COV-2. This is good and important. At the same time there are burgeoning proposals for technological innovation in hazard warnings and dissemination through GIS, AI analytics, mobile applications, etc., (Bisri, 2019; CSIS, 2019; IPCHS, 2019). But while disaster management agencies need to be alert and adaptable to opportunities presented by new technology, there are known concerns regarding the utility of such technology in the varied capability contexts of Low to Mid Income Countries (CSIS, 2019; Thomson, 2019).

The continued limitations of DRR institutions and services in many LMICS locations, mean that by tradition, and for at least the immediate future, self-reliance is at the core of local resilience to disaster. Local communities who experience hazards on a daily basis have a very keen sense of changes, threats, and opportunities within their environment and such local experiences and perspectives are also most frequently the only local historic records available. Greater incorporation of local knowledge would therefore confer benefits to the overall assessment and anticipation of disaster risk as well as very practical insight into appropriate technology and means of early action containment (Thomson et al., 2015), yet substantive efforts to understand local contexts and nuance is often completely lacking.

Incorporating foresight methodology is an important first step in the exploration and assessment of any proposals for more formalised EBH surveillance and mitigation. Improved strategic capability must prioritise investment and partnership, improve capacity to support civil authorities and local communities, and provide deeply practical insight into appropriate technologies. Rather than seeking a top-down technological fix, it is therefore beholden on the actors responsible for EBH surveillance and mitigation that we undertake direct engagement in borderlands to understand the different dynamics of flow and communication.

Foresight approaches are not about prediction, but about understanding existing systems and potential futures by bringing together key stakeholders from different sectors, designations, regions, and communities (Jackson & Chatterjee, 2019; Lauster & Hansen-Casteel, 2018). Specific to improving chokepoint Emergent Biological Hazard EBH surveillance, this by necessity will involve both formal and informal stakeholders who control different political-economic aspects across and between borders. This includes both the legal and illegal flow of goods and people, and the full range of agencies and representatives who traditionally have not been part of disaster risk reduction (DRR) governance or biological threat surveillance.

Incorporating foresight methodology provides practical insight into the necessary hard-soft infrastructure integration that will be required to ensure the full functionality of any new technology. Guiding foresight questions in this context must be culturally adaptive, sensitive, location-specific, and iterative. Only after this can they progress to focused, higher-level questioning that may lead to policy and proposal formation. Such questions can include but are not limited to: current Early Warning, Emergency Services, Customs, and Public Health systems status; national protocols versus on the ground configuration; intergovernmental co-ordination and known structural and operational constraints; levels of agency, sensitivities, capacities, and approval processes; arising lessons and how we consider broader disaster management post-SARS-COV-2; and perceived future needs and trajectories under more diverse drivers such as climate change, landscape destruction, and changing demographics.

Conclusion

The complexity presented by rapidly escalating human-ecological systems fragmentation and global connectivity will increasingly challenge EWS and One Health surveillance programming. To address the threat of cascading disaster impacts, a stronger evidence base is needed to understand the emerging risk profile post-SARS-CoV-2, and to assess the acceptability and scalability of future proposals for technological initiatives for improved multi-hazard surveillance and mitigation across diverse, complex, and networked risk landscapes.

Our overall future disaster risk is stochastic, transboundary, and emergent. In the IPR and similar contexts across the Global South, investment in early warning and early mitigation strategies must therefore strike a balance between addressing the immediate needs of potential and impending hazard impacts, while supporting the efforts required to improve longer-term surveillance and disaster preparedness. Chokepoints between diverse but interconnected human-ecological systems offer strategic opportunities for improving surveillance and containment capability in frequently restricted resource and governance contexts.

Governance, enfranchisement, and participation in frequently contested places will remain important determinants of what surveillance and preparedness recommendations may or may not be viable. But the genuine scale of this strategic investment and how and where it needs to be best implemented has not yet been fully articulated or committed. It is to be remembered that EBHs are potential human security threats no matter where they occur and that this is therefore a weak link, not a strong link, game, meaning that our overall preparedness and resilience is only as strong as the strategic early warning and mitigation capability of all nations.

References

Allen, T. A., Murray, K., Zambrana-Torrelio, C., Morse, S., Rondinini, C., Di Marco, M., Breit, N., Olval, K., Daszak, P. (2017). Global hotspots and correlates of emerging zoonotic diseases. *Nature Communications*, 8(1124).

Ager, A., Burnham, G., Checchi, F., Gayer, M., Grais, R. F., Henkens, M., Massaquoi, M. B. F., Nandy, R., Navarro-Colorado, C., & Spiegel, P. (2014). Strengthening the evidence base for health programming in humanitarian crises. *Science, 345*(6202), 1290–1292.

Albers, H., Lee, K., Rushlow, J., & Zambrana-Torsselo, C. (2020). Disease risk from human–environment interactions: Environment and development economics for joint conservation. *Health Policy, Environmental and Resource Economics, 76*, 929–944.

Baudoin, M. A., Henly-Shepard, S., Fernando, N., Sitati, A., & Zommers, Z. (2016). From top-down to "Community-Centric" approaches to early warning systems: Exploring pathways to improve disaster risk reduction through community participation. *International Journal of Disaster Risk Science, 7*(2), 163–174.

Béné, C., Newsham, A., Davies, M., & Ulrichs, M. (2014), Review article: resilience, poverty, and development, *Journal of International Development*, *623*, 598–623.

Birkmann, J., Chang Seng, D., & Setiadi, N. (2013). Enhancing early warning in the light of migration and environmental shocks. *Environmental Science and Policy, 27*, S76–S88.

Bisri, M. (2019). State of early warning systems in ASEAN. In *ASEAN Risk Monitor and Disaster Management Review (ARMOR)* (pp. 68–94), ASEAN Coordinating Centre for Humanitarian Assistance on Disaster Management.

Black, R., Arnell, N. W., Adger, W. N., Thomas, D., & Geddes, A. (2013). Migration, immobility and displacement outcomes following extreme events. *Environmental Science and Policy, 27*, S32–S43.

Bottrell, D. G., & Schoenly, K. G. (2012). Resurrecting the ghost of green revolutions past: The brown planthopper as a recurring threat to high-yielding rice production in tropical Asia. *Journal of Asia-Pacific Entomology, 15*(1), 122–140.

Burger, A., et al. (2021). Organizing theories for disasters into a complex adaptive system framework. *Urban Science, 5*(3), 61.

Coetzee, C., Van Niekerk, D., & Raju, E. (2016). Disaster resilience and complex adaptive systems theory. *Disaster Prevention and Management: an International Journal, 25*(2), 196–211.

Coetzee, C., & Van Niekerk, D. (2018). Should all disaster risks be reduced? A perspective from the systems concept of the edge of chaos. *Environmental Hazards, 17*(5), 470–481.

Coker, R. J., Hunter, B. M., Rudge, J. W., Liverani, M., & Hanvoravongchai, P. (2011). Emerging infectious diseases in southeast Asia: Regional challenges to control. *The Lancet, 377*(9765), 599–609.

CSIS. (2019). *Can digital health help stop the next epidemic?* Center for Strategic and International Studies.

Cumming, G. S., & Peterson, G. D. (2017, September). Unifying research on social–ecological resilience and collapse. *Trends in Ecology & Evolution, 32*(9). https://doi.org/10.1016/j.tree.2017.06.014695

Cutter, S. L., Barnes, L., Berry, M., Burton, C., Evans, E., Tate, E., & Webb, J. (2008). A place-based model for understanding community resilience to natural disasters. *Global Environmental Change, 18*(4), 598–606.

De Groote, H., Kimenju, S., Munyua, B., Palmas, S., Kassie, M., Bruce, A. (2020). Spread and impact of fall armyworm (Spodoptera frugiperda J.E. Smith) in maize production areas of Kenya. *Agriculture, Ecosystems & Environment, 292*, 106804.

Gall, M., Nguyen, K. H., & Cutter, S. L. (2015). Integrated research on disaster risk: Is it really integrated? *International Journal of Disaster Risk Reduction, 12*, 255–267.

Glantz, M. H. (2003). Guidelines for establishing audits of agricultural-environmental hotspots, *Environment and Natural Resources Series, 15*.

Grol-Prokopczyk, H. (2013). Thai and American doctors on medical ethics: Religion, regulation, and moral reasoning across borders. *Social Science and Medicine, 76*(1), 92–100.

Han, B. A., Kramer, A. M., & Drake, J. M. (2016, July). Global patterns of zoonotic disease in mammals. *Trends in Parasitology, 32*(7). https://doi.org/10.1016/j.pt.2016.04.007565

Hort, K., Sommanustweechai, A., Adisasmito, W., & Gleeson, L. (2017). Stewardship of health security: The challenges of applying the One Health approach. *Public Administration and Development, 39*, 23–33. https://doi.org/10.1002/pad.1826

IPBES. (2020). Workshop on Biodiversity and Pandemics Executive Summary Plenary-approved activities, set out in section 6.1, *Intergovernmental Platform on Biodiversity and Ecosystem Services*, 1–18.

IPCC. (2014). Fifth assessment report to the international panel on climate change. In Y. Hijioka, E. Lin, & J. Pereira (Eds.), *Working Group II: Fifth Assessment Report, Chapter 24: Asia*, Cambridge University Press.

IPCHS. (2019). *State of the Region 2019: Health Security in the Indo-Pacific*, Indo-Pacific Centre for Health Security, Canberra.

Ishida, M. (2008). GMS economic cooperation and its impact on CLMV development. In C. Sotharith (Ed.), *ERIA Research Project Report* (pp. 115–140), IDE-JETRO, Chiba, no. 20074.

Jackson, T., & Chatterjee, K. (2019). *Bracing for the future: A participatory exercise on foresight for food systems in South Asia, February 2019.*

Kanta, K. R., de Sherbinin, A., Jones, B., Bergmann, J., Clement, V., Ober, K., ... Midgley, A. (2018). *Groundswell: Preparing for internal climate migration.* The World Bank.

Keesing, F., Daszak, P., Belden, L., & Dobson, A. (2010). Impacts of biodiversity on the emergence and transmission of infectious diseases. *Nature, 468*, 647–652.

Lauster, M., & Hansen-Casteel, S. (2018). On some fundamental methodological aspects in foresight processes. *European Journal of Futures Research, 6*, 11.

Montgomery, J. M., Woolverton, A., Hedges, S., et al. (2019). Ten years of global disease detection and counting: Program accomplishments and lessons learned in building global health security. *BMC Public Health, 19*, 510.

Morens, D., Daszak, P., Markel, H., & Taubenberger, J. (2020). Pandemic COVID-19 joins history's pandemic legion. *Clinical Science and Epidemiology, 11*(3), e00812-e820.

Nordhagen, S., Calverley, D., Foulds, C., Thom, L. L., & Wang, X. (2012). Credibility in climate change research: a reflexive view. Tyndall Centre for Climate Change Research, Working Paper (Vol. 152), Tyndall Centre for Climate Change Research

Okello A., Welburn S., & Smith, J. (2015). Crossing institutional boundaries: mapping the policy process for improved control of endemic and neglected zoonoses in sub-Saharan Africa. *Health Policy Plan, 30*(6), 804–812.

Piot, P. (2014). Ebola's perfect storm. *Science, 345*(6202), 1221.

Rigaud, K. K., Sherbinin, A. D., Jones, B., Bergmann, J., Clement, V., Ober, K., Schewe, J., Adamo, S., Mccusker, B., Heuser, S., & Midgley, A. (2018). *Groundswell: Preparing for internal climate migration.* The World Bank.

Sen, A. K. (1999). *Development as freedom.* Oxford University Press.

TERN. (2016). *Australia's terrestrial ecosystem research framework strategic framework 2016–2025.* University of Queensland.

Thomson, J., Prior, T., Paddon, M., & Plant, R. (2014). Resilience or transformation? A case study of environmental change, livelihood, & shifting disaster risk thresholds from within Lao PDR. *Conference Proceedings: Climate Change & Social Impacts*, Colombo.

Thomson, J., Mukhiebir, P., Plant, R., & Prior, T. (2015). Expanding our understanding of Early Warning Systems and the role of general system

resilience. *Conference Proceedings: Australian-New Zealand Disaster & Emergency Management Conference*, Gold Coast, Australia.
Thomson, J., Mukhiebir, P., & Plant, R. (2018). The imperatives and challenges of All-Hazards Early Warning Systems. *Conference Proceedings: 8th International Conference on Building Resilience ICBR*. Lisbon.
Thomson, J. (2019). Keynote: Improved early warning: A systems approach for forecasting transboundary multi-hazard disaster events. *Workshop Proceedings: ASEAN Intergovernmental Conference on Transboundary Early Warning Systems (EWS)*, Phnom Penh Cambodia.
UNEP-ILRI. (2020). *Preventing the next pandemic: Zoonotic diseases and how to break the chain of transmission*, United Nations Environment Programme and International Livestock Research Institute, Nairobi, Kenya.
UNFAO. (2020). Desert Locust Bulletin N 505, General situation during October 2020 Forecast until mid-December 2020, 2nd November, Rome.
UNISDR. (2009). *Terminology on disaster risk reduction, briefing note 03* (p. 18). United Nations International Strategy for Disaster Reduction.
UNISDR. (2015). Sendai framework for disaster risk reduction 2015–2030, *Third World Conference on Disaster Risk Reduction*, Sendai, Japan, 14–18 March 2015, March, pp. 1–25.
Walker, B., Carpenter, S. R., Rockstrom, J., Crépin, A., & Peterson, G. D. (2012). 'Drivers, "slow" variables, "fast" variables, shocks, and resilience.', *Ecology and Society, 17*(3), 30–34.
World Bank. (2021). Safeguarding animal, human and ecosystem health: One health at the World Bank. https://www.worldbank.org/en/topic/agriculture/brief/safeguarding-animal-human-and-ecosystem-health-one-health-at-the-world-bank, accessed 18 July 2021.
Zommers, Z., Singh, A., Kelman, I., & Glantz, M. H. (2014). Introduction In Z. Zommers, & S. Singh (Eds.), *Reducing Disaster: Early Warning Systems for Climate Change*, Springer.

CHAPTER 11

The Role of Gardening in Response to Cascading Disaster on Peri-Urban Fringe of Port Vila, Vanuatu

Andrew MacKenzie

Introduction

Rural–urban migration in Pacific Small Island Developing States was once temporary. Villagers would travel to towns to earn cash by working or selling handicraft and fresh produce; then return to their home villages. Since the late twentieth century; with growing formal economies centred on towns and cities, rural–urban migrants have been more likely to permanently settle. These new migrant communities, now second and third generation urban residents, require the provision of services, infrastructure and employment opportunities not previously built into the original urban infrastructure. Too often, government policy in many island nations has not kept up with the shifting demographic patterns and residents'

A. MacKenzie (✉)
University of the South Pacific, Port Vila, SHEFA 1980, Vanuatu
e-mail: andrew.mackenzie@usp.ac.fj

A. Lukasiewicz and T. O'Donnell (eds.), *Complex Disasters*, Disaster Risk, Resilience, Reconstruction and Recovery,
https://doi.org/10.1007/978-981-19-2428-6_11

requirements, due to lack of skills, resources and financial means or political will to anticipate and respond to future urban needs.

Small Island Developing States (SIDS) refer to mainly Pacific, Caribbean and Indian Ocean nations and territories that share similar features such as small populations, geographic remoteness and high transport and transaction costs. They are vulnerable to global economic shocks, biodiversity loss and climate change impacts because communities often lack economic alternatives to exploiting fragile marine and terrestrial ecosystems (UN, 2020). The lack of services and infrastructure is endemic in the rapidly expanding cities of SIDS in the Pacific (Petrou & Connell, 2017). Despite these rapidly growing urban populations, however, the majority of voters still live in rural areas. Political incentives to address urban policy issues are therefore lacking; compounding social and environmental problems, particularly in the context of increasing frequency and intensity of natural disasters. Consequently, issues such as poor housing, unemployment, inadequate transport infrastructure, environmental degradation, the rapid spread of non-communicable diseases and social discontent have become chronic. Worse still, these issues have become increasingly tolerated by policymakers (Petrou & Connell, 2017). These peri-urban communities are the first place that new migrants settle in search of low-cost living and employment. The resulting confluence of population growth, infrastructure deficits and poor health outcomes are most evident on the rural–urban fringe in the Pacific (Baker & Gauri, 2017).

Peri-urban development in the Pacific cities ranges from planned subdivisions to informal settlements, often built on otherwise productive landscapes that support the growing of food and other products catering for the expanding towns and cities (Bekessy et al., 2012). As the Pacific urbanises, these peri-urban landscapes become the sites of spatial contest between the need to produce fresh food and the need to provide housing and infrastructure. This chapter examines how this spatial contest plays out in Port Vila, the capital of the Republic of Vanuatu in the Western Pacific. It examines these changes in the context of the cascading effects of increased frequency and intensity of disasters resulting from climate change.

The relationships between disaster and urban socio-ecological systems (SES) are not well researched in the Pacific (Loïc Le Dé et al., 2018). Research has tended to focus on preparation and emergency response rather than systemic changes to improve resilience to disaster (Donovan,

2017). Urban scholarship has much to offer in relation to understanding how SESs interact with disaster on the rural–urban fringe in the Pacific (Baker & Gauri, 2017). This chapter explores the relationships between hazards, vulnerability and adaptive capacity of communities in the peri-urban context (Pelling, 2011). Rather than focussing on causal and risk-based approaches, it conceptualises disasters as assemblages of relationships between unfolding events, affected communities and the environment. Drawing on the work of Donovan (2017) this chapter places cascading disasters in the context of existing social problems and exposes the levels of institutional lack of knowledge that disasters reveal (Donovan, 2017). It adapts Pelling's (2011) framing of SES stressors to identify how and why the vulnerability of communities may change in response to cascading disasters (Table 11.1). Identifying the SES stressors allows decision-makers to address the factors that can lead to systems failure before it reaches a threshold. Table 11.1 identifies four stressors and the relationship between the built environment and the surrounding gardens that may contribute to them. These stressors on socio-ecological systems result from the nonlinear and interdependent relationship between cascading disasters and anthropogenic processes (Cutter, 2018). For this reason, cascading disasters once considered to be extreme and consequential, are becoming more frequent and chronic for communities in the longer term (Ballard et al., 2020).

In pursuing this line of thinking, we can see that systems become stressed and communities can become vulnerable as a result of cascading disasters, maladaptive practices and the consequences of those actions. Alternatively, communities that engage in adaptive practices, such as supporting gardening, demonstrate higher levels of resilience in the face of the social, environmental and economic stressors.

The rural–urban fringe is a contested space between the productive use of landscapes for growing food and the provision of essential services and infrastructure in context of disasters. On the rural–urban fringe, those that benefit from the infrastructure and housing that create urban sprawl also benefit from the productive agricultural landscapes as sources of food and income (Magee et al., 2016). From an urban policy perspective, limiting urban expansion or displacing productive landscapes in favour of growth can be potentially both beneficial and detrimental to those affected. For this reason, this research focuses on socio-ecological stressors impacting communities and the adaptive actions that individuals take, rather than evaluating the merits of urban policies in SIDS.

Table 11.1 Stressors on peri-urban SES in Port Vila

SES stressor	*Built spaces*	*Gardening*
Lack resources to change practices	Informal settlements in Port Vila are located on land that may be subject to arbitrary decisions of custom land owners	Land for gardening is allocated based on kinship relations and ties to key decision makers in the community. Access to land for gardening is uneven
Lack information	Informal settlements are often built on custom land and gardens are cleared from adjacent bush by opportunistic villagers seeking to supplement their income	During hardship, changes in gardening activity is driven by necessity (in response to hardship) rather than through strategic planning or business development that responds to the market demand for fresh produce
Failure of institutions	Lack of planning, design standards, institutional support and enforcement of leases results in poor outcomes in terms of planning for population growth and service provision on the rural–urban fringe	Lack of coordination between government departments results in land use planning that favours leasing to private interests for development rather than adopting a more strategic approach to preserving productive agricultural peri-urban land for economic development and food security
SES overwhelmed	A combination of natural disasters, poor regulation and arbitrary decision making by custom land owners result in settlements continually being built and rebuilt without lasting improvements in sustainable development outcomes	Sea level rise, urban expansion, erosion, pollution and weak regulation will reduce the land area available for gardening on the rural–urban fringe. Gardening will become less viable as an adaptive action for a smaller percentage of the growing peri-urban communities

Source Developed by Author, adapted from Pelling (2011)

Migration, Disasters and Landscape Change

The changing nature of disasters sets the context for this analysis whereas population increase is a major driver of the stressors. The United Nations Global Compact for Safe Orderly and Regular Migration (GCM) is instructive for considering urban challenges facing Pacific nations experiencing population pressures from migration (Newland, 2018). The GCM acknowledges that the multi-causality of migration interacts with political, economic and demographic drivers. It also states that policy coherence rests on a number of global instruments related to climate change, disaster and environmental governance that provide a structure to analyse these changes. This research uses the Sendai Framework for Disaster Risk Reduction to identify how the cascading effects of disaster will affect socio-ecological systems in three ways.

1. Human alteration of natural systems (through maladaptive practices) will exacerbate environmental decline and community vulnerability over time.
   Pacific SIDS are particularly vulnerable to the effects of climate change (Komugabe-Dixson et al., 2019). In urban areas, this vulnerability is compounded by poor planning and unregulated sprawl into productive landscapes that are often subject to natural disasters, such as flooding, landslides and tsunamis. The lack of spatial planning policy as much as environmental damage places these communities at greater risk.
2. Rural–urban migration will place more people at risk.
   Rural–urban migration in the Pacific is characterised by the breaking of customary ties to home islands and the reconstitution of those relationships in urban settlements (Rawlings, 1999). Internal migrant communities form on the peri-urban fringe; often on custom land and at the behest of the custom land owner who is usually the village chief. Settlements may therefore emerge under the tenuous consent of a powerful individual (McDonnell, 2018). As a result, neither the community nor the government is willing to make long-term investments in infrastructure or services. This results in the uneven pressure on existing services on the peri-urban fringe and the consequence is a greater risk of vulnerability for both the migrant and established communities (Petrou & Connell, 2017).

3. The impacts of disaster will become continuous as recovery times shorten and reducing the resilience of systems to shocks.

The combined effects of poor planning, poor environmental management and increased population pressure on resources result in a high exposure to risk for peri-urban communities. In addition, the cascading effects of disaster over time will become more chronic as recovery times between hazardous events shorten. The nature of recovery will therefore change as socio-ecological systems are altered by these factors.

Port Vila, Vanuatu

Port Vila is home to over 50,000 predominantly Ni-Vanuatu, most of whom have migrated from the 83 inhabited islands of the archipelago over the past forty years seeking jobs in tourism, retail, construction and the public service (VNSO, 2018). Port Vila's population grew an average of 2.2% in the municipal areas and the peri-urban areas at 2.6% per year in the decade between 2010 and 2020 (World Bank, 2020). While the city is growing at a greater rate than the rural areas, three quarters of the Vanuatu's population are from rural villages and predominantly subsist by gardening on customary owned land (Bedford et al., 2017). Under the Vanuatu constitution, all land belongs to Ni-Vanuatu. Custom land constitutes over 90% of land, including all land outside of the urban footprint of Port Vila and other urban settlements in Vanuatu (McDonnell, 2018). Custom land ownership rules vary from island to island; in most cases the village chief has final decision on custom land tenure (McDonnell, 2018). Legally, property rights can be purchased for up to seventy years under lease agreements between the custom land owner and the lessee; resulting in a de-facto private market for land sales (Ballard et al., 2020). On Efate, the most populous island and home to Port Vila; over 50% of custom land is under lease to expatriate investors and wealthy middle-class Ni-Vanuatu (McDonnell, 2018). Leaseholders enjoy exclusive rights to their land, however, in practice, many villagers continue to access leased land to garden, fish or collect timber under explicit or tacit consent of the leaseholder (McDonnell, 2018). The continuing access to leased land is not guaranteed, but tolerated, especially during times of hardship caused by disaster. This combination of customary and contemporary systems of land tenure, in the context of frequent natural disasters, manifests in an opaque system of occupation, development and settlement on the peri-urban fringes of Port Vila (Lindstrom, 2012).

Politically, there is little interest in resolving the problems of rapid urban growth, even though urban planning policy is inadequate, diffuse and poorly enforced (Petrou & Connell, 2017). The Vanuatu government's ambivalence towards urban planning policy allows for a more dispersed power relationship between those with control of over urban landscapes and those exercising power in the absence of legal enforcement (McDonnell, 2018). Likewise, recent agricultural policy has been almost silent on the role of urban agriculture and ambiguous about peri-urban land management for food security, despite these landscapes producing the bulk of Port Vila's fresh food (James, 2018).

Gardening; an Adaptive Action on the Rural–Urban Fringe

Gardening is central to the identity of many Pacific Island people; especially in Vanuatu. Traditional village life is organised around the allocation of small allotments by the chief to members of the village community. The garden is not usually co-located with the house but is within 500 m of the village. The majority of plants are grown for food, construction materials, fibre and medicine (Lebot & Simeoni, 2015). Gardening is one of the key activities that rural–urban migrants continue to practice as long as they have access to land and sufficient resources and time over and above formal employment (James, 2018; Lindstrom, 2012). In times of hardship, brought about by disaster, gardening substitutes incomes, provides additional food and reinforces cultural and social ties (Lindstrom, 2012). In essence gardening is an adaptive action that helps to reduce the vulnerability of the urban poor who live on the rural–urban fringe (Gallent & Shaw, 2007). It also plays an essential role in population health and well-being and has particular importance to the urban economy. The bulk of fresh fruit and vegetables consumed in Port Vila come from the three area councils on the rural–urban fringe (FAOUN, 2015). The three peri-urban local areas in Port Vila, Erakor, Eratap and Mele produce 61% of all vegetable crops grown in the Port Vila (FAOUN, 2015). However, this food sovereignty is not reflected in the Ni-Vanuatu diet due to the prevalence of imported dry goods and processed food, resulting in very high rates of non-communicable diseases, especially in urban areas where the access to gardens is scarcer (James, 2018).

Adaptive Actions on the Rural–Urban Fringe

This chapter examines how gardening is an adaptive action on the expanding urban edge that reduces vulnerability. Donovan (2017) argues that adaptation is about improving livelihoods in the same way that reducing vulnerability and the building of resilient socio-ecological systems improves wellbeing. However, the decisions that individuals make (to improve wellbeing) are dependent on a range of motivations. For example, to build homes for shelter and to garden to provide food and cash crops are not necessarily adaptive or maladaptive actions, but rather individual components of disaster response that have a spatial dimension (Biesbroek et al., 2009).

In urban scholarship, adaptation is conceived as a process that is technical, expert and results in a defensive analysis of urban conditions, rather than a socio-ecological analysis of adaptive potential of urban systems.

In considering the spatial dimensions of adaptation, Pelling (2011) argues that socio-ecological systems fail for four reasons: The community lack resources, forcing them to persist with unsustainable practices; they lack information, undermining self-regulating practices; Institutions fail to adapt; or the speed of change overwhelms the system regardless of the communities' capacity (see Table 11.1).

Research Method

The research reported in this chapter analysed High Resolution Satellite Imagery (HRSI) data on the peri-urban fringe of Port Vila (Fig. 11.1). The Mele study area is a more traditional village in Vanuatu where gardeners are allocated lots outside of the village area by the chief. Normalised Difference Vegetation Index (NDVI) analysis examined changes in land use and land cover (Bhatt & Rao, 2018). Selected study area was analysed over seven years (2014–2021). The research project began in 2014, preceding cyclone Pam, the most consequential and physically damaging event in modern Vanuatu history (Handmer & Iveson, 2017). The most recent data was collected in 2021, twelve months into the national state of emergency declared at the beginning of the COVID-19 pandemic. These two events were not examined in isolation as many other factors and natural disasters directly or indirectly affected the landscape patterns of uses in Port Vila during this time. The observations reported here about the role of gardening as an adaptive action need to

Fig. 11.1 Location of study areas and Port Vila (*Source* Google, 2021)

be considered in the context of SES stressors that, over time, will make it harder for peri-urban communities to continue to turn to gardening following disaster.

Changes in gardening activity were analysed between 2014 and 2021 to determine the changes to landscape patterns adjacent to villages and other urban settlements in the study areas. The selected study areas included peri-urban settlements (formal subdivisions, traditional villages and informal settlements), and surrounding landscapes accessed for gardening. Ground truthing was undertaken using site inspections, drone footage and aerial photography to validate the assumptions about the data and the real-time representation of land cover types. The goal of the analysis was to determine the changes in land use over time as an indicator of the nature and extent of gardening activity in the context of an expanding urban footprint.

Of interest to this chapter are the changes from dense woody vegetation to partially cleared/grassland. This change is most indicative of limited woody vegetation cleared using a bush knife for gardening activity which is the dominant means of preparing land for planting gardens. Unlike commercial horticultural production, the gardeners on Port Vila's urban fringe do not remove all woody vegetation to plant crops. The study area boundaries were determined by topographical (steep slopes) and hydrological (rivers) features. These natural boundaries were selected because they represent natural boundaries limiting access to landscapes for gardening. Landscapes outsides of these boundaries were considered unlikely to be used for gardening.

Changing Landscape Patterns in Response to Disaster

This chapter focuses on findings from the Mele study area to highlight the interacting factors that influence the changes in land use as a result of gardening. Mele is the largest endemic village in Vanuatu. Melemaat, to the northwest of Mele is a migrant village established by Ni-Vanuatu from Maat village on the island of Ambae who migrated following a series of volcanic eruptions and cyclones in the mid-1970s. The Mele chief at the time granted the migrants from Maat permission to settle on Mele village land. Of interest to this chapter are the areas to the west of the Mele creek, north of the villages and south of the hill slopes of Klems hill. Three observations can be made about the changes in patterns over the study period.

The villages have grown in size and density reflecting the population increases and resulting building activity. While Vanuatu remains a predominantly rural population, it is the urban fringe areas that are experiencing the greatest increase in population growth. The Mele study area includes Mele village which is the largest village by population in Vanuatu (VNSO, 2018). The other major settlements are Melemaat, a migrant community originally from Ambae island and Blacksands community from Tanna. These three communities grew between 2014 and 2015 despite the destruction caused by cyclone Pam and the economic impact of the state of emergency. The data shows that Mele, the established village with custom ownership of most of the study area to the west of the Airport grew the most in size and density. This is consistent with the population growth of 2.6% per year for peri-urban communities. Of interest to this research was the impact of this population increase on the demand for landscapes in response to disasters.

The second observation from the data was the increase in forested areas to the east of Mele creek reflecting a change from grazing to forestry and commercial plantations. The properties to the east of Mele creek have been leased for many decades by established farming families. This land is predominantly used for grazing or commercial coconut plantations. The data shows a significant increase in plantations between 2014 and 2021. In contrast to the landscapes to the east of Mele creek, these changes are wholesale changes typical of a commercial operation and in response to other commercial factors beyond the scope of this study. The difference between the changes in landscape patterns between the leased commercial farms to the east of Mele creek and the incremental and fragmented changes to the custom land to the west reveals the difference in how these landscapes are accessed.

The Mele study shows a significant change from forest woodland to semi-cleared gardens between 2018 and 2021 (Figs. 11.2 and 11.3). This would suggest that the thinning of the forest was in part attributed to increased gardening activity and confirmed by field visits. The Mele study area also shows that these landscapes are gradually being displaced by other forms of development such as commercial intensive agricultural production and urban expansion. This can be seen as a continuum

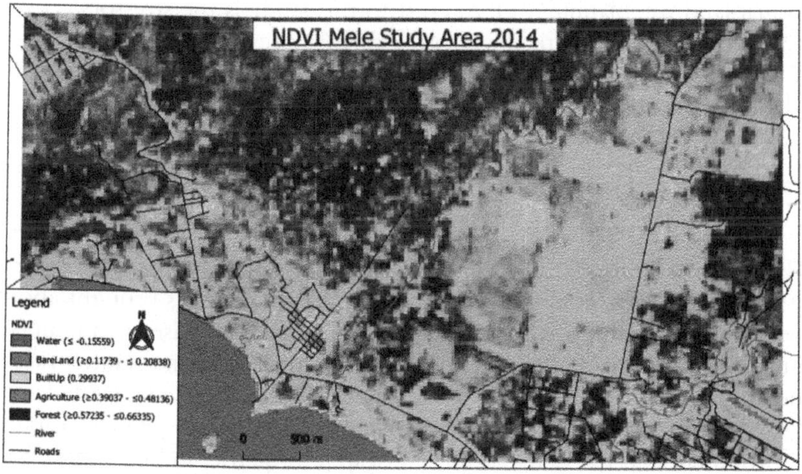

Fig. 11.2 NDVI of Mele study area 2014 before Cyclone Pam

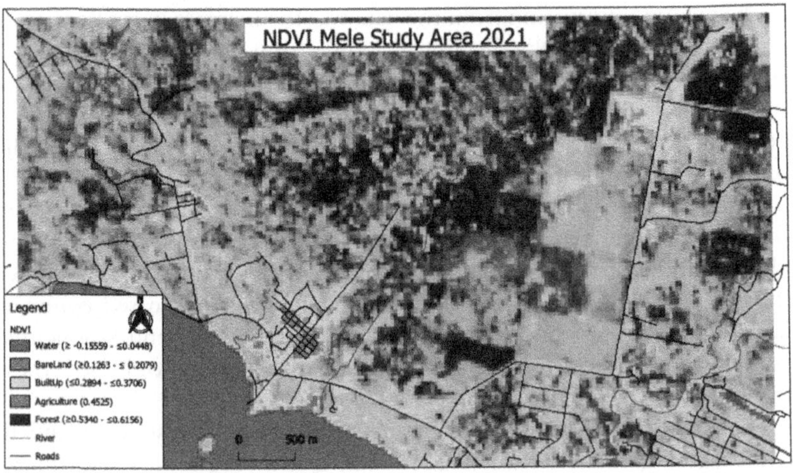

Fig. 11.3 Mele study area 2021 approximately 12 months since the declaration of the state of emergency and borders were closed

of a long-standing practice on the peri–urban fringe of opportunistic gardening alongside commercial agricultural production in the absence of formal land use planning. Of interest to this study is the stressors caused by cascading disasters that will place further pressure on residual landscapes to accommodate poor urban migrants who turn to gardening to adapt during disaster (Pelling, 2011).

Gardening as an Adaptive Action in Response to Cascading Disaster

This research can help us to gain a better understanding of how gardening fits into a broader understanding of the drivers of changes brought about by disaster. Gardening as an adaptive action is well recognised as an important cultural and economic activity in Vanuatu, however its value is rarely incorporated into land-use planning. The Mele study provides evidence of a strong correlation between the size, density and rate of growth of built environment and the demand for landscapes for gardening in response to cascading disasters. Interestingly, while Cyclone Pam was a large consequential event, it hasn't produced the same cascading effects as Covid-19 has in Mele. The reasons for this are a mix of interrelated factors to do with population growth and the rapid increase in poverty caused by the sustained loss of cash employment during the pandemic.

This is evidenced by the increase in gardening appears to be a response to the hardship caused by the loss of alternative forms of income and food as the tourism sector shut down. In contrast, gardening does not appear to be an adaptive response to hardship caused by cyclone Pam. This is perhaps due to a comparatively short time that the economy (and jobs) were impacted. Indeed, the flood of international aid workers and income flowing into Port Vila following Pam would have kept the resorts and other tourism businesses busy. This finding is comparable to the increase in gardening activity detected in other developing nations following a sustained economic shock. Following the collapse of the Soviet Union in 1989, Havana, the capital of Cuba experienced a similar economic shock as exports to the former Soviet Union collapsed (Saunders, 1999). The example of Havana is instructive for the Vanuatu government from a policy perspective. What started as a community-level urban agricultural movement to adapt to hardship became a municipal and then national-level goal to maximise the amount of peri-urban horticultural production (Koont, 2009), so much so that the government encouraged and supported a wide range of gardening and animal husbandry activities and enterprises across the entire city, not just on the urban fringes (Premat, 2009).

While lessons can be learned from the response of the municipal government in Havana, it should be noted that the factors that caused the hardships in Cuba were permanent and in response to a global geopolitical reordering that required a sustained change in Cuban economic policies. The challenge for gardeners in Port Vila is that the hardship caused by the pandemic is viewed as temporary and once the borders open and tourists return, so too will the jobs, the incomes and the livelihoods that depend on paid work rather than gardening. From a land use policy perspective, two problems face these communities. The first is that there is no way of knowing at this stage how long will these communities have to depend on gardening, and secondly, in the absence of robust planning policies will these landscapes be available and suitable for gardening in the event of another future disaster that will require Ni-Vanuatu to turn to gardening to reduce their vulnerability to disaster.

Conclusion

Gardening has, to some extent, buffered these communities (and the Port Vila economy) in times of hardship. However, without adequate attention being paid to the combined stressors of population increase (and the associated infrastructure development that goes with it) and the reduction of suitable land for gardening the SES will be at risk of crossing a threshold in the aftermath of future disasters.

In this context we argue that gardening is an orienting action post disaster and is an important adaptive tool for individuals. The multiple benefits are well documented and the diffuse nature of the peri-urban interface demonstrates that there is a degree of co-existence between the gardening communities and the informal settlements, particularly on the urban edge. The findings show that the land use pressures are more diverse than a simple competition for resources between built and non-built landscapes. When applying Pelling's SES stressors, we argue that gardening will continue to play an important role in reducing the vulnerability of individuals who have access to suitable land, however a combination of factors is contributing to reducing the land available for gardening and this presents a greater threat than one single factor such as urban edge development.

The aim of this chapter has been to understand how gardening, as evidenced by the changes to landscape patterns over time, provides clues about the importance of gardening as an adaptive action. Understanding the role of gardening within a socio-ecological systems perspective raises the prospect that DRR research and practice can gain a better understanding of how gardening fits into a broader SES understanding of the drivers of changes brought about by disaster. The reason that gardening receives little attention from a DRR planning perspective may be because, unlike agriculture or forestry, it is a land use that is for more diffuse, opportunistic and integrated into the expanding urban footprint. Yet for Port Vila, gardeners provide the bulk of fresh produce sold in the markets in town. Gardening provides meaningful alternatives to formal employment in times of hardship, it generates essential fresh produce and makes it available to poor urban residents through community networks that exchange goods using a barter along with cash. The diffuse, fragmented nature of gardening also reveals many of the difficulties for DRR researchers to evaluate the social and economic value of gardening and equally difficult for land use planners to allocate resources towards protecting viable landscapes from urban sprawl.

References

Baker, J., & Gauri, G. (2017). *East Asia and Pacific Cities: Expanding Opportunities for the Urban Poor*. Retrieved from Washington, DC.

Ballard, C., McDonnell, S., & Calandra, M. (2020). Confronting the naturalness of disaster in the Pacific. *Anthropological Forum*, 1–14. https://doi.org/10.1080/00664677.2020.1729698

Bedford, S., Siméoni, P., & Lebot, V. (2017). The anthropogenic transformation of an island landscape: Evidence for agricultural development revealed by LiDAR on the island of Efate, Central Vanuatu, South-West Pacific. *Archaeology in Oceania*, 53. https://doi.org/10.1002/arco.5137

Bekessy, S. A., White, M., Gordon, A., Moilanen, A., McCarthy, M. A., & Wintle, B. A. (2012). Transparent planning for biodiversity and development in the urban fringe. *Landscape and Urban Planning, 108*(2), 140–149. https://doi.org/10.1016/j.landurbplan.2012.09.001

Bhatt, C., & Rao, G. S. (2018). HAND (height above nearest drainage) tool and satellite-based geospatial analysis of Hyderabad (India) urban floods, September 2016. *Arabian Journal of Geosciences, 11*(19), 1–14.

Biesbroek, G. R., Swart, R. J., & van der Knaap, W. G. M. (2009). The mitigation–adaptation dichotomy and the role of spatial planning. *Habitat International, 33*(3), 230–237. https://doi.org/10.1016/j.habitatint.2008.10.001

Cutter, S. L. (2018). Compound, cascading, or complex disasters: What's in a name? *Environment: Science and Policy for Sustainable Development*, *60*(6), 16–25. https://doi.org/10.1080/00139157.2018.1517518

Donovan, A. (2017). Geopower: Reflections on the critical geography of disasters. *Progress in Human Geography, 41*(1), 44–67. https://doi.org/10.1177/0309132515627020

FAOUN. (2015). *Agriculture sector of Vanuatu severely damaged by tropical Cyclone Pam*. Retrieved from https://search-proquest-com.virtual.anu.edu.au/docview/1773557697?accountid=8330

Gallent, N., & Shaw, D. (2007). Spatial planning, area action plans and the rural-urban fringe. *Journal of Environmental Planning and Management, 50*, 617–638.

Handmer, J., & Iveson, H. (2017). Cyclone Pam in Vanuatu: Learning from the low death toll. *Australian Journal of Emergency Management, 32*(2), 60–65.

James, S. 2018. Food Security in Port Vila, Vanuatu. Port Vila, Vanuatu.

Komugabe-Dixson, A. F., de Ville, N. S. E., Trundle, A., & McEvoy, D. (2019). Environmental change, urbanisation, and socio-ecological resilience in the

Pacific: Community narratives from Port Vila, Vanuatu. *Ecosystem Services, 39*, 100973. https://doi.org/10.1016/j.ecoser.2019.100973

Koont, S. (2009). The urban agriculture of Havana. *Monthly Review, 60*(8), 44–63.

Lebot, V., & Simeoni, P. (2015). Community food security: Resilience and vulnerability in Vanuatu. *Human Ecology, 43*(6), 827–842.

Lindstrom, L. (2012). Vanuatu migrant lives in village and town. *Ethnology: An International Journal of Cultural and Social Anthropology, 50*(1), 1–15.

Loïc Le Dé, R., Leone, F., & Gilbert, D. (2018). Sustainable livelihoods and effectiveness of disaster responses: A case study of tropical cyclone pam in Vanuatu. *Natural Hazards, 91*(3), 1203–1221. http://dx.doi.org.virtual.anu.edu.au/https://doi.org/10.1007/s11069-018-3174-6

Magee, A. D., Verdon-Kidd, D., Kiem, A. S., & Royle, S. A. (2016). Tropical cyclone perceptions, impacts and adaptation in the southwest pacific: An urban perspective from Fiji, Vanuatu and Tonga. *Natural Hazards and Earth System Sciences, 16*(5), 1091–1105. https://doi.org/10.5194/nhess-16-1091-2016

McDonnell, S. (2018). Selling "sites of desire": Paradise in reality television, tourism, and real estate promotion in Vanuatu. *Contemporary Pacific, 30*(2), 413–436.

Newland, K. (2018). The global compact for safe, orderly and regular migration: An unlikely achievement. *International Journal of Refugee Law, 30*(4), 657–660.

Pelling, M. (2011). *Adaptation to climate change*. Routledge.

Petrou, K., & Connell, J. (2017). Rural-urban migrants, translocal communities and the myth of return migration in Vanuatu: The case of Paama. *Journal De La Société Des Océanistes, 144–145*, 51–62.

Premat, A. (2009). State power, private plots and the greening of Havana's urban agriculture movement. *City & Society, 21*(1), 28–57. https://doi.org/10.1111/j.1548-744X.2009.01014.x

Rawlings, G. E. (1999). Foundations of urbanisation: Port Vila Town and Pango Village, Vanuatu. *Oceania, 70*(1), 72–86. https://doi.org/10.1002/j.1834-4461.1999.tb02990.x

Saunders, J. (1999). Seizing the reins: Green city projects in Tbilisi, Russia, Luton, UK, Havana, Cuba, and Berkeley, US. *New Internationalist, 313*, 28–29.

UN. (2020). *World Social Report 2020: Inequality in a Rapidly Changing World*. UN.

VNSO. (2018). *Household Income and Expenditure Survey 2017*. Retrieved from Vanuatu Government, https://vnso.gov.vu/index.php/en/statistics-report/survey-reports/household-income-expenditure-survey

World Bank. (2020). *World Bank rankings 2020*. Retrieved from World Bank.

PART IV

Protracted Disasters and their Impacts on Recovery

CHAPTER 12

Public Hazard Awareness for Culturally and Linguistically Diverse Communities During Protracted Events: A Case Study of Japan's Pandemic Response

David Green and Justin Whitney

Introduction

Japan is a country known for disasters, be they natural, man-made or fictional, as portrayed in popular culture. International observers are familiar with dramatic worldwide coverage of the 11 March 2011 'triple disaster', the 1995 Great Hanshin-Awaji Earthquake, and Godzilla, a gigantic beast that has repeatedly clawed its way out of the sea, destroying everything in its path, since 1954. Thanks to this extensive

D. Green · J. Whitney (✉)
Graduate School of Law, Nagoya University, Nagoya, Japan
e-mail: justin.whitney@outlook.com

D. Green
e-mail: david.green@law.nagoya-u.ac.jp

A. Lukasiewicz and T. O'Donnell (eds.), *Complex Disasters*, Disaster Risk, Resilience, Reconstruction and Recovery,
https://doi.org/10.1007/978-981-19-2428-6_12

experience with disaster, Japan is a country that is generally considered well-prepared for sudden onset events. Regularly scheduled disaster drills, disaster training programs and education initiatives targeting neighbourhood associations, schools and places of work are both widespread and common.

However, whereas most disaster events in Japan, such as earthquakes, tsunami, typhoons and the occasional radioactive monster come quickly and require an immediate and short-term response before shifting to longer-term recovery activities, the corona virus has represented a new and different challenge to the government. Not only is COVID-19 a protracted event requiring a prolonged response effort that places ongoing demands on the nation's healthcare resources, but the government has also had to reconsider its approaches to public messaging. Messaging approaches have traditionally focused on simple, predetermined mitigation strategies, such as tethering furniture to walls to reduce injury following an earthquake, guidance on preparing emergency survival kits and instruction on appropriate evacuation plans and procedures. The static nature of the content of these messages has lent itself to translation into multiple languages, which the Japanese government has used to communicate hazard risk to its growing foreign-born population.

While effective in guiding community behaviour before, during and immediately after sudden onset disasters such as those common in Japan, this approach bears little relevance to a protracted hazard like the COVID-19 pandemic. In contrast to the above-mentioned strategies that are relatively static in their messaging, the Japanese government, like those of other nations, must continually update official messaging as knowledge of the virus and its accompanying response evolves, and as people's behaviour and attitudes towards the virus change over time. This is a challenge that is further complicated by the need to effectively compete with conflicting narratives and misinformation—sources to which culturally and linguistically diverse (CALD) populations are seen to be particularly susceptible.

COVID-19 thus represents a novel challenge for Japan in addressing a major protracted event. Japan is far from alone in facing this challenge; Australia, a nation considered one of the traditional countries of immigration, has openly and repeatedly acknowledged that more needs to be done to effectively convey health messaging to its CALD groups (Butcher, 2020; Kelly et al., 2020). This was particularly the case in Melbourne, following widespread outbreaks of COVID-19 in several public housing towers, populated largely by people from CALD backgrounds in July 2020 (Wild et al., 2021). Like Japan, Australia is regularly forced to confront significant natural hazards, such as bushfires, floods and cyclones. However, Australia's multicultural policies tend to pay only cursory attention to hazard awareness and preparedness, and are likewise largely oriented around rapid-onset events. By contrast, hazard-related matters act as a central driver of Japan's multicultural policies, where municipal 'multicultural coexistence plans' often prioritise disaster awareness for CALD residents (MIC, 2020b). The way in which Japan has endeavoured to adapt its existing CALD outreach strategies to communicating COVID-19 directives to its migrant communities may thus provide lessons from which other countries, like Australia, can draw.

While the Japanese government may exhibit high levels of preparation for rapid-onset disasters, it was much less prepared for a protracted event like the corona virus. For example, although the government can mandate closures and fine non-compliant businesses as of February 2021 (Japan Times, 2021), it lacks the legislative powers to mandate a full lockdown. Policies to stop the spread of COVID-19 have included the ridiculed 'Abenomask' response in which every household in Japan was to receive two reusable masks (Mainichi Daily News, 2020), and finger pointing is common between the national and local governments across successive waves of sharply increasing infection rates followed by 'state of emergency' declarations that have been of questionable effectiveness (Nikkei Asia, 2021a, 2021b). Although Japan was lauded for its COVID-19 response in the early onset of the pandemic, the nation succumbed to subsequent waves of the virus, as confused messaging on keeping the population safe abounded. But how has Japan's public messaging been catering to its growing CALD population? While government at all tiers in Japan has actively attempted to engage CALD groups with hazard awareness information, much of it has tended to be translations of the largely static information provided to the Japanese populace as described above. In the current evolving and protracted crisis, how have Japanese local governments adapted their messaging strategies to keep their CALD communities safe?

This chapter is primarily concerned with Japanese government outreach to its CALD community in the wake of the COVID-19 pandemic. We look to answer how a relatively homogenous country like Japan has worked to provide essential information to its CALD community during this unique protracted event. Although all communities are impacted in the event of a disaster, that impact can cascade out from CALD groups, especially if government messaging is lacking. Indeed, as has repeatedly been the case in Australia, there have been notable reports in the Japanese media of COVID-19 spreading first from CALD communities into greater society (Kyoto, 2020; NHK News, 2021). Like in Australia, factors including lifestyle, language barriers and financial difficulties are all considered causes for the emergence of clusters in Japanese foreign communities (Osaki, 2020). By examining information outreach to CALD communities in Japan, we aim to shed some light on government information dissemination methodologies, the relative uniformity (or lack thereof) in government messaging, and ultimately look to provide a starting point for discussion on how to improve information outreach to CALD communities in the event of a protracted disaster. Given the similarities in the dilemmas facing Japan and Australia in effectively engaging their respective CALD communities in this protracted event, we consider how Australia might learn from Japan's evolving actions.

Disaster and Japanese Government Response

For the purpose of this chapter, we consider Japan's foreign population to be synonymous with its culturally and linguistically diverse population. Japanese society has traditionally had a high level of ethnic homogeneity, historically low numbers of immigrants and strict citizenship laws, making immigrants and immigrant communities distinct minorities. Although Japan has a contingent of former colonial subjects, mainly from Korea, that entered the country before World War II and continue to maintain foreign nationality, the vast majority of Japan's CALD population entered the country after the revision to the 1990 Immigration Control and Refugee Act (Green, 2017). Most CALD residents are thus relative newcomers to Japan. As of 2020, Japan's foreign population stood at 2.8 million people, or 2.3% of Japan's total population (MOJ, 2021). While this is a small proportion of the overall Japanese populace, it is steadily growing and likely to expand further as Japan deals with a rapidly ageing population.

Foreigners being caught up in disasters in Japan is not uncommon. The 1995 Great Hanshin-Awaji earthquake, which claimed an estimated 6131 lives, saw a disproportionately high number of foreigners killed and injured, with some estimates indicating a death rate approximately four times higher than Japanese residents (Cabinet Office, 2000; Sasaki,

1995; Takezawa, 2000). Anecdotal evidence from more recent disasters such as the 2016 Kumamoto earthquake, Typhoon Jebi in September 2018 and Super Typhoon Hagibis in October 2019, suggests that foreign tourists and non-Japanese residents continue to be disproportionately affected by hazards (Baird, 2019; Japan Times, 2019; Kikuchi & Osumi, 2016; Yoshida, 2016). This greater vulnerability of foreign nationals has not gone unnoticed by the Japanese government. One significant example of government acknowledgement of this vulnerability gap was the 2006 circulation of a notification calling for all subnational governments to develop Regional Multicultural Coexistence Promotion Plans by the Ministry of Internal Affairs and Communications.

Although development of this notification finds its roots in the activities following the 1995 earthquake, it was not until after the Niigata Chuetsu earthquake in October 2004 that much of the post-disaster multilingual support for foreign nationals became formalised. In 2005, the Council of Local Authorities for International Relations (CLAIR), an affiliated organisation of the Ministry of Internal Affairs and Communications (MIC), developed multilingual emergency management information, tools and resources for interpreters and translators. In the same year, the International Office of the MIC established the 'Study Group on the Promotion of Multicultural Coexistence'. After meeting seven times over the ensuing 12-months, this group devised the overarching Regional Multicultural Coexistence Promotion Plan scheme to be recommended to local governments. With natural disasters being a key driver of this initiative, disaster prevention strategies such as emergency drills and training, and ensuring access to disaster-related information through multilingual support formed one of the six specific measures highlighted within the notification (MIC, 2006).

Thus, in just over ten years after the Great Hanshin-Awaji earthquake, a model to systemise much of the ad-hoc work that had been carried out to support the foreign residents in the aftermath of that and subsequent disasters was finally deployed nationally (MIC, 2006; Tamura, 2017). All of Japan's regional prefectural governments, most of its largest cities, and many other local governments developed multicultural plans in the ensuing years, with many also integrating multicultural policy into their local emergency prevention plans. There has also been visible growth in government messaging, particularly in the lead up to the delayed 2020 Tokyo Olympics. At the national level, the *Jōhō Nanmin Zero* ('zero information refugees') strategy was launched in December

2016 to ensure information was adequately delivered to foreign residents and tourists during times of emergency (MIC, 2020a), and a raft of local and prefectural government initiatives ranging from training volunteer interpreters, including involving foreign residents in disaster training drills, and creating multilingual disaster preparedness manuals, posters and pamphlets, among other efforts have been developed.

Notwithstanding these activities, it remains a common practice in Japan to group foreign residents with other vulnerable populations in policy documents (Cabinet Office, 2013; Tokyo Metropolitan Government, 2020). While national-level efforts to coordinate and facilitate the development of Multicultural Coexistence Promotion Plans have seen a significant number of plans developed by local governments, the very nature of the locally led approach to their development has also resulted in significant variation in how local governments have fulfilled the nationally driven mandate. Local responsibility also brings with it limitations in resources and skills.

A common thread exists, however, across all tiers of government, insofar as much of the effort in developing awareness materials centres around translation of Japanese language materials. Despite the almost complete focus of the Japanese government engagement of CALD communities being on the provision of translation and interpretation services, post-event studies to ascertain the uptake and effectiveness of such materials have thus far not been conducted. Timeliness, accuracy, understandability and trustworthiness are all critical elements of crisis communication. With increasing global human mobility, translation is also becoming an essential component to government messaging (O'Brien, 2019, 315). It is crucial to note however, that the mere translation of Japanese language materials does not guarantee a message will be clear or understood by target audiences. O'Brien (2019, 307) points out, for example, that translation without cultural adaptation might not be sufficient.

Risk communications expert Peter Sandman (2007) argues that a key component of risk communication is 'precaution advocacy'—the act of alerting insufficiently concerned people to serious hazards. Sandman suggests it is crucial that people's concern (outrage and fear) be appropriately raised to motivate them to take precautions. Given the correlation between higher hazard experience and the adoption of hazard precaution strategies (Lindell et al., 2009, 1086), the likely diversity in such life experiences among those born overseas and between these groups and the native Japanese will no doubt add to the complexity associated with arousing sufficient concern—even fear—to motivate these communities to prepare. Without an understanding of what might motivate CALD residents to engage with hazard awareness materials, what their traditional sources of this information might be, and how these sources might be leveraged or brought into alignment with messaging designed to protect the broader community, it will be impossible for local governments to ensure the safety of their residents, both foreign and native. As we have seen with the case of the COVID-19 pandemic, this is especially true during times of emergency when the event is protracted and continually changing.

Mandated social distancing and nationwide lockdowns in the midst of the COVID-19 pandemic have resulted in a surge in the use of internet-based communication, as individuals seek to communicate with friends and family, work from home, continue their education, enjoy digital entertainment and obtain information. Japan is not alone in its increased reliance on internet-mediated communication and service delivery. This has extended to the delivery of government services, particularly in the largest cities hardest hit by corona virus. Indeed, while internet-based communication has been the most common and convenient approach for both practitioners and foreign residents for some time, it is now undoubtedly the first and primary means of communication.

The ubiquity and reach of the Internet may enable many people, including foreign residents, to be well informed regardless of their level of language proficiency. However, the global reach of this technology also poses significant challenges for delivering consistent messaging to all who may be affected. Research has found, for example, that in Australia, individuals who speak a language other than English at home are among those displaying higher levels of hesitancy regarding COVID-19 vaccinations (Biddle et al., 2021). This challenge may be linked to research suggesting individuals born overseas are more likely to be active online when consuming news compared to those not born overseas (Rodrigues, 2017), thus exacerbating misinformation being spread on social media platforms, including among culturally diverse groups (Kassam & Hsu, 2021; Xiao et al., 2021). It is clear that governments must better harness the same information technology, to ensure that a single source of truth is maintained and can cut through the plethora of information available on these platforms.

Hypothesis

As we can see from the above discussion, most outreach to CALD community residents in Japan originates at the local level, with considerably less involvement from the national government. Because we are mainly concerned with government outreach to CALD residents during the corona virus pandemic, we place our focus exclusively on this period. Since we have established that outreach appears to be initiated at the local level, at least when considering previous disasters, our first hypothesis (H1) stipulates that local governments distribute COVID-19 information to CALD communities in a disparate way, where some communities provide a higher level of information than others. A unified provision of information would help to ensure that a single government narrative is maintained and that all CALD communities receive consistent messaging as the COVID-19 pandemic evolves. However, because local governments initiate most CALD community outreach, we can reasonably expect inconsistent government messaging, likely adding to confusion over how to respond to the corona virus.

In attempting to provide some explanation for this hypothesis, we further speculate (H2) that community information provision is driven by local CALD community size and concentration. We would expect, in other words, that municipalities with larger populations and concentrations of foreign residents would be more likely to engage in active information dissemination campaigns targeting them. Communities with smaller CALD populations or with fewer local government resources would have less CALD community-targeting information available through official channels and likely have to rely on outside information sources. As the pandemic continues, problems associated with inconsistencies between the information available to CALD groups and the circumstances specific to that community are likely to worsen.

Methodology

In attempting to assess government emergency and health-related messaging and outreach to CALD residents in Japan during the pandemic, we look to internet-based communication targeting foreign residents. As noted above, the use of the internet in information dissemination has grown considerably in recent years. Websites are not only the most commonly used avenue for outreach to CALD residents (MIC, 2017), but internet-based communication has an added advantage for empirical research as it also provides a uniform metric that can be used to compare city efforts across Japan.

Our emphasis in this study is on outreach from Japan's ordinance-designated cities. Ordinance-designated cities have the distinction of being divided into administrative subsections and have a higher degree of autonomy from regional prefectural governments compared to smaller cities, towns and villages. Most importantly for this study, ordinance-designated cities represent the largest 21 of Japan's 1724 municipalities, including the Tokyo metropolitan region, and housed 42% of Japan's foreign population as of 2020 (MOJ, 2021). In looking at local government outreach efforts, we are able to capture a large portion of foreign population residency by focusing on these 21 cities. Because ordinance-designated cities house a large proportion of Japan's foreign population, they also tend to be the most active in offering services to them. We can assume that other municipalities with smaller foreign populations will be less active, an assumption supported by the annually conducted nationwide survey on the status of local government development of plans and guidelines for the promotion of multicultural coexistence. This survey shows that as of April 2020, all 21 cities in the analysis had formulated such plans, with less than 40% of the remaining municipalities having followed suit (MIC, 2020c). Consequently, we believe this study is representative of Japan's more active cities in CALD community outreach.

To assess local government outreach, we reviewed the websites for all 21 cities, as well as their corresponding international centre websites, looking for information relating to the corona virus. City websites represent direct communication from the local government to residents, while international centres are public entities receiving funding from local governments, tasked with promoting internationalisation at the regional and local levels. These centres often work with their respective municipal governments to address the various needs of local CALD communities. While city websites often have foreign language components targeting CALD residents, international centres regularly house the expertise for engaging with CALD residents and thus tend to have websites that provide more detailed information on residing in Japan in a multilingual format. Consequently, international centres provide an avenue for foreign residents with lower levels of Japanese proficiency in particular to access information.

Our assessment consisted of conducting 'outreach checks' primarily by visiting city and international centre English language websites. Although extensive COVID-19 information is often readily available on Japanese language city websites, CALD residents can have difficulty accessing and understanding such materials. English is the most common language of outreach after Japanese, while some cities may provide information in other languages as well as simplified Japanese. We do not account for information in other languages in this study, although we do note if materials are provided in simplified Japanese.

City and international centre websites were accessed between December 2020 and February 2021, a period when COVID-19 infections were on the rise and a state of emergency was declared across Japan's major urban areas, making it an ideal time for local governments to provide important information regarding the virus to residents. It is worth noting that we opted to exclude information related to vaccine procurement in our data collection, as Japan only started the initial phases of vaccine rollout in late February 2021. Many cities had not yet prepared public vaccine plans in English or Japanese at the time of data collection.

After a collection of websites was made for each city, we conducted a content analysis of the various websites to determine common traits for outreach across the municipalities, settling on 21 indicators. Most indicators are binary, where we simply note whether the type of outreach was addressed. Some websites could be more or less detailed, where cities may have a small number of sites with detailed information, or may have a larger number of more narrowly focused sites. After running the relevant tallies, the 21 indicators were combined into four broad categories based on the type of information represented. These categories include medical information, financial information, technical information and general information. By classifying the indicators in such a way, we are able to determine the broad categories cities concentrate their efforts on. A discussion of the results follows.

Results

Table 12.1 provides a list of all variables used and their description, while Table 12.2 gives the descriptive statistics for each variable. Looking first at general trends in terms of the number of outreach websites in English,

Table 12.1 Variable labels

Variable Name	*Description*
Websites	Number of websites found addressing COVID-19, either city of international center
Symptoms	1 = Explanation on what to do if experiencing symptoms, 0 = otherwise
Family care	1 = Explanation on how to care for family members if they have COVID-19, 0 = otherwise
Prevention	1 = Explanation of measures to prevent contacting COVID-19, 0 = otherwise
Infection	1 = Explanation of infection severity levels, 0 = otherwise
PCR testing	1 = Information on the availability of testing in the city or on the testing process, 0 = otherwise
Financial relief	1 = Information on how to obtain financial relief if job is negatively affected by COVID-19, 0 = otherwise
Company support	1 = Information on how companies or businesses can obtain financial relief if negatively affected by COVID-19, 0 = otherwise
Cash payment	1 = Information on cash payment given to all residents, including foreign residents, in June 2020, 0 = otherwise
Housing benefits	1 = Information on extension of housing benefits to people who lost their job or business due to COVID-19, 0 = otherwise
Simple Japanese	1 = Information available in simplified Japanese, 0 = otherwise
Updates	1 = The city has regular, visible COVID-19 related updates, 0 = otherwise
Confirmed cases	1 = The city has updated listings of confirmed cases, 0 = otherwise
Consultations	1 = Information available of COVID-19 related consultations, either by phone, email or in person, 0 = otherwise
School closure	1 = Information on school closures due to COVID-19, 0 = otherwise
Government closure	1 = Information on government facilities closures due to COVID-19, 0 = otherwise
Events	1 = Information on holding events, including restrictions and rules, 0 = otherwise
Evacuation	1 = Revised emergency evacuation information and recommendations in the wake of COVID-19, 0 = otherwise

(continued)

Table 12.1 (continued)

Variable Name	*Description*
Immigration	1 = Information on visa extensions or birth reporting exceptions granted due to COVID-19, 0 = otherwise
App	1 = Information on registering for or using the COCOA application, the government's official COVID-19 tracker, 0 = otherwise
Prefecture	1 = City or international center webpage links to prefecture or national government sites, 0 = otherwise
Pathway	1 = City or international center homepage has a clear and accessible pathway to COVID-19 information in English, 0 = otherwise
Total	Composite outreach score
Foreign population	Total foreign population, 2019
Total population	Total population, 2019
Foreign percentage	Percentage of foreign population, 2019
Medical info	Composite of Symptoms, Family Care, Prevention, Infection, Confirmed Cases and PCR Testing variables. 0 = minimum, 6 = maximum
Financial info	Composite of Financial Relief, Company Support, Cash Payment and Housing Benefits variables. 0 = minimum, 4 = maximum
Technical info	Composite of Simple Japanese, App, Prefecture, Pathway and Machine Translation variables. 0 = minimum, 4 = maximum
General info	Composite of Updates, Consultations, School Closure, Government Closure, Events, Evacuation and Immigration variables. 0 = minimum, 7 = maximum

Source Authors' construct

cities range from a low of four websites (Kawasaki) to a maximum of 45 dedicated sites (Hiroshima), with an average of 13.43 websites targeting CALD residents. The most common outreach we found across cities was providing consultations (90%), giving COVID-19 prevention tips (90%) and offering financial relief information (71%). The least common means of outreach included changes to emergency evacuation protocols (10%) and information on government office closures (14%).

From our composite outreach score, we can see that cities range from a low of two points in Saitama, to a high of 20 points in Hiroshima. The average outreach score across ordinance-designated cities was 9.38 out of 20. Figure 12.1 lists the outreach scores for all cities. These results may be somewhat surprising and unexpected, where a city with a relatively larger

Table 12.2 Descriptive statistics

Variable name	*Mean*	*Standard dev*	*Min*	*Max*	*Obs*
Websites	13.43	8.42	4	45	21
Symptoms	0.33	0.48	0	1	21
Family care	0.19	0.40	0	1	21
Prevention	0.90	0.30	0	1	21
Infection	0.43	0.51	0	1	21
PCR testing	0.24	0.44	0	1	21
Financial relief	0.71	0.46	0	1	21
Company support	0.43	0.51	0	1	21
Cash payment	0.43	0.51	0	1	21
Housing benefits	0.52	0.51	0	1	21
Simple Japanese	0.48	0.51	0	1	21
Updates	0.62	0.50	0	1	21
Confirmed cases	0.43	0.51	0	1	21
Consultations	0.90	0.30	0	1	21
School closure	0.19	0.40	0	1	21
Government closure	0.14	0.36	0	1	21
Events	0.33	0.48	0	1	21
Evacuation	0.10	0.30	0	1	21
Immigration	0.38	0.50	0	1	21
App	0.33	0.48	0	1	21
Prefecture	0.81	0.40	0	1	21
Pathway	0.81	0.40	0	1	21
Total	9.38	4.28	2	20	21
Foreign population	57,921.57	105,126.3	5916	489,857	21
Total population	1,760,676	1,930,101	702,395	9,486,618	21
Foreign percentage	2.45	1.29	0.74	5.29	21
Medical info	2.52	1.50	0	6	21
Financial info	2.10	1.51	0	4	21
Technical info	2.95	1.20	0	4	21
General info	2.67	1.56	0	7	21

Source Authors' construct

foreign concentration like Saitama ranks disproportionately lower than a city like Hiroshima, a smaller city with a comparatively smaller foreign concentration. The Tokyo Metropolitan area ranks in the middle in spite of easily having the largest foreign population and highest concentration of foreign residents in the country next to Osaka.

Due to the small sample size, we were unfortunately unable to obtain meaningful results from running regression analyses. However, one factor

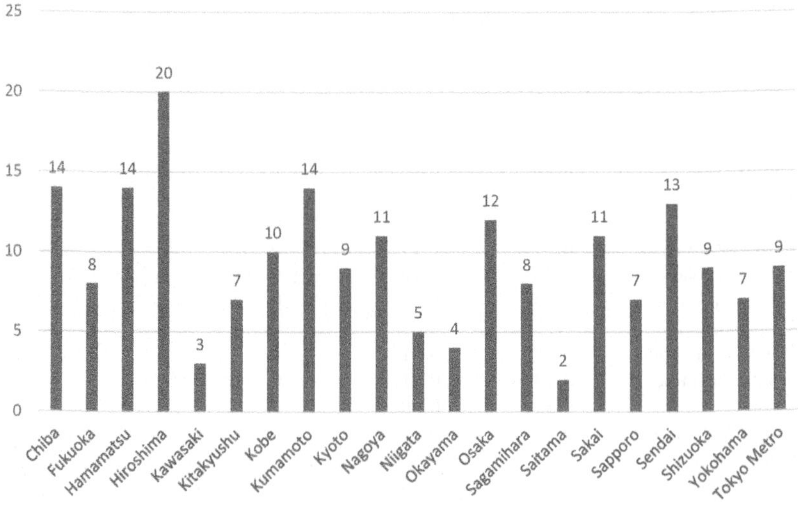

Fig. 12.1 Outreach scores by ordinance designated city (*Source* Authors' construct)

we were able to test for was whether foreign population size had an effect on the level of outreach. Although a regression provided no significant results, there was a correlation between foreign percentage and the composite outreach score, significant at the 90% confidence level. One can perhaps assume that there is at least a weak association between foreign population concentration and the level of outreach Japan's largest cities pursue for COVID-19.

Figure 12.2 illustrates the composite rankings across our four broad categories of medical, financial, technical and general information provision. Here we can see that most cities rank towards the middle in information provision, although there are a few outliers that provide both more and less information respectively. Hiroshima, for example, is the highest-rated city in this survey and is exceptionally good at providing medical and general information. Tokyo, on the other hand, receives average marks in technical and general information, scores high on medical information and provides no discernible financial information. That Tokyo scores well on medical outreach may come as little surprise given that Tokyo was the epicentre for COVID-19 outbreaks in Japan during this period, although a lack of financial information may

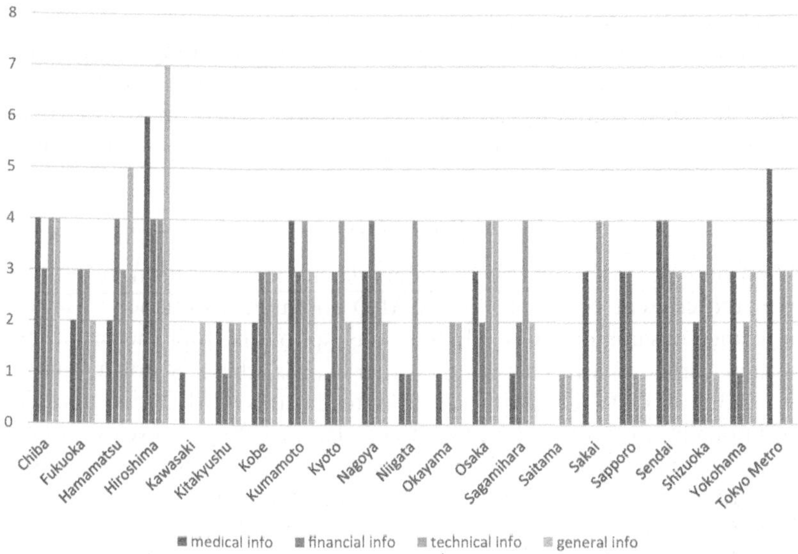

Fig. 12.2 City outreach by type (*Source* Authors' construct)

be somewhat unexpected. Likewise, Saitama and Kawasaki, both cities in the greater Tokyo region with large foreign populations, provide surprisingly little outreach across the four categories covered here. This finding is especially striking, as Kawasaki in particular is regarded as being one of the most active municipalities in Japan in reaching out to its CALD population (Green, 2013). Chiba and Yokohama, also major cities in the Tokyo region, appear to have relatively higher levels of outreach.

In looking at the different categories of outreach, most cities score between two and three in providing medical information, receive either a zero or three in financial information provision, a three or four in technical information and a two or three in general information provision. We can summarise that providing technical information is actually the most common occurrence across cities, mainly because they often have a clear pathway to COVID-19 information on their homepages and they often link to additional prefecture or national government information. Financial information seems to be hit-and-miss, where cities are likely to either provide no information for people experiencing financial stress or opt to provide more extensive information. Medical and general information tend to fall towards the middle, where most cities provide some level of medical information, most often information on preventing the spread of the corona virus. For general information, COVID-related consultations were the most common category of provision, where most cities gave some information on how to contact government facilities with inquiries.

Discussion

Based on our results we find considerable city-level variation in both the level and content of corona virus CALD community outreach in Japan's ordinance-designated cities. These results uphold both of the hypotheses posited earlier in the chapter. In other words, there is not a uniform provision of COVID-19 information to Japan's CALD communities. Additionally, we find evidence of a correlation between foreign population concentration and the level of information provision, an association likely to be strengthened with a larger sample size. If local governments do indeed prioritise outreach to communities with large concentrations of CALD residents, areas with fewer CALD residents are at risk of being overlooked for such support. At the same time, if sheer numbers of CALD residents motivate city outreach, questions over the vulnerability of various individuals or groups within CALD communities are likely being ignored. A more nuanced approach to outreach and messaging would consider variability within CALD groups, possibly either through survey research or group engagement activities.

Stepping back somewhat, it is very likely that Japan was unprepared for a protracted event like the corona virus. Messaging protocols were largely not in place, particularly those targeting CALD residents. This lack of preparation was most likely driven by Japan's legacy of sudden onset hazards and the messaging strategies developed to cope with them. The longer 'campaign' style of messaging that has become synonymous with COVID-19 is a newer phenomenon in Japan, one that has had to develop over time.

Protracted events such as COVID-19 add an additional layer of complexity for government in responding to disaster, where official messaging must adapt and change as the understanding of the hazard changes. Inconsistent application of these messages across different jurisdictions can lead to an inconsistent and ineffective understanding of the hazard and appropriate coping behaviours, which in turn can undermine public trust in government. For CALD communities, inconsistent or unclear official messaging can lead to greater reliance on other sources of information, most likely friends, family and social media, often in their native language, which may not be consistent with the directives of the host country. As we demonstrate, there appears to be inconsistent messaging to CALD communities across local governments in Japan's largest cities. While a range of causal factors has been attributed to the observed COVID-19 clusters emerging within CALD groups in Japan (Osaki, 2020), this inconsistency in messaging has the potential to erode trust in official sources of information and drive individuals to seek their information elsewhere. Should this be the case, it is likely to interact with lifestyle, language and financial related factors, further compounding the vulnerability of CALD groups.

Conspicuously lacking in the Japanese context is coordination from the national government on COVID-related messaging, especially when considering CALD communities. Although Japan's CALD population is a small proportion of its overall populace, in a protracted pandemic minority groups receiving insufficient or inconsistent messaging run the risk of being a new source of virus spread. Japan may perhaps benefit from considering the Australian approach of leveraging community groups, where the government provides funds and expertise for CALD groups to design and implement their own messaging (Australian Government Department of Health, 2021). Australia's approach is not without its own issues, for example the stigma surrounding COVID and the socio-economic consequences of having a confirmed infection have impeded testing in CALD communities (Burnet Institute, 2021). However, collaboration between government and diverse community groups in Australia is at a much higher level compared to Japan.

Some limitations to this study should also be addressed. For one, this study only considered English language outreach to CALD communities. While outreach in English was the most common language after Japanese, the vast majority of foreign residents in Japan are not native English speakers. There is some question as to how effective English language communication is, although it does represent a baseline of dissemination priorities. It is also possible that while most information was available in English aside from Japanese, information provision in other languages covered areas not addressed in English. By the same token, this study does not compare English language outreach to the materials available in Japanese. While we anecdotally note that considerably more material is available in Japanese, future research could potentially compare the level of difference between Japanese language outreach and materials provided to residents in other languages. Finally, while data was collected over a three-month period, with the changing nature of the corona virus it is possible that cities may change the available information and levels of outreach. Our study represents a snapshot in the middle of an ongoing protracted event, but is not a comprehensive project. It is possible that official messaging drastically improved, for example, which we are unable to capture in this instance. However, our results do demonstrate the relative disparities in English language information provision across Japan's largest cities as they worked to contain the spread of COVID-19.

Conclusion

This chapter has looked at practices of local government information dissemination targeting culturally and linguistically diverse communities in Japan during the COVID-19 pandemic. Even before the virus erupted, cities demonstrated disparate levels of English language outreach to foreign residents. We find that this tendency was fully present as the government attempted to provide essential information to foreign residents in the midst of the virus outbreak. Levels of outreach vary considerably across Japan's largest cities, where outreach appears to be motivated, at least in part, by the concentration of CALD residents in the municipality.

In terms of outreach initiatives, cities most commonly provided technical information to CALD residents, including linking to regional and national information centres and Japan's contact-tracing application. Medical and general information fell more towards the middle, where most cities provided some level of medical information, most often on preventing the spread of corona virus. Most cities also provided information on setting up consultations. Least consistent was financial information, where cities either provided no information to people experiencing financial stress as a result of the corona virus, or gave extensive information on the various financial assistance schemes available.

Most apparent with this study is a lack of national government guidance in messaging to CALD residents. Individual cities often do not have the time or resources to coordinate information campaign responses, particularly as they juggle a range of competing priorities in the midst of a protracted event. To foster greater CALD community trust in government, government should consider both coordinating responses at the national level, empowering local CALD communities with local-level information dissemination, and further facilitating communication between local CALD communities and government.

Cultural and linguistic differences can present significant barriers when attempting to communicate with migrant populations and minority groups, increasing their vulnerability to emergency events. The protracted nature of a pandemic only makes these challenges greater. As most governments have limited resources, messaging to both the majority and CALD communities must be carefully considered. While local government first-hand knowledge is crucial in addressing community needs in times of emergency, national guidance on messaging could help to provide a consistent and unified voice from government. For an ethnically homogenous country like Japan, which has only relatively recently started coming to terms with an increasingly diverse population, greater emphasis should be placed on balancing current local government initiatives with national coordination and CALD group engagement and consultation.

While Australia has been grappling with multicultural policies for a much longer period of time, a key difference in Japan's approach is the centrality of hazard awareness and emergency preparedness and response. Indeed, these concepts are a key driver of the inception, creation and ongoing development of Japan's policies associated with engaging its CALD population. While generally a peripheral consideration for the Australian emergency management sector, as hazards increasingly interact with the sprawling population of Australia, the likelihood of CALD groups being impacted in a protracted event will also increase. This has indeed been seen in not only in COVID-19 clusters, but also in the recent floods, catastrophic fires and tropical cyclones that have impacted many parts of Australia. Like Japan, it is clearly time for emergency management and hazard awareness to be recognised as an important component of Australia's multicultural policy frameworks. Australia has well-developed emergency management and multicultural policy domains. How these two domains can be successfully integrated might be something Australia can learn from Japan's experience.

Acknowledgements The authors would like to acknowledge the support of the Japan Society for the Promotion of Science and the Australia Japan Foundation for funding the research underpinning this chapter.

References

Australian Government Department of Health. (2021). COVID-19 Vaccine Campaign to Support Culturally and Linguistically Diverse Audiences, https://www.health.gov.au/ministers/the-hon-greg-hunt-mp/media/covid-19-vaccine-campaign-to-support-culturally-and-linguistically-diverse-audiences.%20Accessed%2028%20May%202021

Baird, C. (2019, September 5). Accounts of Those Who Were Left Stranded by Typhoon Jebi in Japan. *Japan Times*, https://www.japantimes.co.jp/news/2018/09/05/national/accounts-left-stranded-typhoon-jebi-japan (Accessed 27 May 2021).

Biddle, N., Edwards, B., Gray, M. & Sollis, K. (2021). Change in Vaccine Willingness in Australia: August 2020 to January 2021. Preprint, https://doi.org/10.1101/2021.02.17.21251957

Burnet Institute. (2021). Shame, Stigma Barriers to COVID-19 Testing for Young and Culturally Diverse, Report Finds, https://www.burnet.edu.au/news/1408_shame_stigma_barriers_to_covid_19_testing_for_young_and_culturally_diverse_report_finds. Accessed 26 August 2021.

Butcher, N. (2020). Better Engaging Culturally Diverse Communities during COVID-19. *Pursuit*. 4 September 2020, https://pursuit.unimelb.edu.au/articles/better-engaging-culturally-diverse-communities-during-covid-19 (Accessed 25 September 2021).

Cabinet Office of Japan. (2013). The Fifth Study Group on Evacuation Support for Persons Requiring Assistance in the Event of a Disaster [*Saigaiji yō Eengosha no Hinan Shien ni kansuru Kentōkai Dai Go-kai*]. Document 4: Necessary Measures for Different Groups of People Requiring Assistance During a Disaster [*Saigaiji yō Engosha-tō no Tokusei-Goto ni Hitsuyōna Taiō ni Tsuite*]. http://www.bousai.go.jp/taisaku/hisaisyagyousei/youengosya/h24_kentoukai/5/index.html (Accessed 27 May 2021).

Cabinet Office of Japan, Disaster Management Section. (2000). Information Materials on Lessons Learned from the Great Hanshin-Awaji Earthquake [*Hanshin Awaji Daishinsai Kyōkun Jōhō Shiryōshū*]. https://ndlonline.ndl.go.jp/detail/R300000004-I8829027-00. (Accessed 28 May 2021).

Green, D. (2013). Local Foreign Suffrage in Kawasaki City: The changing state of voting rights in Japan. *Electronic Journal of Contemporary Japanese Studies, 13*(1).

Green, D. (2017). As its population ages, Japan quietly turns to immigration. *Migration Policy Institute* (blog), https://www.migrationpolicy.org/article/its-population-ages-japan-quietly-turns-immigration.%20Accessed%2028%20May%202021

Japan Times. (2019, September 14). Lessons from the Latest Typhoon. https://www.japantimes.co.jp/opinion/2019/09/14/editorials/lessons-latest-typhoon/ (Accessed 27 May 2021).

Japan Times. (2021, February 2). Upper House passes bills allowing fines for defying antivirus steps. https://www.japantimes.co.jp/news/2021/02/03/national/bills-fines-antivirus-measure/ (Accessed 24 August 2021).

Kassam, N., & Hsu, J. (2021). Being Chinese in Australia: Public opinion in Chinese communities. Lowy Institute, https://interactives.lowyinstitute.org/features/chinese-communities/Lowy%20Institute%20%E2%80%93%20Being%20Chinese%20in%20Australia.pdf (Accessed 27 May 2021).

Kelly, D., Shaw, K., & Porter, L. (2020). Melbourne Tower Lockdowns Unfairly Target Already Vulnerable Public Housing Residents. *The Conversation*. 6 July 2020. http://theconversation.com/melbourne-tower-lockdowns-unfairly-target-already-vulnerable-public-housing-residents-142041 (Accessed 25 September 2021).

Kikuchi, D., & Osumi, M. (2016, April 18). Embassies Working to Help Non-Japanese Quake Survivors. *Japan Times*. https://www.japantimes.co.jp/news/2016/04/18/national/embassies-working-to-help-non-japanese-quake-survivors/ (Accessed 27 May 2021).

Kyoto, S. (2020, May 8). Why Announce 'Japanese Nationality' or 'Foreign Nationality' of People Infected by Corona Virus? Experts Warn that Kyoto Prefecture is Inviting Discrimination. [*Korona Kansen-sha 'Nihon Kokuseki' ka 'Gaikokuseki' ka, Naze Happyō? 'Sabetsu Izanau' to Senmonka, Kyōtofu no Taiō ni Keishō*]. https://www.kyoto-np.co.jp/articles/-/238771 (Accessed 28 May 2021).

Lindell, M., Arlikatti, S., & Prater, C. (2009). Why People do what they do to Protect Against Earthquake Risk: Perceptions of Hazard Adjustment Attributes. *Risk Analysis, 29*(8), 1072–1088.

Mainichi Daily News. (2020, April 16). Don't Fit: PM Abe's Cloth Face Masks Hit by Backlash from Nursing Home Workers. https://mainichi.jp/english/articles/20200416/p2a/00m/0na/010000c (Accessed 28 May 2021).

MIC – Ministry of Internal Affairs and Communications. (2006). Study Group Report on the Promotion of Multicultural Coexistence: Promotion of Multicultural Coexistence in Local Communities [*Tabunkakyōsei no Suishin ni kansuru Kenkyūkai Hōkokusho: Chiiki ni okeru Tabunkakyōsei no Suishin ni tsuite*]. https://www.soumu.go.jp/kokusai/pdf/sonota_b5.pdf (Accessed 27 May 2021).

MIC – Ministry of Internal Affairs and Communications. (2017). Questionnaire on How to Deal with Foreign Residents in the Event of a Disaster [*Saigai-ji no Gaikokujin Jūmin he no Taiō ni Kansuru Ankēto*]. https://www.soumu.go.jp/main_content/000488934.pdf%20 (Accessed 28 May 2021).

MIC – Ministry of Internal Affairs and Communications. (2020a). Zero Information Refugee Project [*Jōhō Nanmin Zero Purojekuto*]. https://www.soumu.go.jp/menu_seisaku/kokumin/jyohonanminzero/index.html (Accessed 27 May 2021).

MIC – Ministry of Internal Affairs and Communications. (2020b). Revision of Regional Multicultural Coexistence Promotion Plans [*Chiiki ni okeru Tabunkakyōsei Suishin Puran no Kaitei*]. https://www.soumu.go.jp/main_content/000718717.pdf (Accessed 25 September 2021).

MIC – Ministry of Internal Affairs and Communications. (2020c). Development Status of Guidelines and Plans for Promoting Multicultural Coexistence [*Tabunkakyōsei no Suishin ni Kakaru Shishin Keikaku no Sakutei Jōkyō*] https://www.soumu.go.jp/main_content/000692492.pdf (Accessed 28 May 2021).

MOJ – Ministry of Justice, Immigration Services Agency. (2021). Statistics on Foreign Residents [Zairyū gaikokujin tōkei]. http://www.moj.go.jp/isa/policies/statistics/toukei_ichiran_touroku.html (Accessed 25 May 2021).

NHK News. (2021, May 22). Do Not Eat with Foreigners, Warns Health Centre Infection Prevention Memo [*'Gaikokujin to Shokuji Shinai yō ni' Kansen Kobō Keihatsu Bunsho ni Hokenjo ga Kisai*]. https://www3.nhk.or.jp/news/html/20210522/k10013044921000.htm (Accessed 28 May 2021).

Nikkei, A. (2021a, January 2). Suga vs Koike: Bitter Political Rivalry Hampers Japan's COVID Response. https://asia.nikkei.com/Politics/Suga-vs-Koike-Bitter-political-rivalry-hampers-Japan-s-COVID-response (Accessed 28 May 2021a).

Nikkei, A. (2021b, January 4). Japan Debates State of Emergency while Tokyo Cries for Action. https://asia.nikkei.com/Spotlight/Coronavirus/Japan-debates-state-of-emergency-while-Tokyo-cries-for-action (Accessed 27 May 2021b).

O'Brien, S. (2019). Translation technology and disaster management. In M. O'Hagan (Ed.), *The Routledge Handbook of Translation and Technology* (pp. 304–318). Routledge.

Osaki, T. (2020). Japan Faces Balancing Act over Virus Clusters among Foreign Nationals. *The Japan Times*, 22 November 2020. https://www.japantimes.co.jp/news/2020/11/22/national/social-issues/coronavirus-clusters-foreign-nationals-japan-discrimination/ (Accessed 25 September 2021).

Rodrigues, U. (2017). Australians Born Overseas Prefer the Online World for their News. *The Conversation*. https://theconversation.com/australians-born-overseas-prefer-the-online-world-for-their-news-84355 (Accessed 27 May 2021).

Sandman, P. M. (2007, May 3). What kind of risk communication does pandemic preparedness require? *Center for Infectious Disease Research and Policy* (blog), https://www.cidrap.umn.edu/news-perspective/2007/05/what-kind-risk-communication-does-pandemic-preparedness-require (Accessed 28 May 2021).

Sasaki, K. (1995). The Great Hanshin-Awaji earthquake and problems of foreign residents [*Hanshin Awaji Daishinsai to Gaikokujin Mondai*]. *The Business Review [keiei Kenkyū], 255*, 105–114.

Takezawa, Y. (2000). The Great Hanshin-Awaji earthquake and town-making towards multiculturalism. *Zinbun, 34*(2), 87–99.

Tamura, T. (2017). Disaster Response in the Age of Multicultural Coexistence [*Tabunkakyōsei no Jidai ni okeru Saigai-ji Taiō*]. *Institute of Scientific Approaches for Fire and Disaster* [*Shōbō Bōsai no Kagaku*]. *Special Feature: Foreigners and Disaster Prevention, 130*, 7–11.

Tokyo Metropolitan Government. (2020). Support for people requiring special attention during a disaster [*Saigai-ji yō Hairyo-sha he no Shien*]. Tokyo Disaster Prevention Homepage. https://www.bousai.metro.tokyo.lg.jp/bousai/1000027/1000303.html (Accessed 27 May 2021).

Wild, A., Kunstler, B. Goodwin, D., Onyala, S., Zhang, L., Kufi, M., ... Skouteris, H. (2021). Communicating COVID-19 health information to culturally and linguistically diverse communities: Insights from a participatory research collaboration. *Public Health Research & Practice, 31*(1), 38–51.

Xiao, B., Aualiitia, T., Salim, N., & Yang, S. (2021, March 4). Completely Untrue: Experts Debunk Misinformation That MRNA Vaccines 'genetically Modify' Humans. *ABC News* (blog). https://www.abc.net.au/news/2021-03-04/covid-19-vaccine-misinformation-cald-communities/13186936 (Accessed 27 May 2021).

Yoshida, R. (2016, April 16). Kumamoto International Center Offering Shelter, Translation Services to Foreigners. *Japan Times*. https://www.japantimes.co.jp/news/2016/04/16/national/kumamoto-international-center-offering-shelter-translation-services-foreigners/ (Accessed 27 May 2021).

CHAPTER 13

Earthquakes, Tsunami and Climate Change: Customary Management and Adaptation

Skye Turner-Walker

Introduction

Ongoing earthquakes occurred in and around the Haruku Strait during 2019 and 2020 (and less frequently ongoing into 2021). The most severe in the region in recent living memory, the ongoing earthquakes had major consequences for the area, with the small island of Haruku hard hit. In this case study of the Island of Haruku, the rapid onset of the seismological disaster was complicated by the compounding, cascading and protracted impacts of climate change (slow-onset and ongoing impacts of drought, floods, unpredictable rainy and dry season periods and crop damage) and multiple disasters (actual and anticipated) over time. Adjusting and adapting to the seismological disaster events was influenced by adaptation to other types of disaster arising from climate change

S. Turner-Walker (✉)
Fenner School of Environment & Society, Australian National University, Canberra, Australia
e-mail: skye.turner-walker@anu.edu.au

A. Lukasiewicz and T. O'Donnell (eds.), *Complex Disasters*, Disaster Risk, Resilience, Reconstruction and Recovery,
https://doi.org/10.1007/978-981-19-2428-6_13

impacts and events. This chapter unpacks the information sources and fears factored into local communities' decision-making when responding to earthquakes and the threat of a tsunami.

The interviews drawn on for this case study sought to elicit community experiences and views. Interviews were undertaken during 2019 and 2020 when earthquakes and a series of aftershocks struck the Central Maluku region in Indonesia's north-eastern islands. The interviews for this chapter chart the fear and gradually the recovery and resilience as the community of Haruku village on Haruku Island found ways to adjust and adapt to life in the months following, when earthquakes would happen often daily, and frequently multiple times a day. The chapter investigates the role of climate change adaptation preparations in influencing disaster response, recovery and resilience. Twenty-six households were interviewed within the village prior, during and after the earthquake events. The first interviews focused on climate change adaptation, yet when earthquakes began in the middle of the research data collection phase, the initial research on climate change quickly became sidelined for the risk and immediate concerns from earthquakes and the potential for a tsunami leading discussions toward resilience to environmental changes, impacts and risks from multiple causes, and not only climate change.

For this research, the definition of community refers to a group that operates as a network of relations, social system interactions or unit of belonging and identity usually bounded within a geographic location (Wilson, 2012a, 2012b). In the case used in this research, the term community applies to a geographically bounded entity (small island village) that shares commonalities concerning social ties and networks, a commonality of belonging and place-bound identity and location-centred production and exchange in determining resource governance decision-making. Interviews were undertaken within the village community in Indonesian language interspersed with Ambon and Haruku dialects, which were then translated to English. These interviews were undertaken as individual interviews and household member groups or focus group discussions with participants selected based on a rolling recruitment strategy and their willingness to engage with the research. Many of the interviews were held in the aftermath of the earthquakes in evacuation camps and temporary village relocation arrangements. The research used a grounded theory approach for coding and developing themes from the data collected in these interviews, choosing grounded theory given its constructivist approach that can be used to advance critical inquiry and social justice research, as well as being particularly beneficial in researching situations fraught with contradictions, injustices and relentless troubles (Charmaz, 2020; Martin & Turner, 1986; Myers, 2009) such as disaster contexts.

In Indonesia's Central Maluku, on the island of Haruku, slow-onset and ongoing disaster impacts arising from climate change have been of great concern for many years for the local Indigenous residents, or *Adat Masyarakat*. Haruku is a small island in Maluku (the Moluccas) found between Ambon and Seram Islands. Haruku village, known as *Negeri* Haruku, sits alongside the village of Sameth on the southwestern coast of Haruku Island, straddling the other side of the Haruku Strait from the Port of Teluhu, 9 km away on the neighbouring Ambon Island of Central Maluku. High tides, erratic wind patterns and sea currents, storms, cyclones, floods, coastal inundation, erosion, increased heat and dry periods, and changing rain patterns, have been of significant concern during the last decade. Damaging significant tracts of coastal ecosystems protected under customary practice (such as the endangered Maleo bird habitat, mangroves, and coastal fish populations). The small islands of Central Maluku, including Haruku, have a marine-influenced climate and face frequent isolation during storms, as the islands are accessible only by boat. Storms impact heavily on fishing-dependent livelihoods by restricting fishers who use only small boats for going out to sea (due to economic, customary or cultural-ecological reasons). Several additional factors have amplified this impact on fishing livelihoods. A drop in fish-catch numbers due to climate-related changes in ocean temperature and currents, for example, alters the use and reduces the reliability of local knowledge and practice of customary methods reliant on precise signals and the timing of ocean currents for tracking fish locations throughout the year. Significant overfishing intensifies losses from these climate change impacts, meaning that fish are no longer easily caught close to shore. With fish no longer easily caught close to the shoreline, fishers dependent on small boats are driven to fish further out at sea. These impacts overlap and amplify one another, putting fishers at greater risk to climate change impacts from storms, cyclone impacts and heightened waves (more easily and frequently caught out due to greater seasonal variability), as well as elevated risk from earthquakes and tsunami.

Earthquakes and Tsunami

During 2019 and continuing throughout 2020 and into 2021, Haruku experienced numerous earthquakes. They began on September 26, 2019, when a 6.8 magnitude earthquake occurred along the 42-km long Fault-line that runs through the Strait between Ambon Island and Haruku Island (BMKG, 2019). A further 3089 immediate earthquakes and aftershocks from this earthquake were recorded (BMKG, 2019), and compounding the initial earthquake damage to buildings and infrastructure across Haruku. The earthquakes were at varying intensities, sometimes occurring not only daily but multiple times per day. While no immediate deaths were recorded on the island of Haruku, one person injured during the first earthquake reportedly died some days later, and many had minor injuries or developed gastroenteritis and flu. Throughout the greater Maluku region a total of 148,619 displaced residents were spread throughout the Central Maluku Regency, West Seram and Ambon City, with 6355 houses damaged (Nuraini, 2019). Within Haruku, many of the houses in Haruku village were damaged, and electricity was cut off for a prolonged period. Water pipelines that pumped the water supply to the village from the spring in the mountains were impaired, while roads and the main bridge cracked and the local church steeple broke. With the water pipelines damaged, agriculture was marred without the ability to irrigate, while fishing livelihoods also suffered as ongoing fears of tsunami prevented fishers from returning to the ocean during the continuing earthquakes. Communication and electrical infrastructure were severely damaged, while after fleeing to multiple locations dispersed in the nearby hills, families were left without communication to contact and find one another with phone signals down and mobile phone batteries out of charge.

Most people on land fled to the hills of Haruku Island in fear of a tsunami, losing contact with family members in the process. Opting to relocate to higher-ground evacuation areas, most of the community moved to makeshift accommodation, setting up tarpaulins and tents. However, the conditions also meant that health implications arose due to both flooding of the site, coupled with the lack of sanitation, clean water and highly transmissible close living quarters.

The initial earthquake sparked panic, spurring most residents to flee from collapsed houses in anticipation of further collapse from aftershocks. The panic expressed in interviews with respondents describes the moments of the first earthquake, with heightened reactions. "We left class. My teacher fainted!" was described by some high school students, while others portrayed the conditions in the sandy earthed beachside village and the potential for liquefaction: "*the sand churned, moved, [and] raised as though it were tossed! The motorbike lifted, things fell, including my mother, while drinking tea. The kitchen cupboards crashed, and the roof fell in.*" Many people who were in boats travelling or fishing between Haruku Island and the main island of Ambon in the Strait where the epicentre occurred, were shocked by the sound of the strike-slip fault earthquake (Triyono, 2019). The neighbouring port to the island in Tulehu and the infrastructure for the boats ferrying between Haruku and the main island of Ambon were flattened, with buildings on the water's edge collapsing into the sea. Despite tsunamis being unlikely to occur from strike-slip earthquakes, and a *tidak berpotensi tsunami* (no potential for tsunami) status issued through the *Badan Meteorologi, Klimatologi, dan Geofisika* (BMKG), Indonesia's Agency for Meteorology, Climatology and Geophysics early warning system and app (Triyono, 2019), the panic was intensified by the pervasive threat of a possible tsunami. With infrastructure being damaged, communications with the outside world were severed, including for early warning systems which rely on telecommunications infrastructure and electricity. Most people on land fled to the hills of Haruku Island in fear and anticipation of a tsunami, while others stayed camped outside of their houses—motorbikes facing towards the hills, keys in the ignition, ready to flee. People grouped in four main clusters over the hills, unable to contact family members and loved ones with phone networks and electricity down. When telecommunications were restored, electricity remained off, and most mobile phones had long since run out of battery and died, so messages were conveyed between the four different camps for families to find each other. When the threat of a tsunami arrived it was the knowledge imbedded in the past that furthered the fear and worry brought by hoax messaging on WhatsApp and social media, compouned by the memories from media of the recent past tsunami event in Palu that had people on edge. "*Tsunami has been in Maluku for hundreds of years. So, because of this understanding from hundreds of years ago, we anticipate this.*"

The residual fear of a tsunami permeated into all facets of life for months. Information, particularly for the younger generations, had come via official mobile phone-based early warning system applications that provide information on earthquakes and tsunami that tended to be overridden by a mass of social media posts on apps such asFacebook and WhatsApp. Social media app groups and forwarded messages also came with a number of hoax messages and fake news articles, which spread in the days following (and preceding) the initial earthquake. The earthquake timing coincided with mysterious mass fish deaths in the weeks leading up to the earthquake, with dead fish of all kinds washing up throughout the Strait, and on several beaches on Ambon Island, throughout South Leitimur District, Nusaniwe District, Baguala District and Salahutu District, in Central Maluku Regency. Starting two weeks prior to the earthquake, as thousands of fish were found dead washed up on beaches, local residents interviewed for the local newspaper said, "No one wants to take the fish. Because apart from being afraid that the fish will be contaminated with poison, there are also those who think that the incident is a sign" (Sugiyarto, 2019). The mass fish deaths on several beaches caused speculation that the deaths indicated an earthquake and tsunami (Prabowo, 2019). While the Ambon Fish Quarantine Centre took samples of fish concluding that they had not died from disease (Patty & Ika, 2019), Indonesia's BMKG and *Lembaga Ilmu Pengetahuan Indonesia* (LIPI) the Indonesian Institute of Sciences, concluded that these fish deaths had not occurred due to underwater explosions. Instead they posited that the mass death of fish in several coastal areas on Ambon Island was caused by volcanic activity under the sea. However, this theory went unproven, as no evidence was found (Jemadu, 2019a). Research conducted by some government and academic institutes suggested that the mass fish deaths were due to the movement of water masses or "upwelling" (Prabowo, 2019), a process whereby deep, cold water, rises towards the surface (NOAA, 2021c).

Ultimately, the research did not rule out the possibility of seismic activity occurring in the ocean. Nevertheless, the BMKG and LIPI were quick to attempt to dispel the link between the mass fish deaths throughout the Haruku Strait, resulting from water temperature changes rather than gasses and seismic activity or tsunami-related reasons (Jemadu, 2019a, 2019b; Tubaka, 2019). The *Laporan Isu Hoaks Harian Direktorat Jenderal Aplikasi Informatika Kementerian Komunikasi dan Informatika* ("Daily Hoax Issue Report" from the Directorate General of Information Applications, under the Ministry of Communication and Information) also declared this to be false (hoax) information. Media releases were given to dispel any suggestions that the mass fish deaths on Ambon coast were a sign of earthquake and tsunami (Prabowo,

2019). The role media played significantly amplifed the "felt" presence of tsunami in Indonesia, alongside hoax messages that followed. False media has notoriously followed most large earthquakes in Indonesia, claiming predictions of tsunami, which has meant that the BMKG dedicates a significant public relations resourcing and media outreach to dispelling "hoax" information as it spreads virally on social media. Hoax messages, for example, stated that a much larger earthquake would occur with a tsunami coming (Din, 2019; Fadhil, 2019; Iqbal, 2019). So, when the hoax news and messages coincided with traditional or historical signals used for indicating seismic and tsunami activity, locals say they did not hesitate to react pre-emptively. The timing of the mass fish deaths started two weeks prior to the earthquake on September 26, 2019, and continued during this period, sparking anxiety "*When the dead fish occur on top of the water, we worry again about a tsunami*." According to interview respondents, Haruku locals preferred to invest their trust in the local knowledge of past experiences, which links these signals to tsunami and seismic events. Most people elected for the safer option of higher ground "*just in case*" and particularly for any less-mobile family members including elderly, those with any disabilities and children.

Overall, government reliance and trust are low within the island's community due to a perceived lack of support and resourcing to the island to date. This level of distrust is combineswith perceptions that the national governance structure applied for local levels of district and village government undermine the traditional village structures and are lacking legitimacy at local levels, in combination with the years of regional conflict. In practice, management of land and marine resources and social-cultural systems in the island are a hybrid system of local institutions. Local governance institutions in Haruku village, while primarily customary, are mixed with religious governance (Protestant Church inherited from Dutch colonisation) and the national administrative system (adapted predominately from structures in Java). Haruku is one of the villages in the Maluku islands that still maintains and practices the tradition of *Sasi*. *Sasi* is the customary law found throughout *Adat Masyarakat* areas in Maluku. *Sasi* is present throughout most of Maluku from the northern Islands of Halmahera, Ternate, Buru, Seram, Ambon and Haruku, to the south and southeast of Banda, Kei and Aru, as well as throughout areas of Papua. Practised in Haruku Island, *Sasi* governs and enforces protection of natural resources by guarding ecosystems through regulations, boundary controls, allocations of sustainable use periods or protection time period controls (Mantjoro, 1996; Soselisa, 2019). The *Kewang*, a selected number of people responsible for implementing customary *Sasi* and act as environmental stewards. They are responsible for controlling environmental management of all resources in the village domain, especially during the *Sasi* (control or protection time periods) (Soselisa, 2019). The Kewang are also elected to discuss village economy, the coordination of village livelihoods and economy, and infringements of *Sasi* regulations, and boundary controls with neighbouring villages, including the patrols of (common) natural resources, both marine and terrestrial (Mantjoro, 1996; Zerner, 2014). In the case of Haruku, the Kewang customary institution is involved. Kewang act as a go-between for residents with several local governments as well as non-government and civil society organisations, including climate change adaptation and disaster resilience programming.

The role of the *Kewang* became crucial in Haruku, in propagating a peace dialogue and de-escalating conflict during the Maluku conflict, and have as a result become crucial mediums for the village to engage with local and district government. Maluku's conflict erupted during the post-Suharto period from 1999 to 2004/2005, and although widely attributed predominately to religious clashes between neighbouring villages, the causes are reportedly much more complex and politically driven. Local hearsay and reports from respondent interviews suggest that the role of agitators had underlying political motivations to cause destabilisation in the region for political means. The various movements advocating self-determination for the Malukus and independence from Indonesia have continued in the decades following, contributing to a sense of need for self-sufficiency, and independence from government resourcing, with pervasively little trust in being able to access government resources, found throughout the community. This same sense and lack of trust also extends to early warning systems (EWS), with respondents expressing very little confidence in the EWS for tsunamis throughout Indonesia. EWS failed several times during 2018 alone, resulting in thousands of deaths in Palu and hundreds of deaths in Southern Sumatra and Western Java, both of which were affected by tsunamis the year prior to the earthquake in Haruku with neither given tsunami warnings. In fact, in Palu, a tsunami warning was lifted 36 min after it was first raised, to a "no potential for tsunami" status. While in Sumatra and Java, there was no tsunami warning issued when an underwater volcano (*Anak Krakatau*) erupted, triggering tsunami on both the Sumatra and Java Islands and killing hundreds of people caught unaware. According to the National Board for Disaster Management's Nugroho, "*Indonesia does not yet have a tsunami early warning system for those caused by underwater landslides and volcanic eruptions. The current early warning system is for earthquake activity*" (Griffiths, 2018).

In a place where trust in the information and support from government agencies remains low, the fear of tsunami continued each time fish unexpectedly were found dead en masse on beaches (Nuraini, 2019). Local confidence on what sources of information should count as reliable information to base decision-making upon came through multiple sources, but most linked to accounts from the past. Indonesia's most horrifying tsunamis of the last few decades left their mark in the level of knowledge and sensitisation around tsunami occurrence, particularly in pre-empting tsunami events. For example, the tsunami in Aceh, along with Palu, was responsible for retrospection on the traditional knowledge of earthquakes and tsunamis in the area. The tsunami in Aceh reached up to nine metres in height in parts, and inundated areas up to three to four kilometres inland (Borrero, 2005). Interview respondents in Haruku noted that their pre-emptive reaction to a potential tsunami, came about in trepidation with the knowledge of the dangers of remaining in dwellings located directly at wave height on the coast learnt in part from the media memory of the tsunami in Indonesia's Aceh province in 2004. While almost exactly one year prior, on September 28, 2018, a strike-slip earthquake occurred in Palu, Indonesia, when a 7.5 magnitude earthquake struck in Central Sulawesi, Indonesia, resulting in a 1.5-m tsunami (BMKG, 2018). The earthquake caused a devastating tsunami (Sassa & Takagawa, 2019), accompanied with extensive liquefaction in the coastal areas that likely resulted in a gravity flow of liquefied soil mass to cause tsunami. This event left an imprint through the stories and media on the effects of the earthquake in nearby Palu, Sulawesi, to the neighbouring Central Maluku region. Having heard of the liquefaction of sand in Palu during the 2018 earthquake and having seen some of the images shared by Palu's residents and the tsunami damage that followed, Haruku village residents were anxious that they could experience something similar. Residents understand the similarities between Palu and their own village geography, with Haruku's entire settlement built on the flat sandy coastal plain flanked by mountains and foothills.

> *We stay vigilant about each earthquake because of the potential for a tsunami and particularly since the houses in the village are at the beach height. While the earthquakes continue, everyone chooses to stay, or at least have their children and elders stay on the hills and mountain where it is safer against the tsunami and the liquefaction.*

The parallels, according to interview respondents, with the location of Palu's houses throughout Palu's coastal plain, and the extensive liquefaction despite the earthquake being a strike-slip fault (which are unlikely to produce tsunamis) left great concern in Haruku when met with the earthquakes. Coupled with the tsunami warning being called off in Palu before it hit, uncertainty lingered for coastal dwellers in the region. Many perceived the earthquakes as a religious sign and thought that the constant earthquakes and destruction of houses were in punishment.

> *Psychologically, the earthquake has left a large imprint on us. Some keep imagining the same earthquake again. Some people worry that the earthquakes are punishment and that if they do not repent for their sins, there will be more, or there will be worse. Some have been sprinkling holy water on land and trees to try to bless or protect these. Some have been donating more to the church for protection.*

Earthquakes, Tsunami and Climate Change: Cascading and Protracted Impacts, Response and Adaptation

Climate change impacts had major influence on the earthquake recovery and resilience during ongoing earthquakes, which were notably compounded by the slow-onset and ongoing climate change impacts of drought, floods, unpredictable rainy and dry season periods and crop damage. Preparations in adapting to climate change thereby implicated the overall response to seismological disaster, and the gradual recovery and resilience. Maintaining and increasing resilience to climate change has centred on customary cultural and resource land management practices that guard local ecosystem integrity. Aside from living with climate change day to day and responding through local *Sasi* practices and diversification, Haruku communities have been engaged in climate change adaptation and resilience programmes. Building upon the process of adaptation established through customary management, Haruku became targeted to take part in novel adaptation programmes with community-centric approaches under the umbrella of the national adaptation action activities. Most recently programmes focused on climate change adaptation and disaster resilience, have centred community discussion around preparedness for climate disaster events.

The Haruku programme focused on building evacuation routes for flooding, were used instead for tsunami preparation. However, the learning and disaster risk reduction and preparation aspects have contributed to the local response to earthquake and perceived likelihood of tsunami impacts. Experiences with floods, earthquakes and tsunamis have occurred throughout history between the generations in Maluku, so these histories of experience were used as the basis for broaching discussion on climate change.

Training included evacuation scenarios for flooding, "*for earthquakes we often used to talk, but for flooding... we've made a [response] team.*" *Program Kerja Masyarakat-Kelompok Masyarakat* (POKMAS), the Community Work Program Community Group formed for flooding evacuation preparation, were reignited to build temporary water and sanitation infrastructure for the evacuation camps, as training undertaken for a flood disaster event was reappropriated for the earthquake and tsunami evacuation. Some members rapidly established toilet and water/sanitation spots for washing, which became crucial as water-sanitation related health issues and flu overtook to become one of the largest problems during the relocation. Although the training was for flooding risk, the training and group organisation was utilised in helping with earthquake evacuation and was used in the end for setting up tarpaulins and tent accommodation. Haruku programming on climate change adaptation included activities under climate change adaptation, such as investments in food security. These were implemented through local home vegetable gardening and permaculture, by the local organisation the *Lembaga Penelitian dan Pengabdian Masyarakat* (LPPM), or Institute for Research and Community Service under USAID funding. Efforts sought to improve locally sourced supplies, develop community adaptive capacity and provide knowledge about permaculture and land utilisation. LPPM focused on food security issues for small islands due to storms and the reliance on limited options for transport by sea during severe weather.

At the same time, traditional knowledge passed down from ancestors and, more recently, grandparents, kept record that earthquakes and tsunamis have a long history to the area, and previous settlements had been located on higher ground for several reasons. Haruku has its own tsunami stories from the past, told by the generations before of the Banda Islands. The oldest in the community remember the stories of tsunami impacts in the past as told through the memories passed on from their parents, while others heed the tales. The memory of the tsunami that followed the 7.6 SR earthquake in the Banda Sea in 1950 is still alive in the stories of those on Ambon Island and Haruku (Ulum, 2019). On Sunday, October 8, 1950, before noon, there was a tsunami in Ambon (at the time they were referred to locally as what translates as "***water up and down***" rather than the term used today from the Japanese word tsunami). The tsunami reached three *Negeri* (lands) in Central Maluku, of Hutumuri, Small Hative and Galala. In Haruku, stretching further back, knowledge has also been passed down of a massive earthquake in the Banda Sea on the morning of November 26, 1852. This earthquake is likely to have been at least 7.5—8.8, producing violent shaking that lasted for 5 min (NOAA, 2021a; Okal & Reymond, 2003). Many buildings throughout Banda were destroyed, while Papenberg (Boijang) mountain (the 250 m high volcano that towers over Banda Neira) partially collapsed and large cracks in the ground appeared. Strong aftershocks followed and the earthquake was felt throughout the islands of Laot, Buru, Seram, and potentially also as far as Batjan and Ternate with damage also caused in the islands of Ai and Rozengain. In Haruku, the walls of the Aboru church and the Fort Zelandia on the foreshore of Haruku village were cracked by the earthquake. The tsunami following reached up to 8m in height and slammed into Banda Neira, Sapura, Seram and Haruku islands. The tsunami is said to have caused major damage throughout all of these islands, washing away boats, many villages and their residents (NOAA, 2021b). Tsunami in Central Maluku has been recorded even further back, making tsunami a regular inter-generational occurrence, with the first record stretching back to February 17, 1674. This earthquake and subsequent tsunami are said to have claimed the lives of around 2500 people. Historically, the village of Haruku was located at high ground far from the coast, with the movement to larger coastal settlements occurring due to conversion to Islam and Christianity and during the Dutch-colonial period. However, the hills have offered safety over time to both coastal disasters and conflict. A key example is the limited loss of life experienced during the 7.6 earthquake in the Banda Sea in 1950 that sparked the Ambon tsunami throughout Central MalukuMaluku in three

Negeri. During the conflict between the Indonesian National Army and the Republic of South Maluku (RMS), locals had spotted the TNI navy in the Strait and abandoned coastal settlements for safety from conflict when the tsunami washed housing away, but with little casualties (Ulum, 2019). Some families in Haruku village reported that they had relocated to the coast during the post-Suharto conflict after houses were destroyed or burnt. Interviewees believe that for Haruku, being settled up in the hills was for safety reasons such as being protected from tsunami and earthquakes and potentially having other practical reasons such as being cooler and closer to agroforestry food and freshwater springs providing protection from various eras of conflict.

The histories of tsunamis in local knowledge permeate as a reliable information source and inform short- and long-term response decision-making, providing solutions and ways to adapt and respond to earthquake and tsunami, rather than official information sources such as early warning systems. At the same time, little official government assistance or resourcing was given and official government assistance only came a few days after the initial large earthquake, being distributed to the Haruku District Office three days following the earthquake. Reportedly government teams arrived to check the number of injuries and houses and infrastructure damaged. The community was given disaster care packages (called *sembako*), which referred to packets of instant noodles and bottles of drinking water. Some respondents reported being assisted with tarpaulins for evacuation shelter, while others reported not receiving any such assistance. Respondents interviewed were overall not very happy with the response they received from the local government as people found the *sembako* unhelpful. As their houses had collapsed, they had no way to cook the noodles, while local foods remained plentiful anyhow and drinking water supplies still came from natural springs that were undamaged. Respondents reported wanting help with emergency accommodation and assistance in fixing urgent infrastructures such as water pipelines and electricity.

> *Government response didn't come. No one came. For a long long time. And when they did come in their teams, they brought useless things.. things that cause more work for the local population. Endless supplies of Indomie and bottled water. What are we supposed to do with the rubbish? Not much use when you have no way of cooking but have a whole forest full of ready food, and ocean and rivers full of fish, if you only had the supplies to cook again.*

Your access to a drinking spring remains, but suddenly you have the supply of plastic drinking bottles with no landfill to put the plastic in.

Initial shock and fear gradually evolved into recovery and long-term adaptation and resilience to ongoing earthquakes that stretched over a year and beyond, with the ever-present threat of tsunami shifting from imminent danger to background insecurity. At the same time, respondents reported the ways of adjusting and adapting to life in the months during the earthquakes through which climate change and adaptation preparations played a substantial role in coping with rapid onset seismological disaster and the protracted impacts of these together.

In the context of multiple present-day disaster events, Haruku's residents reported looking to the past for their resilience and safety for the future. Historically, the village was located at higher ground, and while looking to the past to be safer in the future there were also many practical aspects to the evacuation relocations. For example, staying further up the mountains means also that children are closer to school and do not have to walk so far, each day each way. The location is more convenient as the school is located closer to the evacuation camps rather than their homes on the beach area, and families feel safer having their children nearby if there is an emergency. The area is also cooler and closer to water springs for drinking water while the pipe infrastructure remains damaged. For many locals, they see no point to return to their homes on the beach foreshore. Being too scared to take up fishing again, the beachside village is no longer conveniently located for fishing or for boat trips to bring supplies from Ambon as most have decided to avoid the ocean. In the recovery during the months following the first earthquake in September 2019 and following year into 2021, many residents decided to stay longer term in the evacuation sites. Spending daytime to work on lower grounds but choosing to spend nights sleeping at the camps on the hills, in case of another earthquake or a tsunami. Others returning to the evacuation location to be with elderly, ill or disabled family members and children. Most households insist that the most vulnerable in their families, the elderly, disabled and children, remain in the evacuation sites permanently for safety in case another earthquake came that everyone needed to react to (tsunami or not) quickly. The higher ground location also offers safety from several climate disaster impacts. Sloping and still highly forested, the area protects from potential landslides and coincides with safe zones mapped out for flood and landslide evacuation. The ocean breeze reaches the hill heights offering a reprieve from the heat, as does the higher level of vegetation and established trees.

In the same way that the community turned to the experience of the past for dealing with earthquakes and tsunami, interviewees spoke of customary stewardship and traditional knowledge to protect from climate change impacts and disaster events. In interviews, respondents reported that they no longer wished to stay in houses that use concrete or "*beton*" (concrete blocks), the materials most houses have been constructed in the village during the last few decades. At the same time, those that choose to return to their houses prefer to sleep outside at night or adjacent to their houses in tents for fear that their concrete constructed houses will collapse if there is another earthquake. Instead preferring to build back their houses with safer materials for earthquakes—with lighter roofs and flexible (allowing some movement) structures, such as their traditional types of housing: "*It's better to build houses out of wood, so that there is less danger if they collapse, or if there is liquefaction.*" Many decided to permanently relocate to higher ground—where they started building temporary versions of accommodation, as well as moving businesses such as local shops to the new location, starting household vegetable gardens and improving water supplies.

For Masyarakat Adat in Haruku, maintaining as well as increasing resilience in broad terms, including to climate change and disasters, centres on customary cultural and resource land management practices that guard local ecosystem integrity, as well as offer a diversity of options for

livelihood security. Without receiving any government assistance—such as support for tarpaulins, temporary accommodation materials, reconstruction of pipes for water supplies or assistance to reconstruct housing in the longer term, the village was essentially self-sufficient in their disaster response. Electricity remained off for weeks, and ultimately people were responsible for helping themselves and each other in the aftermath of the earthquake. They provided their own materials to make the temporary accommodation. The meals and food they provided themselves and shared with neighbours or anyone in need. Haruku island has robust local customs for environmental protection under *Sasi*, which employ methods of timing and space in order to regulate access to resources and territories to maintain a sustainable ecological, livelihood and conservation equilibrium (e.g., through customary forest conservation, fish stock sustainability practices). Customary rules mean that resources are used most efficiently, usefully and maintained within ecological bounds (sustainably). For example, the young fruit in the forest should not be taken so that they have the opportunity to first grow and produce seed. The systems that build resilience of the environmental systems in and around Haruku village, ensure the resilience of food and social systems.

> *Everyone shares the produce if anyone is in need. We always have produce such as banana and cassava, vegetables, and various edible leaves. So we can survive. Accessing water is more difficult as we currently have to carry it from the village to the evacuation site (this is different depending on the location) ...If we can't fish, and the clove harvests fail, then we still have kasbi [cassava] ... kasbi can be sold too. We can also sell vegetables, like melinjo. If people are unable to go to the sea, such as because the earthquakes under water continue, then people here return to the fields.*

Slow-onset and ongoing disasters arising from climate change have been of great concern for many years for *Adat Masyarakat* in *Negeri* Haruku. Ways to respond to climate change have primarily involved responses based on the complex local knowledge of the ecological system that everyone in Haruku lives within, at least to some degree. Within the community and continuing the traditions of *Sasi*, the impacts of climate change were felt, as residents observed the changes around them and changes to *Sasi* systems. Climate change impacts have altered how to interpret customary law and stewardship of their land and marine area. For the majority dependent on natural resource-based livelihoods, people relied on their ability almost entirely to read their environment and the intricate ecosystem signals and adapting strategies to change for the bulk of their livelihoods, for example, through the use of diversification as continuing customary systems. When interviewed both prior to the earthquakes and following, interviewees widely reported a changing climate. Haruku is also widely known for their customary marine management practices, or sasi laut or marine customary management (Novaczek et al., 2001). According to interviews with the *Kewang*, *Sasi* is ever-changing and evolving through its application by the community in the Central Maluku region, with a constant alteration or adaptation appropriate to the lifestyle and beliefs integrated between traditional and modern lives that people lead. Changes in the traditional lifestyle of

Haruku people raises a constant awareness and balance between individual ownership rights, where individual ownership confirms the bond between the owner and nature or objects owned, and that which is shared or part of "the commons." These bonds are influenced by the beliefs of people who view that nature has a soul, so humans must act cautiously and with respect towards nature. This is the basic concept of environmental stewardship that is contained in the tradition of *Sasi*.

> *We operate between these. If we do not receive food aid or are unable to buy food from outside, eating here is still no problem at all. We have staples based on sago - bapeda, as well as kasbi and others. Kasbi isn't affected by pests, variable rains, storms or waves, or earthquakes. No one goes hungry also because we also all share. We have double livelihoods that we can swap between if the conditions are not right to go to the sea, because of either climate change or earthquakes and the threat of tsunami. We also have the staple foods from the sago plant, so we do not go without.*

Despite *Sasi* customary systems ensuring that diverse ecological food systems are sustained, the impacts of the seismological disaster have been compounded by the slow-onset and ongoing climate change impacts of drought, floods, unpredictable rainy and dry season periods and crop damage. Agrarian livelihoods were affected by the damaged water pipes, which rendered irrigation systems no longer functional and instead dependent on laborious bucket collections taken from far down the hill slopes. Crop loss from increased heat periods and dry spells kept occurring, while flooding made living conditions in tents difficult and compounded water-sanitation issues. Broken pipes from earthquakes meant that the new gardens for food made closer to the evacuation sites had difficulty being maintained with varying seasonal patterns and increased heat. Conversely, rain variability and flooding caused further havoc at the evacuation sites. Health implications arose due to the flooding of evacuation sites, coupled with the lack of sanitation and clean water and crowded conditions at the sites. Livelihoods are often supplemented by fishing for tuna, which can bring in large sums if successful (although success has been rapidly declining, due to climate change related influences and overfishing). However the repeat earthquakes, which happen in the channel between the island of Haruku and the main island of Ambon, had left fishers petrified and stopped them from returning to the ocean: "*The earthquake just continues. While we do not pursue our livelihoods because of the earthquakes, our income is dwindling.*" Thus fishing, which is usually used to supplement household incomes and food, ceased following the earthquake. Initial earthquakes had corresponded with the timing of the wave season, when the waves are considered too big for speed boats to go out in the ocean, which in turn corresponds with the timing when people are busy with the clove season harvests. Interviewees reported that the waves in recent years had also become stronger. As an island, storms and an increase in big waves meant not only the ability to catch fish to supplement household food or for fishing livelihoods was affected, but getting supplies from the outside was also regularly more difficult.

Respondents report that not just due to climate change but also overfishing, fish numbers are dwindling. In the past, fishing could be done by the shore by grabbing fish and therefore did not require people to have boats or to be able to travel out far from the shore where storms could impact them, or risk during ongoing earthquake activity while far out at sea. "*It used to be that we went from the shore up to 50 m into the sea was packed with fish immediately. If it is now from the morning sometimes until the night, we have to fish.*"

Most families also supplement income with clove plantations of varying sizes as cash crops to harvest. Coincidentally the islanders were lucky that the timing of the initial earthquake was the same time as the clove harvest season. Instead of fishing many relied on harvesting their cloves, as clove harvest time is also when families tend to have a bit more cash from harvest. However, as earthquakes continued to prevent fishers from returning to the oceans, and clove season passed, livelihood incomes continued to decrease. Many people also abandoned their harvests (and lost much needed income) when they fled, reluctant to return. Instead of collecting, harvesting vegetables and fruits or fishing for food, at this time they were instead able to buy in food and food stocks (like rice) until cash flows from cloves harvests and fishing gradually dwindled. Compounding these issues is the impacts of climate change, which often presents new problems with seasonal variation and increased heat and dry periods, new plant diseases and pests. Agricultural patterns had been affected, not just in yields and planting patterns but also for processing. "*We do not know any more when the rain will come or when it will be dry.*" At the same time, rainfall variation affects the harvesting as the crops cannot be dried. Clove harvests are the island's primary cash crop and are often used to send children to school or (such as in the case of earthquakes) build or rebuild housing. Clove harvesting is dependent on drying in full sun, according to the season of harvest. The drying process needs full sun without any clouds; otherwise, the quality of the clove is affected. Climate change has affected the drying process, compared with years previously, driving yields down. Nevertheless, despite the decreased yields, which would typically mean cloves would be in higher demand, following the earthquake, the price for clove harvests plunged to IDR 50,000 (USD3.50) per kilogram, with some collectors buying from farmers at only IDR 40,000 (USD2.80) per kilogram. Reduced selling prices happened even though clove harvests for the year were relatively scarce as people were so desperate to sell in the aftermath of the earthquake that the market could buy from farmers at a meagre price taking full advantage of farmers economic needs in recovery to the 2019 earthquake.

With the main cash crop affected, alongside fishing incomes following the series of earthquakes, climate change and disaster impacts have compounded to impact household incomes detrimentally. Having a flow-on effect as to whether households can afford schooling for children, or whether household members are driven to seek work off the island (e.g., labour contracting, or work on fishing trawlers drive people to leave either to Ambon, Jakarta, Papua or eastern regions for fishing). Cash coping assistance also comes from outside remittance being sent from family members, to be able to fund things such as housing construction and reconstruction. Interviewees estimate that around half of the local population of Haruku, do not work in Haruku but stay outside in areas such as Jakarta, West Papua or Ambon. With family members working in an array of professions (e.g., government or as teachers) or as labourers on construction projects or fishing trawlers. These remittances became vital for buying materials from the Ambon main island to rebuild housing and supplement incomes with livelihoods lost. However, it means that during post-earthquake evacuation, many able-bodied people left the island for work elsewhere, while elderly and children and those tending for them stayed behind.

Conclusion

Traditional knowledge, social media and hoax news, sensitisation to tsunami events, government and NGO programming, all entwined as agents of information in roles shaping the initial shock and fear, and gradually the recovery and resilience as a small island community. Where the community has responded to disaster and adapted to ongoing disaster impacts, climate change has destabilised the response, leading to compounding, cascading and protracted impacts. Yet preparations in adapting to climate change have also aided the response and resilience to seismological disaster by providing options that can be "re-adapted" for climate change to disaster contexts; such as home gardening for food security, flood evacuation mapping, local response team organisations and flood crisis management. While *Sasi adat* practices that have proven vital to climate change adaptation, by having a diverse range of livelihood and food security options, and the buffering of intact ecological system integrity—have provided continued safety to the community also in the aftermath of earthquake disaster. In such a way, this case study has provided an example of how climate change adaptation preparations spur disaster preparedness and response resilience, and vice versa. However, both climate change adaptation and disaster response and resilience have benefited from ecosystem-based approaches, entrenched in local knowledge and protection systems under customary *Sasi*.

References

BMKG. (2018). Gempabumi Tektonik M=7.7 Kabupaten Donggala, Sulawesi Tengah pada hari Jumat, 28 September 2018, Berpotensi Tsunami | BMKG. BMKG | Badan Meteorologi, Klimatologi, dan Geofisika. https://www.bmkg.go.id/press-release/?p=gempabumi-tektonik-m7-7-kabupaten-donggala-sulawesi-tengah-pada-hari-jumat-28-september-2018-berpotensi-tsunami&tag=press-release&lang=ID. Accessed 3 October 2021.

BMKG. (2019). Gempabumi Tektonik M 6,8 Mengguncang Kota Ambon, Tidak Berpotensi Tsunami | BMKG. BMKG | Badan Meteorologi, Klimatologi, dan Geofisika. https://www.bmkg.go.id/berita/?p=gempabumi-tektonik-m-68-mengguncang-kota-ambon-tidak-berpotensi-tsunami&tag=gempabumi&lang=ID. Accessed 3 October 2021.

Borrero, J. C. (2005). Field data and satellite imagery of tsunami effects in Banda Aceh. *Science, 308*(5728), 1596–1596. https://doi.org/10.1126/science.1110957

Charmaz, K. (2020). With constructivist grounded theory you can't hide: Social justice research and critical inquiry in the public sphere. *Qualitative Inquiry, 26*(2), 165–176. https://doi.org/10.1177/1077800419879081

Din. (2019). BMKG Konfirmasi Hoaks Gempa Besar dan Tsunami Ambon. *Teknologi*. https://www.cnnindonesia.com/teknologi/20190927142115-192-434647/bmkg-konfirmasi-hoaks-gempa-besar-dan-tsunami-ambon. Accessed 3 October 2021.

Fadhil, H. (2019). BMKG: Isu Viral akan Ada Gempa Besar dan Tsunami di Ambon Hoax. *Detiknews*. https://news.detik.com/berita/d-4724430/bmkg-isu-viral-akan-ada-gempa-besar-dan-tsunami-di-ambon-hoax. Accessed 3 October 2021.

Griffiths, J. (2018). How Indonesia's tsunami warning system failed its citizens again. CNN. https://www.cnn.com/2018/12/24/asia/indonesia-tsunami-disaster-year-intl/index.html. Accessed 3 October 2021.

Iqbal, M. (2019). Viral Isu Gempa Besar & Tsunami Maluku, BMKG: Hoax! *CNBC Indonesia*. https://www.cnbcindonesia.com/news/20190927174535-4-102793/viral-isu-gempa-besar-tsunami-maluku-bmkg-hoax. Accessed 3 October 2021.

Jemadu, L. (2019a, September 16). Ikan Mati Massal di Pantai Ambon, LIPI Terjunkan Peneliti. suara.com. https://www.suara.com/tekno/2019a/09/16/213959/ikan-mati-massal-di-pantai-ambon-lipi-terjunkan-peneliti. Accessed 3 October 2021.

Jemadu, L. (2019b, September 16). Ikan Mati Massal di Pantai Ambon Pertanda Tsunami? BMKG: Hoaks. suara.com. https://www.suara.com/tekno/2019b/09/16/171315/ikan-mati-massal-di-pantai-ambon-pertanda-tsunami-bmkg-hoaks. Accessed 3 October 2021.

Latief, H., Kodijat, A., Ismoyo, D., Bustamam, B., Adyasari, D., Nurbandika, N., & Rahayu, H. (2016). Air Turun naik di tiga Negeri mengingat tsunami Ambon 1950 di Hutumuti, Hative kecil dan Galala (Water up throughout three Negeri regions: Remembering the 1950 Ambon tusnami in Hutumuri, Small Hative and Galala). United Nations Educational, Scientific, and Cultural Organization (UNESCO) Office Jakarta – Indian Ocean Tsunami Information Centre (IOTIC). http://iotic.ioc-unesco.org/resources/publications-and-reports/69/publication. Accessed 3 October 2021.

Mantjoro, E. (1996). Traditional management of communal-property resources: The practice of the Sasi system. *Ocean & Coastal Management, 32*(1), 17–37. https://doi.org/10.1016/S0964-5691(96)00013-0

Martin, P. Y., & Turner, B. A. (1986). Grounded theory and organizational research. *THe Journal of Applied Behavioral Science*. https://doi.org/10.1177/002188638602200207

Myers, M. D. (2009). *Qualitative research in business and management*. SAGE Publications Ltd. https://uk.sagepub.com/en-gb/asi/qualitative-research-in-business-and-management/book244733. Accessed 3 October 2021.

NOAA. (2021a). National Centre for Environmental Information (NCEI) Hazard Earthquake Information. National Centres for Environmental Information. https://www.ngdc.noaa.gov/hazel/view/hazards/earthquake/event-more-info/1934. Accessed 3 October 2021.

NOAA. (2021b). NCEI Hazard Earthquake Information. National Centres for Environmental Information. https://www.ngdc.noaa.gov/hazel/view/hazards/earthquake/event-more-info/1934. Accessed 3 October 2021.

NOAA. (2021c). What is upwelling? National Ocean Service, National Oceanic and Atmospheric Administration (NOAA), US. https://oceanservice.noaa.gov/facts/upwelling.html. Accessed 3 October 2021.

Novaczek, I., Harkes, I. H. T., Sopacua, J., & Tatuhey, M. D. D. (2001). Institutional analysis of Sasi Laut in Maluku, Indonesia. ICLARM, Penang, MY. https://idl-bnc-idrc.dspacedirect.org/handle/10625/30324. Accessed 3 October 2021.

Nuraini, D. (2019, October 14). BNPB: Ratusan Ikan Terdampar di Maluku Tak Terkait Aktivitas Kegempaan | Ekonomi. Bisnis.Com. https://ekonomi.bisnis.com/read/20191014/99/1158862/bnpb-ratusan-ikan-terdampar-di-maluku-tak-terkait-aktivitas-kegempaan. Accessed 3 October 2021.

Okal, E. A., & Reymond, D. (2003). M 8.5—141 km WNW of Tual, Indonesia. In The mechanism of great Banda Sea earthquake of 1 February 1938: applying the method of preliminary determination of focal mechanism to a historical event. *Earth and Planetary Science Letters*, *216*(1), 1–15, https://earthquake.usgs.gov/earthquakes/eventpage/official19380201190422_30/executive. Accessed 3 October 2021.

Patty, R. R., & Ika, A. (2019, September 16). Misteri Penyebab Ribuan Ikan Mati Mendadak di Ambon Terungkap Halaman all (The Mystery of the Cause of Thousands of Fish Die Suddenly in Ambon Revealed). KOMPAS.com. https://regional.kompas.com/read/2019/09/16/17015561/misteri-penyebab-ribuan-ikan-mati-mendadak-di-ambon-terungkap. Accessed 3 October 2021.

Prabowo, P. H. (2019, September 23). Hoaks, ikan mati massal di pantai Ambon pertanda tsunami. Antara News. https://www.antaranews.com/berita/1076582/hoaks-ikan-mati-massal-di-pantai-ambon-pertanda-tsunami. Accessed 3 October 2021.

Sassa, S., & Takagawa, T. (2019). Liquefied gravity flow-induced tsunami: First evidence and comparison from the 2018 Indonesia Sulawesi earthquake and tsunami disasters. *Landslides, 16*(1), 195–200. https://doi.org/10.1007/s10346-018-1114-x

Soselisa, H. L. (2019). Sasi Lompa: A critical review of the contribution of local practices to sustainable marine resource management in Central Maluku, Indonesia. *IOP Conference Series: Earth and Environmental Science, 339*, 012013. https://doi.org/10.1088/1755-1315/339/1/012013

Sugiyarto. (2019, September 15). Ribuan Ikan Mati Mendadak dan Terdampar di Sejumlah Pantai di Ambon, Penyebabnya Masih Misterius (Thousands of fish died suddenly and washed up on a number of beaches in Ambon, the cause is still mysterious). Tribunnews.Com. https://www.tribunnews.com/regional/2019/09/15/ribuan-ikan-mati-mendadak-dan-terdampar-di-sejumlah-pantai-di-ambon-penyebabnya-masih-misterius. Accessed 3 October 2021.

Triyono, R. (2019). Gempa magnitudo 6,8 guncang Ambon, 20 orang meninggal dunia. BBC News Indonesia. https://www.bbc.com/indonesia/indonesia-49834676. Accessed 3 October 2021.

Tubaka, N. (2019, September 18). Ratusan Ikan Terdampar dan Mati Misterius di Maluku, Ada Apa? Mongabay.Co.Id. https://www.mongabay.co.id/2019/09/18/ratusan-ikan-terdampar-dan-mati-misterius-di-maluku-ada-apa/. Accessed 3 October 2021.

Ulum, M. (2019). Gempa Ambon, Ingatan Tsunami dan Kisah Unik Setelahnya. https://papua.bisnis.com/read/20191007/414/1156483/gempa-ambon-ingatan-tsunami-dan-kisah-unik-setelahnya. Accessed 3 October 2021.

Wilson, G. A. (2012a). Community resilience and environmental transitions. *Routledge*. https://doi.org/10.4324/9780203144916

Wilson, G. A. (2012b). Community resilience, globalization, and transitional pathways of decision-making. *Geoforum, 43*(6), 1218–1231. https://doi.org/10.1016/j.geoforum.2012.03.008

Zerner, C. (2014). Through a Green Lens: The Construction of Customary Environmental Law and Community in Indonesia's Maluku Islands. *Law & Society Review*, *28*(5), 1079–1122, https://doi.org/10.18574/9780814789339-044

CHAPTER 14

Kinship as Cultural Citizenship in Post-disaster Housing Reconstruction: Narratives from Fisherfolk in the Philippines

Pamela Gloria Cajilig

Introduction and Context

Isla Sasa, an estuarine island in Manila Bay, is perpetually enveloped in a bustle of housing reconstruction activity. Whilst walking around the four-hectare island, one needs to sidestep concrete mixes, builders hauling bamboo-and-palm leaves from the wharf, and children playing amidst stacks of hollow blocks as parents keep a watchful eye. Several hazards threaten life on Isla Sasa and therefore shape housing reconstruction efforts. The island, home to 3000 residents who are mostly fisherfolk, is prone to daily flooding caused by land subsidence as well as sea-level rise in Manila Bay at four times the global annual average of

P. G. Cajilig (✉)
School of Architecture and Urban Design, RMIT University, Melbourne, Australia
e-mail: s3668872@student.rmit.edu.au

A. Lukasiewicz and T. O'Donnell (eds.), *Complex Disasters*, Disaster Risk, Resilience, Reconstruction and Recovery,
https://doi.org/10.1007/978-981-19-2428-6_14

3.2 mm (Ng, 2020). Isla Sasa is also typhoon-prone due to the Philippines' location within the Pacific typhoon belt. The country is visited by an average of twenty typhoons annually (Asian Disaster Reduction Center, 2019). Residents and community leaders agree that the strongest recent typhoon in the island was Typhoon Glenda (Rammasun) in 2014, in which entire houses were lost to the wind and sea. As of writing, many inhabitants are currently rebuilding after super typhoons Rolly (Goni) and Ulysses (Vamco) destroyed large swathes of the country in November 2020. According to the residents, living with the frequent flooding and typhoons in Isla Sasa means accepting that one has to rebuild their home every ten to fifteen years despite limited resources. Rebuilding after disaster is a costly and daunting challenge for fisherfolk who live in one of the world's most hazardous landmasses (Bankoff, 2015) and who are one of the poorest vulnerable sectors in the Philippines (Philippine Statistics Authority, 2020).

The significance of kinship networks to post-disaster housing reconstruction in Isla Sasa became clear during my initial walks around the island with Sally, Elena, and Lilia, a group of residents who are also community leaders and who are aware of my doctoral research topic on housing and disaster. Whilst walking, the three women helpfully pointed out houses that they believed to be disaster resilient; these were mostly two- or three-storey homes made of concrete. The recognition of these houses was often accompanied by narratives of how certain homeowners were able to build more resilient houses with the help of their kin. On the other hand, my companions attributed the dereliction of other houses to either broken or lost ties between homeowners and their relatives.

The importance of kinship to disaster recovery also surfaced during focus groups discussions I co-facilitated in their *barangay* as part of a collaboration between universities in the Philippines and in the United States. During my conversations with homeowners whose houses were destroyed by Typhoon Glenda, I therefore made it a point to ask about the relationships that made rebuilding possible. Who among their kin have been essential to rebuilding their homes? How are these relationships created and maintained? What are the consequences of relying on kinship to reduce disaster risks? And, why has it been necessary to rely on such relationships to overcome catastrophe?

The last question of *why* Isla Sasa inhabitants turn to their kin during disaster draws from those who analyse kinship not only as an element of social organisation but as a negotiation of the efforts of the state and allied actors to help as well as control citizens affected by disaster (Carruth, 2018; Remes, 2016). The discussion illustrates how mobilising kinship networks for assistance throughout several disasters manifests the incapability and hesitation of administrative units to address the question of long-term adaptation on the island despite international and domestic policy frameworks that mandate the localisation of all phases of disaster risk reduction. Kinship emerges as a manifestation of post-disaster housing reconstruction construed as a private burden in the context of citizenship regimes that comprise the inadequacies of state-sanctioned efforts in disaster preparedness, mitigation, and recovery as well as encourage competition for limited post-disaster housing reconstruction resources.

Consequently, I frame kinship within post-disaster housing reconstruction as a mode of citizenship practice to underscore its political and relational nature. Kinship is an element of social organisation and an economic resource (Oracion, 2015) *as well as* a political act used by the vulnerable to navigate structures of power that define the terms of belonging and exclusion within efforts to surmount catastrophe. Specifically, I frame the mobilisation of kinship networks as the *cultural citizenship* of certain Isla Sasa inhabitants who continue to struggle with rebuilding their homes since Typhoon Glenda in 2014. I draw from Ong et al. (1996) in defining "cultural citizenship" as the beliefs, values, and practices that arise from negotiating the processes of subjectification enforced by the state and actors vested with the power to act on the state's behalf. To grasp kinship as cultural citizenship, I first identify the *citizenship regimes* (Vink, 2017) that shape experiences of belonging and exclusion in post-disaster housing reconstruction at Isla Sasa: the regime of localism, the response regime, and, the *palakasan* system, a competition for power that goes beyond the rule of law and that is informed by local definitions of good leadership (Agpalo, 1999; Lopez, 2020; Pye & Pye, 1995). I then proceed with discussing two household-level cases of cultural citizenship within Isla Sasa that nuance the possibilities and limitations of relying on kinship to negotiate these intersecting regimes within a context of protracted and overlapping disasters.

The decentring of blood relations in this chapter echoes anthropological work about non-Eurocentric perspectives of kinship (Franklin & Mckinnon, 2001). The reinvention of "*kinship*" (a widely studied subject within classical anthropology) as an epistemic device has been critical of the imposition of Eurocentric ontologies on other cultures (Schneider, 1984) as embodied by earlier anthropological studies on the subject by, for example, Evans-Pritchard (1940), Levi-Strauss (1969), Morgan (1877), Malinowski (1922), and Radcliffe-Brown (1941). Critiques of these initial studies are informed by concerns around social justice, including feminism and civil rights; these critiques argue against evolutionary and structural-functionalist perspectives that assume cultural stability, latent biologism, essentialised notions of sex and gender, as well as hard distinctions between nature and culture (Collier & Yanagisako, 1987; Franklin & McKinnon, 2001; MacCormack & Strathern, 1980). Carsten (2000) thus brings forward the term "relatedness"—alongside or in opposition to "kinship"—to (1) breach the analytic divide between nature and culture in the study of kinship; and (2) to include "indigenous idioms of being related" instead of depending on predetermined and Eurocentric definitions (p. 4). Carsten's broad approach reflects the analysis of social organisation in the Philippines by Torres (1985). The latter observes that social ties in the Philippines are based on extended families from both maternal and paternal roots, as well as residence; neighbours therefore are also part of one's kindred. In the Philippines, the family is a fluid and dynamic contraption as children are encouraged to establish relationships with a specific combination of relatives to generate their own unique "*family*", by, for example, maintaining good relations with their godparents (Pye & Pye, 1995).

Finally, this chapter is a contribution to theory about post-disaster housing reconstruction. The field is one of the least understood aspects of disaster risk reduction (Chang, 2010); it requires significant resources and is fraught with complexity that is rooted in operational and political risks (Barakat, 2003; Barenstein & Pittet, 2007; Bilau et al., 2015; Charlesworth & Ahmed, 2015; Takasaki, 2011). Many studies of post-disaster housing reconstruction focus on programmes that are financed and implemented by officialised institutions (e.g., Global Shelter Cluster, 2018; Lyons et al., 2010). However, the majority of those affected by disaster are forced to rebuild and recover through their own efforts (Bankoff, 2015; Flinn et al., 2017). As such, nuancing the field requires investigation of the informal resources used by disaster-affected groups as well as an exploration of themes that are on the fringes of disaster management research yet, on the contrary, are central to the lived experience of catastrophe—such as kinship.

Methods and Techniques

Ethnographic research in this study involved a combination of face-to-face and remote qualitative research techniques because of travel restrictions of the Philippine COVID-19 quarantine. Before the quarantine, I co-facilitated three focus group discussions with the island's elected officials, civic group officers and volunteers. The structure of the discussions was based on the toolkit for the analysis of the disaster resilience of communities by McCaul and Mitsidou (2016), which includes such topics as hazards and capacities relevant to disaster management. I also conducted thirteen informal interviews with homeowners about the socio-material aspects of island in relation to disasters. Upon the implementation of the COVID-19 quarantine, I conducted twenty-eight more remote interviews with homeowners and adult residents as well as an interview with a municipal official. These interviews focused on the resources and processes used to remake their homes and risk management practices given frequent exposure to hydrometeorological hazards. The interviews were arranged with the help of a community contact with multiple considerations in mind, including tide levels, wind conditions, presence of typhoons, and mobile signal strength, among others.

I now turn to the three citizenship regimes that inform kinship as cultural citizenship within post-disaster housing reconstruction in Isla Sasa.

Citizenship Regimes in Philippine Disaster Risk Reduction and Management

Disaster risk reduction policies are "technologies of citizenship" used by governments to regulate and, in so doing, subjectify citizens whilst also seeking to address their concerns (Cruikshank, 1999). These forms of subjectification are manifested through intertwined citizenship regimes that are mobilised in Philippine disaster risk reduction and management (DRRM). I trace kinship as cultural citizenship in Isla Sasa firstly to the *regime of localism* that is informed by the devolution of authority in Philippine local governance, including disaster governance. However, inequalities within and among local government units as well as limited resources and skills result in localisms that reinforce a *response regime* within disaster management. The pressure on local governments to address the full spectrum of disaster risk reduction challenges with scarce resources in turn prompts participation in a competition for power called the *palakasan* system (Agpalo, 1999), a citizenship regime that encourages the pursuit of personal contacts to advance self-interest (Pye & Pye, 1995).

The Regime of Localism

The emphasis on the "local" within policy frameworks arises from the need to reconcile conflicting policy areas and align the multi-sectoral work necessitated by such supralocal forces as globalisation (Bailey & Pill, 2015) and climate change. The regime of localism in Philippine disaster management stems from both domestic and international policy frameworks that institutionalise the "*empowerment*" of local communities (Bailey & Pill, 2015) as a risk reduction and development strategy. This regime is also rooted in the centrality of hazards in everyday life and the archipelago's long history of informal leadership at the local level for the collective benefit of communities (Bankoff, 2015, pp. 430–431). Localism in the form of devolution—that is, "the transfer of authority from a central government to smaller units of government"—was institutionalised by the Aquino administration that followed the end of the twenty-year Marcos dictatorship in 1986; specifically, the 1991 Local Government Code decentralised national government functions to the country's 1600 municipalities organised by the Department of Interior and Local Government (Heinonen, 2000). This localism is further reflected by two laws that underscore the importance of participation of all levels of government and of marginalised communities in strengthening disaster resilience: the Climate Change Act (CCA) and Philippine Disaster Risk Reduction and Management Act (DRRM Law) (Piepora et al., 2019). The localism in these domestic policy frameworks (Section 14 of the CCA and Section 2 of the DRRM Law) align with the emphasis on participation of local governments and communities in such international frameworks as the Paris Agreement which aims to limit global warming to preferably 1.5 °C (Article 14; United Nations Framework Convention on Climate Change, 2021); the Sendai Framework for Disaster Risk Reduction 2015–2030 (Priority Two; Tozier de la Poterie & Baudoin, 2015); and the Grand Bargain formed during the 2016 World Humanitarian Summit (Inter-Agency Standing Committee, n.d.).

Localism has broad ideological appeal; it is promoted by progressives who value participation as a force for social transformation and by proponents of market-driven approaches to development that subscribe to the narrative of triumph over adversity through self-help (Bailey & Pill, 2015). Localism is also attractive to governments that are characterised by "*fragility*", or the lack of will or ability to provide for constituents; this fragility may stem from limited resources that cannot fully address scale of disasters, or it may be rooted in apprehensions over the political ramifications of taking responsibility for the effects of disaster (Hilhorst et al., 2017; Mukherji, 2017). There are, however, concerns regarding the misuse and abuse of localised approaches to disaster management. Localised methods for disaster risk reduction may shroud and reproduce structures of power and may merely function as sources of income and status among practitioners (Lyons et al., 2010; Stiefel & Wolfe, 2011; Titz et al., 2018). Lastly, Piepora and colleagues (2019) note that financial constraints inhibit the capacity of local government units to effectively address disaster: the poorer and more vulnerable local government units have less funding available to invest in disaster management. In a focus group with local leaders on the island, a councillor brought up the council's failure to materialise a wide range of disaster management plans due to lack of resources, thus forcing the barangay to become dependent on higher levels of government and civic groups:

> *We* [the council] *are asked to make plans [by the municipal and provincial governments] but when it comes to the budget, there's none. For example, after disaster, there really is no livelihood here. We want to give food and relief but there's no money. So, what we just do, we just request from the municipal government, the provincial government, and other NGOs that can help us.*

The Response Regime

The DRRM Law outlines the expected contributions of local government units (via their respective Local Disaster Risk Reduction Management Offices) as well as avenues for disaster management financing. Section 2 requires all local government units, including barangays, to contribute to the four pillars of disaster risk reduction of the National Disaster Risk Reduction Management Plan: (1) Disaster Prevention and Mitigation; (2) Disaster Preparedness; (3) Disaster Response; and (4) Disaster Rehabilitation and Recovery. Section 21 of the DRRM Law stipulates the maintenance of a Local Disaster Risk Reduction Management Fund (LDRRMF) consisting of not less than 5% of revenue from regular resources. The fund may be used for a range of DRRM activities, including preparedness training, purchase of rescue equipment, and post-disaster activities. The DRRM Law also states that of the amount appropriated for LDRRMF, 30% needs to be budgeted as a Quick Response Fund for relief and recovery programmes. However, the use of estimated revenues as the basis for computing the Quick Response Fund disadvantages LGUs whose constituents are comprised by the most marginalised sectors in the Philippines, including fisherfolk (Philippine Statistics Authority, 2020; Piepora et al., 2019). The regime of localism thus lends itself to disaster management practice focused on response, particularly immediate relief, whilst long-term disaster risk reduction remains an enormous challenge (Alcayna et al., 2016; Piepora et al., 2019). The overemphasis on relief is confirmed by Brucal et al. (2020) who analysed Philippine disaster spending using local government financial data: relief comprises the largest portion of local government spending on disaster.

The focus on relief at the expense of foregoing long-term disaster risk reduction measures needs to be understood in the context of disaster recovery challenges in the Philippines. The country's planning and implementation of recovery programmes are impeded by recurrent disasters, limited and inefficient disaster risk financing, political manoeuvring, lack of data, lack of transparency, and lack of coordination between the national government and local governments (Alcayna et al., 2016; The World Bank, 2020; Villacin, 2017). The stakes of being unable to address long-term recovery are high. In 2013, a difficult year that is associated with the destruction of Typhoon Haiyan, Filipino families survived by reducing spending on health and education by an estimate of 25%; spending on transportation and communication dipped by 35% whilst spending on nutritious food dropped by 30% (Jha et al., 2018). Long-term evaluations of post-disaster needs reveal further gaps and challenges in recovery: the lack of attention to health and well-being, including mental health; lack of housing tenure; prolonged settlement in high-risk zones; and the exacerbation of vulnerability through political processes (Alcayna et al., 2016).

The inadequacies of the response regime in disaster risk reduction reported in scholarly and industry publications (e.g., Alcayna et al., 2016; Brucal et al., 2020; Piepora et al., 2019) were reflected during focus groups and interviews with community stakeholders tasked to implement relief distribution in Isla Sasa. The Barangay Secretary mentioned that 5% of Isla Sasa's revenue only amounts to P135,000 ($2727). This amount is only good for some preparedness activities and minimal relief distribution for the barangay's three thousand constituents at P15 ($0.30) per person. The Barangay officers' contributions remain concentrated on early warning, evacuation, and administrative support for the relief distribution of higher levels of local government, private individuals, and civic society organisations. Indeed, barangay officers and residents mentioned that the municipal and provincial governments could always be depended on to generously provide grocery supplies (such as rice, coffee, and instant noodles) to all families in the island as a response to frequent disasters; householders interviewed said they feel the support of the government the most during relief distribution. However, both Barangay officers and residents clarified that whilst they appreciate regular relief distribution, the community has long been in dire need of support for livelihood rehabilitation and housing reconstruction in light of the community's experiences of daily flooding and typhoons.

Meanwhile, the municipal government of Isla Sasa has its own constraints. The push for "Sustainable Settlements" in the town's 2011–2020 Comprehensive Land Use Plan includes the provision of environmentally sustainable housing designs and materials as well as the retreat and resettlement of its most vulnerable and low-lying barangays, including Isla Sasa (Barisky et al., 2014). However, the Municipal Planning and Development Officer (MPDO) during our interview stated that the town revisited this adaptation and mitigation strategy in their 2020 disaster risk reduction planning session. The suspension of these plans was the outcome of the Local Disaster Risk Reduction Management Council's realisation that the town's 2020 disaster risk reduction budget of P17,000,000 ($339,620) for around 133,000 constituents can realistically only cover response. Even though the municipal government is responsible for preparedness, mitigation, and recovery, their limited funding implies that the resources for addressing challenges beyond disaster response requires support from higher levels of government. According to the MPDO:

> *If you look at it, the bulk of our efforts are still in response. As for preparedness, the equipment, those needed for the long-term—we cannot do it. We need to source from the national government.*

The Palakasan System

Agpalo (1999) compares and contrasts the Philippine *pangulo* (head or leader) Regime (PPR) with the British Parliamentary Regime (BPR) and the American Presidential Regime (APR). These three regimes differ in, among other things, their prioritisation of the three values of the French Revolution. The BPR tends to most emphasise the value of liberty whereas the APR places greater importance on equality; the PPR tends to underscore fraternity. Moreover, the socio-historically informed unfolding of political culture in the Philippines has led to the emphasis on the fraternal value of *pagdamay* (solidarity) as a cornerstone of leadership within the PPR. The PPR rationalises patronage to leaders in support of their pagdamay; it also legitimises the mobilisation of resources to materialise a leader's pagdamay, even if this may go against the law (Lopez, 2019). Indeed, the pangulo regime is underpinned by *palakasan* (from the root word lakas meaning "power" or "strength"), or "competition for power wherein whomever holds the most lakas succeeds" (Lopez, 2019, p. 90). Palakasan assumes that the ultimate expression of power is immunity to the law; leaders and those affected by disaster may participate in palakasan by forming alliances in order to re-direct scarce short-term resources in the spirit of *pagdamay* (Ballano, 2020; Pye & Pye, 1995, p. 124).

The presence of *palakasan* as a disaster citizenship regime in Isla Sasa was implied during a focus group discussion with barangay officers that was imbued by tension stemming from inconsistencies in the housing damage assessment after Typhoon Glenda and the cash assistance received thereafter. Three Barangay officers mentioned that many people on the island (including one of them) did not receive any assistance package from the government despite damage to their homes. This assertion was countered by two other officers, who said that conversely, many people who received shelter assistance (including one of the officers who complained) exaggerated their housing damage to receive more support even though they are relatively well-off or can finance housing repairs through other means. Meanwhile, a fisher and occasional construction worker, directly attributed the unevenness within the distribution of livelihood rehabilitation assistance after Typhoon Ulysses in 2020 to palakasan.

> *When you are malakas (powerfully allied) with nanunungkulan (incumbents), you will be able to get fishing nets. If you are not malakas with nanunungkulan, you won't be able to.*

I will next discuss how those affected by Typhoon Glenda practice their cultural citizenship (their negotiation of these three intersecting regimes) by turning to their kin and neighbours for support to overcome the longer-term challenges of disaster recovery. I maintain that the establishment and mobilisation of kinship networks and reciprocal relations has been crucial in competing for limited housing reconstruction and livelihood rehabilitation resources for long-term disaster recovery and, by extension, in negotiating forms of exclusion in the distribution of various forms of assistance.

Kinship as Cultural Citizenship

Case 1: Daniel and Jenny

Daniel (the fisher who commented on inconsistent livelihood rehabilitation assistance in Isla Sasa) and his wife Jenny (an occasional fish vendor) and their two daughters used to live in a bamboo-and-palm house on stilts that stood over the river at the Northern side of the island. During Super Typhoon Ondoy in 2009 (Ketsana), their home was completely destroyed but not washed away. The couple only received P5000 ($100) as housing reconstruction assistance from the government whereas their neighbours, whose concrete house remained intact, received P10,000 ($200). During Super Typhoon Glenda (2014), they could only watch as big waves washed away their entire home towards the sea. They were consequently assessed by the government as having a "fully damaged" home. During separate interviews, they each recalled receiving P3000 ($60) worth of cash assistance from the government as well as some groceries whereas other households interviewed received P5000 ($100) cash assistance even though their homes were classified as "partially damaged". Other householders mentioned that they were also included in the government's distribution of plywood, however, Daniel and Jenny said that they were not part of the "beneficiary" list for housing materials. Indeed, these inconsistencies may be due to the challenge of recalling events that occurred several years ago. However, Jenny and Daniel insisted that they have repeatedly been excluded from receiving reconstruction and livelihood assistance throughout several typhoons and incidences of severe flood because of their lack of political connections within the barangay. "*Those who make the lists, those who assess the damage, they are the problem*", said Daniel. Rather than voice out their concerns about being repeatedly excluded from beneficiary lists to officials, the couple prefer to keep their sentiments to themselves and their immediate circle. "*We just try to disregard our tampo (roughly, resentment) towards the government [the barangay]. We just laugh it off*" Daniel stated. The couple also do not want to come across as rude to the *barangay* officials by bypassing them and directly asking for assistance from the municipal government because this might further lessen their chances of being included in beneficiary lists.

Instead, Daniel and Jenny have relied on immediate family members, in-laws, and childhood friends for cash assistance, clothing donations, materials, and labour after the various disasters that affected their family. Daniel's parents offered a small plot of land next to them for the couple to resettle after Typhoon Glenda. For their disaster recovery needs, the couple also reached out to political connections outside of the barangay that were formed through ritual kinship. When Daniel and Jenny were married (and as is the case in many Filipino families), Daniel's father invited three of his *kababata* (childhood friends) to become the couple's *ninong sa kasal* (godfathers at the wedding) and thus their advisers and protectors. This is in line with Kikuchi's (1992) observations regarding the cultivation of solidarity through ritual kinship in Philippine society as a way of overcoming the lack of social unity and intergenerational lack of social protection. Indeed, Selma, Daniel's mother, mentioned that when she and her husband were in the process of selecting Daniel and Jenny's godparents, they prioritised one's capacity for helpfulness (*pagiging matulungin*) based on financial resources as their primary criteria. Furthermore, Kikuchi maintained that parents living in poverty choose members of the political and economic elite as their children's godparents to guarantee the former's social mobility; meanwhile, powerful individuals who have political ambitions are keen on developing ritual kinship with the working class.

Daniel and Jenny's three *ninong* are the former police chief of the town, a former barangay captain, and the town's current Vice Mayor. Daniel approaches the first two for smaller forms of cash assistance for the repair of his fish corral after typhoons or after a big flood. When the couple were overlooked during the distribution of shelter supplies after Typhoon Glenda, they approached the Vice Mayor, who was then the town's mayor. This *ninong* owned a franchise of a popular burger chain in town. Jenny consequently wrote a handwritten letter of request and then went to the burger joint hoping to chance upon their *ninong*. "*I did not feel hiya [ashamed] anymore. I found strength from within*", Jenny described how she overcame feeling too embarrassed to ask for help and her anxiety over the possibility of being rejected by someone with higher stature. The mayor finally arrived and gave Jenny P3000 ($60) as a form of *pagdamay* (solidarity). The couple used the money to purchase galvanised iron sheets for their home's roofing. In 2009, after Typhoon Ondoy, the same *ninong* also demonstrated pagdamay by giving the couple P2000 ($50) also for fixing the roof of their old house by the river, which was blown away by Typhoon Ondoy (Ketsana). The couple was also excluded then from the government's distribution of cash assistance. Jenny believes the amount was lower because the *ninong* must have used up personal funds during a failed bid for an electoral post during the 2007 Philippine General Elections.

Daniel and Jenny show appreciation by bringing their *ninong* fresh fish and shrimp from their *buklad* (fish corral). Selma said that their extended family reciprocates the assistance received by the couple through attending birthday parties and other special occasions of the *ninong* when invited. The extended family also vote for *ninong* who run for public office and share their good deeds to other constituents. However, Selma stresses that one cannot ask for help from *ninong* (and others who are in a position to extend personal assistance) too often because of *hiya*, that is, feeling social pain for failing to meet reciprocal expectations (Bulatao, 1964; Torres, 1985).

Case 2: Linda

Linda, an orphan and a widow in her sixties, moved to the island in the 1970s from her home province of Cebu when she married her late husband, an Isla Sasa local. She finished her undergraduate degree in Social Work as a scholar at the age of fifty-five. She is one of the few residents of her generation in Isla Sasa who finished college. Linda has no political connections in the barangay through blood ties or ritual kinship.

However, because of her degree, she manages to grow her clout by volunteering for community-level organisations (in which she is often elected as an officer) sponsored by the government and civic organisations. She also volunteers to conduct interviews for the Philippine census and for needs assessment in relation to various social services. Through her volunteer community work, Linda developed her contacts in the municipal and provincial governments. This network became helpful in the aftermath of Typhoon Glenda, when she became impatient with the slow response of the barangay officers. Linda recalled,

> They [the barangay officers] don't know my technique. Whenever there's a disaster, I take pictures, because that's the order of the provincial government. Now, when they see the pictures, they know that there really is a calamity.

Linda sent the pictures she took of the damaged homes on the island to a provincial government contact. She knew that bypassing the barangay officers would spark controversy, but she asserted that she had nothing to fear because the damage depicted on the images is undeniable. Linda's provincial contact deployed a needs assessment team which evaluated her home as "partially damaged" and, as such, she received P5000 ($100) in cash from the government. Meanwhile, her son's family received P10,000 ($200) because their house was assessed as "fully damaged". Other homeowners reported receiving similar amounts. Some, like Daniel and Jenny in Case 1, reported receiving less despite their home being assessed as "fully damaged". Still, others with partially damaged homes said they received livelihood assistance but were not given housing assistance.

Linda and her son's family used the money to purchase bamboo poles and plywood to rebuild. Linda has a tiny shanty on stilts for cooking and sleeping constructed within the ruined concrete walls of her home—the only structures left standing after Glenda. This assistance received by Linda was not enough to completely rebuild her home, especially since her only sources of income are the minimal allowances she receives for volunteering. In 2021, seven years after Typhoon Glenda, Linda's old concrete home remains in disrepair. She continues to cook and sleep inside the bamboo shanty but spends most of the day with neighbours who have larger houses. Linda still hopes to rebuild her home, however, starting with obtaining filling material for raising the ground level to reduce the risk of flooding. In 2020, she purchased thirty sacks worth of filling material. The filling typically costs P300 per sack ($6), but she was able to obtain this for only half the price. She attributes this bargain to her good reciprocal relations with seven of the community's *mananambak* (labourers who source and transport filling material). She does this by including certain *mananambak* families in needs assessment interviews and beneficiary lists (upon their request) in various community programmes, such as feeding programmes for school-age children. According to Linda, "*They [the mananambak] say, "Pick me, pick me". Always they say, "Pick me, interview me". I move fast. I am able to help because I can get a lot of things*".

Linda insists that she does not expect anything in return by including her neighbours in beneficiary lists or by showing them how they can get themselves on such lists; she believes this is the best way for her to show *pagdamay* as someone who is more aware of citizen rights (because of her social work background) compared to the rest of the community. Here, Linda appears to be adhering to the principle of *delicadeza*, that is, the need to be discrete in the exercise of power; *delicadeza* is the fine line between the acceptable and the contemptible that those in power must tread (Pye & Pye, 1995). Whilst *hiya* prevents those in need from asking too much, *delicadeza* prevents the powerful from abusing their position by expecting favours in return or by only favouring certain people.

Linda's *pagdamay* for the mananambak families and her other neighbours have been reciprocated in the form of food, cash, and housing materials especially after typhoons and intense tidal flooding on the island when daily resources tend to be scarce despite municipal and provincial government efforts to provide food relief. However, what may be construed as *pagdamay* and inclusion by Linda and the *mananambak* may be understood as *palakasan* and exclusion by those outside Linda's circle. During my participation in a post-typhoon relief effort in Isla Sasa, I learned that Linda's efforts in community volunteering (and the efforts of other community volunteers) are often the subject of resentment and gossip. This is likely because acts of volunteering raise anxieties about *palakasan* and its exclusionary nature. As such, residents who feel excluded from circles of *pagdamay* within the barangay may appeal to the sense of *pagdamay* of powerful people outside the barangay, (as Daniel and Jenny did) and, in the process, also inadvertently participate in *palakasan*.

The Ongoing Struggle to Recover

As of writing, the COVID-19 pandemic has compounded the challenges of poverty alleviation brought about by disasters driven by "natural" hazards. The pandemic has reversed the steady decline in the country's poverty in recent years, with 2.7 million more people living below the poverty line (The World Bank, 2020). Livelihood rehabilitation in Isla Sasa has become a greater struggle due to the repeated destruction of fish corrals during the series of typhoons in November 2020, the unprecedented tidal flooding levels on the island in June 2021 as indicated by tide calendars, and more typhoons from July to September 2021 (Arceo, 2021; CNN Philippines, 2021; Reliefweb, 2020, 2021). Unfortunately, residents have been unable to pursue alternative forms of livelihood due to COVID-19 quarantine restrictions. "People need income and food. For now, house repairs can wait", said an Isla Sasa community health worker, thus hinting at an even more protracted housing reconstruction process. The remark also validates recommendations to bolster social protection and disaster preparedness measures by strengthening programmes for food and subsistence (The World Bank, 2020).

Whilst Jenny and Daniel scramble for funds amidst pandemic restrictions to rebuild their family fish corral that was damaged in November 2020, the couple continues to grieve for the peaceful and dignified life they lost along with their former home after Typhoon Glenda. They still hope to save for a concrete house that would remain stable amidst the strong waves of the annual Southwest monsoon, but more pressing concerns abound. The couple prioritises the education of their young daughters so they can eventually escape the poverty of Isla Sasa. Meanwhile, Linda still lives in the fragile shanty within the muck, stench, and ruins of her old house. As experienced by other households on the island, the typhoons and flooding from late 2020 to mid-2021 have completely washed away the *panambak* she obtained through her efforts in cultivating a reciprocal relationship with the *mananambak*. She is embarrassed by the state of her home, but she is thankful that, at the very least, she has a roof over her head whilst she sleeps.

Concluding Reflections and Recommendations

I borrowed from Ong et al. (1996) and Vink (2017) to define cultural citizenship in this chapter as the beliefs and practices that emerge out of negotiating citizenship regimes that identify a polity's criteria for belonging as well as members' rights and responsibilities. The discussion identified three interrelated citizenship regimes that shape post-disaster housing reconstruction in Isla Sasa based on localism, response, and the *palakasan* system that encourages the formation of alliances to deploy power beyond the rule of law and in the spirit of solidarity. These regimes collectively recognise the challenges of citizens with "partially damaged" and "fully damaged" homes considering state-led and private efforts to provide construction materials and cash assistance for shelter in the immediate aftermath of catastrophe, yet these fail to address the permanence and accrual of hazards, risks, and vulnerabilities of those affected. The inhabitant's homes and livelihoods have remained vulnerable to flooding and to numerous typhoons after Typhoon Glenda in 2014. Meanwhile, the COVID-19 pandemic not only threatens their health and well-being but also limits their agency in identifying alternative sources of income, and, ultimately, in marshalling resources to make their homes more disaster resilient.

The narratives in this chapter corroborate Pye and Pye's observations regarding the ease with which social processes merge into political life. They emphasise how reliance on kinship networks, though useful in overcoming challenges related to the access to and availability of recovery assistance in the immediate aftermath of disaster, engenders a localism that sustains the pervasiveness of top-down and exclusionary approaches to disaster management and governance (Ponce de Leon, 2021). Case 1 (Daniel and Jenny) shows that those who want to regularly seek assistance from powerful *ninong* are expected to maintain paternalistic relations by, for example, reciprocating through pledging their votes and participating in electoral campaigns. The availability of kinship as post-disaster resources also entangles with the couple's limitations and unwillingness to engage with *barangay* officials to voice out their frequent exclusion from beneficiary lists. Meanwhile, those who fail to nurture reciprocal relations with Linda in Case 2 also potentially risk exclusion from the beneficiary lists she develops after conducting needs assessments on behalf of the state and civic organisations. The turn to kinship has therefore driven the notion of disaster recovery as a private burden and, consequently, downplays the critical role of the public realm in disaster recovery (Finn, 2020). In this case, the mobilisation of kinship does little to address the need for sustained recovery, thus contributing to the protracted and overlapping effects of disaster.

The narratives from *barangay* and municipal officers reveal how disaster risk reduction has been dominated by response efforts in the context of rapid-onset disasters, thus reproducing a view of disasters as exceptional events (Neal, 1997). State-led efforts are limited to immediate relief after typhoons. They fail to address the island's flooding due to land subsidence and sea-level rise (a slow-onset and perpetual disaster), the question of retreat as a long-term flood adaptation pathway, and the overall issue of sustainable dwelling. This highlights the need for domestic and international policy frameworks to revisit the idea of "disaster" as an abnormal and uniformly experienced event to better reflect the lived experiences of the vulnerable and, ultimately, to address the disaster-poverty nexus (Jha et al., 2010; Khan et al., 2008). The research of Bankoff (2007) on the centrality of hazards to lifeworlds in the Philippines, the work of those who question the idea of dividing disasters into distinct phases (Neal, 1997), and literature that underscores the significance of merging reconstruction with planning (Barakat, 2003; Larkham, 2018) are key to this reframing. Meanwhile, disaster management in the Philippines needs to recognise the possibilities and limitations of local capacities, including resource-sharing through kinship networks (Gaillard et al., 2019). Such programmes need to be informed by an understanding of the short- and long-term socio-economic and political effects of sharing across various types of relationships and resources (e.g., Takasaki, 2011); this raises the need for comparative and longitudinal research on the subject. Furthermore, this study focuses on the perspectives of residents and *barangay*- and municipal-level leaders because higher-level government officials were not responsive to requests for remote interviews to accommodate COVID-19 travel limitations and safety protocols. As such, further research into subject needs to examine perspectives of citizenship within disaster risk reduction from higher levels of governance. Finally, disaster risk reduction and management in the Philippines needs to be sensitive to vulnerabilities and inequalities across local government units (Piepora et al., 2019) and provide a level of support that is commensurate to the extreme hazards and poverty that characterise specific administrative districts. Disasters, therefore, are not just environmental events; they are socially, politically, culturally, and ethically complex phenomena.

References

Agpalo. (1999). The Philippine pangulo regime. *Philippine Political Science, 20*(43), 45–60.

Alcayna, T., Bollettino, V., Dy, P., & Vinck, P. (2016). Resilience and disaster trends in the Philippines: Opportunities for national and local capacity building. *PLOS Currents Disasters*. https://doi.org/10.1371/currents.dis.4a0bc960866e53bd6357ac135d740846

Aldrich, D. P. (2017). The importance of social capital in building community resilience. In W. Yan & W. Galloway (Eds.), *Rethinking resilience, adaptation and transformation in a time of change* (pp. 357–364). Springer International Publishing.

Arceo, A. (n.d.). *Tropical Depression Emong slightly intensifies over Philippine Sea*. Rappler. Retrieved 8 September 2021, from https://www.rappler.com/nation/weather/tropical-depression-emong-lpa-pagasa-forecast-july-4-2021-5pm

Asian Disaster Reduction Center. 2019. *Information on disaster risk reduction of the member countries*. Asian Disaster Reduction Center. https://www.adrc.asia/nationinformation.php?NationCode=608&Lang=en

Bailey, N., & Pill, M. (2015). Can the state empower communities through localism? An evaluation of recent approaches to neighbourhood governance in England. *Environment and Planning c: Government and Policy, 33*(2), 289–304. https://doi.org/10.1068/c12331r

Ballano, V. (2020). Legal pluralism, cultural values, and post-disaster housing project in a Philippine municipality (SSRN Scholarly Paper ID 3721048). *Social Science Research Network*. https://doi.org/10.2139/ssrn.3721048

Bankoff, G. (2007). Living with risk; coping with disasters. *Education about Asia, 12*(2), 4.

Bankoff, G. (2015). "Lahat para sa lahat" (everything to everybody): Consensual leadership, social capital and disaster risk reduction in a Filipino community. *Disaster Prevention and Management, 24*(4), 430–447.

Barakat, S. (2003). *Housing reconstruction after conflict and disaster*. Overseas Development Institute.

Barenstein, J. D., & Pittet, D. (n.d.) *Post-disaster housing reconstruction: Current trends and sustainable alternatives for tsunami-affected communities in coastal Tamil Nadu* (p. 16). Institute for Applied Sustainability to the Built Environment, University of Applied Sciences of Southern Switzerland.

Barisky, T., Carter, C., & Crego-Liz, E. (2014). *Towards a local climate adaptation plan*. University of British Columbia. https://issuu.com/christopherj.carter/docs/9.20.14_lccap_hagonoy.compressed

Bilau, A. A., Witt, E., & Lill, I. (2015). A framework for managing post-disaster housing reconstruction. *Procedia Economics and Finance, 21*, 313–320.

Bowles, S., & Gintis, H. (2002). Social capital and community governance. *The Economic Journal, 112*(483), F419–F436. https://doi.org/10.1111/1468-0297.00077

Brucal, A., Roezer, V., Dookie, D. S., Byrnes, R., Ravago, M.-L. V., Cruz, F., & Narisma, G. 2020. *Disaster impacts and financing: Local insights from the Philippines* (p. 17). Grantham Research Institute on Climate Change and the Environment and Centre for Climate Change Economics and Policy.

Bulatao, J. (1964). Hiya. *Philippine Studies, 12*(3), 424–438.

Carruth, L. (2018). Kinship, nomadism, and humanitarian aid among Somalis in Ethiopia. *Disasters, 42*(1), 149–168.

Carsten, J. (2000). Introduction: Cultures of relatedness. In J. Carsten (Ed.), *Cultures of relatedness: New approaches to the study of kinship* (Illustrated edition). Cambridge University Press.

Chang, S. E. (2010). Urban disaster recovery: A measurement framework and its application to the 1995 Kobe earthquake. *Disasters, 34*(2), 303–327.

Charlesworth, E., & Ahmed, I. (2015). *Sustainable housing reconstruction: Designing resilient housing after natural disasters*. Routledge.

CNN Philippines. (2021, September 8). *Tropical Storm Jolina maintains strength, Signal No. 1 up in Metro Manila* [News]. CNN Philippines. https://cnnphilippines.com/news/2021/9/8/weather-updates-September-8.html

Collier, J. F., & Yanagisako, S. J. 1987. *Gender and kinship: Essays toward a unified analysis*. Stanford University Press.

Cruikshank, B. (1999). *The will to empower: Democratic citizens and other subjects*. Cornell University Press.

Eadie, P., & Su, Y. (2018). Post disaster social capital: Trust, equity, bayanihan, and Typhoon Yolanda. *Disaster Prevention and Management: an International Journal, 27*(3), 334–345.

Evans-Pritchard, E. E. (1940). *The Nuer: A description of the modes of livelihood and political institutions of a Nilotic people*. Clarendon Press.

Finn, D. (2020). Streets, sidewalks and Covid-19: Reimaging New York city's public realm as a tool for crisis management. *Journal of Extreme Events, 7*(04), 2150006. https://doi.org/10.1142/S2345737621500068

Flinn, B., Schofield, H., & Morel, L. M. (2017). The case for self-recovery. *Forced Migration Review, 55*, 12–14.

Franklin, S., & McKinnon, S. (Eds.). (2001). *Relative values: Reconfiguring kinship studies*. Duke University Press.

Gaillard, J. C., Cadag, J. R. D., & Rampengan, M. M. F. (2019). People's capacities in facing hazards and disasters: An overview. *Natural Hazards, 95*(3), 863–876.

Global Shelter Cluster. (2018). *The state of humanitarian shelter and settlements 2018—Beyond the better shed: Prioritizing people* [Annual Report]. Global Shelter Cluster.

Heinonen, T. (2000). Conceptualizing and applying a minimum basic needs approach in southern Philippines. *Health Policy and Planning, 15*(2), 194–206. https://doi.org/10.1093/heapol/15.2.194

Hilhorst, D., van der Haar, G., & Weijs, B. (2017). Facing fragilities: The socially embedded nature of socio-economic recovery. In D. Hilhorst & G. van der Haar (Eds.), *People, aid and institutions in socio-economic recovery: Facing fragilities* (pp. 1–15). Routledge.

Inter-Agency Standing Committee. (n.d.) *About the grand bargain*. Inter-Agency Standing Committee, https://interagencystandingcommittee.org/about-the-grand-bargain

Jha, A. K., Barenstein, J. D., Phelps, P. M., Pittet, D., & Sena, S. (2010). *Safer homes, stronger communities: A handbook for reconstructing after disaster*. The World Bank and The Global Facility for Disaster Reduction and Recovery (GFDRR).

Jha, S., Martinez, A., Jr., Quising, P., Ardaniel, Z., & Wang, L. (2018). Natural disasters, public spending, and creative destruction: A case study of the Philippines. *SSRN Electronic Journal*. https://doi.org/10.2139/ssrn.3204166

Khan, H., Vasilescu, L. G., & Khan, A. (2008). Disaster management cycle—A theoretical approach. *Management and Marketing Journal, 6*(1), 43–50.

Kikuchi, Y. (1992). *Uncrystallized Philippine society: A social anthropological analysis*. Cellar Book Shop.

Larkham, P. (2018). Disasters: Recovery, re-planning, reconstruction, and resilience. In C. Hein (Ed.), *The Routledge handbook of planning history* (1st ed., pp. 428–440). Routledge.

Levi-Strauss, C. (1969). *The elementary structures of kinship* (Revised ed. edition). Beacon Press.

Lopez, R. (2020). Internationalizing 'pagdamay' and 'palakasan': A Philippine perspective on Duterte foreign policy. In F. A. Cruz & N. M. Adiong (Eds.), *International studies in the Philippines: Mapping new frontiers in theory and practice*. Routledge.

Lyons, M., Schilderman, T., & Boano, C. (Eds.). (2010). *Building back better: Delivering people-centred housing reconstruction at scale*. Practical Action Pub.

MacCormack, C., & Strathern, M. (Eds.). (1980). *Nature, culture and gender* (Illustrated edition). Cambridge University Press.

Malinowski, B. (1922). *Argonauts of the Western Pacific*. Routledge and Sons.

McCaul, B., & Mitsidou, A. (2016). *Analysis of the resilience of communities to disasters toolkit: User guidance manual*. GOAL International. https://www.goalglobal.org/wp-content/uploads/2019/11/ARC-D-Toolkit-User-Manual-2016.pdf

Morgan, L. H. (1877). *Ancient society*. Charles H. Kerry and Company.

Mukherji, A. (2017). Post-disaster housing recovery. In *Oxford research encyclopedia of natural hazard science* (Vol. 1). Oxford University Press.

Neal, D. (1997). Reconsidering phases of disaster. *International Journal of Mass Emergencies, 15*(2), 239–264.

Ng, D. (2020, March 14). *Why Manila is at risk of becoming an underwater city—CNA* [News]. Channel News Asia, https://www.channelnewsasia.com/news/cnainsider/why-manila-risks-becoming-underwater-city-climate-change-12537632

Ong, A., Dominguez, V. R., Friedman, J., Schiller, N.G., Stolcke, V., Wu, D. Y. H., & Ying, H. (1996). Cultural citizenship as subject-making: Immigrants negotiate racial and cultural boundaries in the United States. *Current Anthropology, 37*(5), 737–762. JSTOR.

Oracion, E. (2015). Kinship networks and resiliency to flooding of Pagatban riverside communities in Negros Oriental. *Philippine Sociological Review, 63*, 27–51.

Philippine Statistics Authority. (2020, June 3). *Farmers, fisherfolks, individuals residing in rural areas and children posted the highest poverty incidences among the basic sectors* [Government]. https://psa.gov.ph/content/farmers-fisherfolks-individuals-residing-rural-areas-and-children-posted-highest-poverty

Piepora, Z., Belarga, A., & Alindogan, A. (2019). At a glance. In M. C. Alejanria & W. Smith (Eds.), *Disaster archipelago: Locating vulnerability and resilience in the Philippines* (pp. 1–20). Lexington Books, http://web.a.ebscohost.com.ezproxy.lib.rmit.edu.au/ehost/ebookviewer/ebook/bmxlYmtfXzIyOTUxOTZfX0FO0?sid=1f9ec6b4-fdf3-48de-88c5-77ebe7c1ff83%40sdc-v-sessmgr03&vld=0&format=EB&rld=1

Ponce de Leon, I. Z. (2021). The purok system of San Francisco, Camotes: A communication perspective of community-based haiyan response. *International Journal of Disaster Risk Reduction*, 102379.

Pye, L. W., & Pye, M. W. (1995). *Asian power and politics: The cultural dimensions of authority*. Belknap Press.

Radcliffe-Brown, A. R. (1941). The study of kinship systems. *The Journal of the Royal Anthropological Institute of Great Britain and Ireland, 71*(1/2), 1–18.

Reliefweb. (2020). *Philippines: Typhoon Vamco (Ulysses) snapshot (as of 12 November 2020)—Philippines*. ReliefWeb, https://reliefweb.int/report/philippines/philippines-typhoon-vamco-ulysses-snapshot-12-november-2020

Reliefweb. (2021, July 26). *DSWD DROMIC Report #4 on the effects of Southwest Monsoon enhanced by Typhoon "Fabian" as of 25 July 2021, 6PM—Philippines*. ReliefWeb, https://reliefweb.int/report/philippines/dswd-dromic-report-4-effects-southwest-monsoon-enhanced-typhoon-fabian-25-july

Remes, J. A. C. (2016). *Disaster citizenship: Survivors, solidarity, and power in the Progressive Era*. University of Illinois Press.

Schneider, D. M. (1984). *A critique of the study of kinship*. University of Michigan Press.

Stiefel, M., & Wolfe, M. (2011). The many faces of participation. In A. Cornwall (Ed.), *The participation reader* (pp. 19–30). ZED Books.

Takasaki, Y. (2011). Targeting cyclone relief within the village: Kinship, sharing, and capture. *Economic Development and Cultural Change, 59*(2), 387–416.

The World Bank. (2020). *Building a resilient recovery* (Philippines Economic Update). The World Bank, https://documents1.worldbank.org/curated/en/983051607354214738/pdf/Philippines-Economic-Update-Building-a-Resilient-Recovery.pdf

Titz, A., Cannon, T., & Krüger, F. (2018). Uncovering 'community': Challenging an elusive concept in development and disaster-related work. *Societies, 8*(3), 71.

Torres, A. T. (1985). *Kinship and social relations in Filipino culture*. 13.

Tozier de la Poterie, A., & Baudoin, M.-A. (2015). From Yokohama to Sendai: Approaches to participation in international disaster risk reduction frameworks. *International Journal of Disaster Risk Science, 6*(2), 128–139.

United Nations Framework Convention on Climate Change. (2021). *Key aspects of the Paris Agreement*. United Nations Framework Climate Change. https://unfccc.int/process-and-meetings/the-paris-agreement/the-paris-agreement/key-aspects-of-the-paris-agreement

Villacin, D. T. (2017). *A review of Philippine government disaster financing for recovery and reconstruction* (Discussion Paper Series No. 2017–21, p. 51). Philippine Institute for Development Studies.

Vink, M. (2017). Comparing citizenship regimes. In A. Shachar, R. Bauböck, I. Bloemraad, & M. Vink (Eds.), *The Oxford handbook of citizenship* (pp. 220–244). Oxford University.

CHAPTER 15

The Disconnections that Facilitate Protracted Disasters: Barriers to Adapting to Fire in the Australian Landscape

Zoë D'Arcy

Introduction

It was easy, during the 2019–2020 Australian bushfires, to imagine that the Australian forests and weather had combined with fire to form an enemy with focused and deadly intent. For seven months, bushfires burnt across dry, drought-stricken landscapes on an unprecedented scale. I am a volunteer fire-fighter with the state's Rural Fire Service. Over several months I, along with tens of thousands of colleagues, spent a lot of time on fire trucks trying to contain the fires, and to protect properties from burning. There is no doubt that many women and men showed great courage in doing this work. It was dirty, hard, involved very long hours,

Z. D'Arcy (✉)
School of Architecture and Urban Design, RMIT University, Melbourne, VIC, Australia
e-mail: s3686427@student.rmit.edu.au

A. Lukasiewicz and T. O'Donnell (eds.), *Complex Disasters*, Disaster Risk, Resilience, Reconstruction and Recovery,
https://doi.org/10.1007/978-981-19-2428-6_15

and there was always a very real threat of suddenly being confronted with life-threatening situations.

But the focus given throughout the 2019–2029 bushfires—at all levels of society—to 'fighting' these disastrous fires that were essentially uncontrollable, made me feel deeply uneasy. Fire-fighters were celebrated as 'heroes' pitted against the combined 'enemies' of Australian bushfires and the bush landscape. This narrative simply endorsed current policies, practices, values and knowledge rather than seeking new ways to address the root causes of our unprecedented situation. Indeed, focusing on the 'war on wildfire' seemed to halt conversations about issues that the bushfires had exposed. Issues that included how to address the climate change that had driven the fires; whether Australia's health systems could cope with the physical and mental health problems caused by fire and smoke; whether burnt ecosystems made fragile by decades of unsustainable landuse practices could ever recover; and provisions of 'recovery' services to small towns and people who had been struggling to survive even before the fires occurred. For example, amongst the worst affected by the fires were already-marginalised Aboriginal peoples of south-eastern Australia, who were then further marginalised during bushfire recovery processes (Williamson et al., 2020).

The 2019–2020 bushfires were disastrous, and obviously required an emergency response. However, the fires—as rapid-onset disasters are wont to do—intersected with a range of complex slow-onset enviro-socio-political disasters that were already in place. Slow-onset, lingering situations that largely arose from power imbalances, and that challenge the idea that disasters are only exceptional events, and now part of everyday existence (Anderson et al., 2019). What are the barriers to identifying these protracted disasters as emergencies, and responding to them as we would a rapid-onset emergency? Is it simply that people are disconnected—from each other, from the realities of the Australian landscape, from government decision-making processes, or increasingly, from the provision of essential services?

This chapter explores the discourse of 'disconnections' being a barrier to communities effectively taking an adaptive approach to fire into the future. This discourse came from interviews with practitioners currently working with Australian communities to improve their fire knowledge and practices. Participants came from a range of fire management perspectives, and their responses offer an insightful look into the ongoing complexities of people coexisting with fire in the Australian landscape. The participants' responses contained similarities in identifying where 'disconnections' exist that prevent an adaptive approach to fire in the landscape. These can be categorised as being between people, place, and government. However, there were differing opinions about what adaptive 'connections' between these agents should be. These seemed to stem from differing narratives of the roles these agents 'should' play in emergency and fire management, and what the nature of those relationships should be between agents.

Why an Adaptive Approach to Fire?

Alternative models of fire management to 'war on wildfire', such as 'fire-adaptive management' (Almstedt & Reed, 2013; Bosomworth et al., 2014; Ruane, 2019), or 'co-existing with fire' (Moritz et al., 2014; Roos et al., 2016) are increasingly put forward as approaches that have the potential to examine and address other underlying 'problems'—the drivers behind why bushfires are becoming more frequent and why they are causing so much damage (Tedim et al., 2020). In climate change adaptation literature, a community's adaptive capacities are a good indicator of its abilities to withstand both rapid and slow-onset emergencies (Cinner & Barnes, 2019).

Pelling (2010) argues that adaptation—specifically social and political adaptation to climate change—offers the chance for transformative change. Adaption offers an intervention in current power frameworks and narratives. He warns that merely becoming 'adapt*ed*'—to adapt without critical awareness—means accommodating the existing risks rather than understanding the alternatives that might exist to deal with the root causes of risks (Pelling, 2010). So, becoming 'adapt*ive*' rather than 'adapt*ed*' to fire, involves understanding both the dominant and the counter-narratives of fire management in Australia. These narratives link human and more-than-human actors—people, landscape, and fire—and position them into particular roles in relationship to one another.

The dominant fire narrative has 'heroes' (fire-fighters), 'generals' (operation command), 'villains' (fire and the dry landscape), and victims (local residents and wildlife). Much of the narrative around achieving transformative change through fire-adaptive approaches is around collaborative decision-making approaches that acknowledge and encompass diverse knowledges and values (i.e., Bosomworth et al., 2014; Roos et al., 2016). To examine these narratives about fire, landscape, and people, and therefore the interviewees' discourse of 'disconnections', this chapter explores the concepts of both communities and emergencies using a more-than-human conceptual framework.

Applying More-Than-Human Theory to Communities and Emergencies

Communities as 'Contact Zones'

The term 'communities' in this chapter uses a more-than-human conceptual model to encompass the entangled relationships and encounters between people, the landscape they live within, and fire. In this model, communities are conceptualised as 'contact zones' where there are human and more-than-human encounters. These encounters are influenced by their 'entangled relationships', which are often characterised by power asymmetries.

'Contact zone' is a term that was first coined by decolonisation scholar Mary Louise Pratt to indicate contested spaces where 'cultures, meet, clash and grapple with each other, often in contexts of highly asymmetrical relations of power such as colonialism, slavery, or their aftermaths as they are lived out in many parts of the world today' (Pratt, 1992). Pratt used the concept to examine and contest imperial histories told in travel writing (Pratt, 2019), but it has recently been used in environmental research to explore more-than-human environmental knowledge production and actions within colonised environments (i.e., Haraway, 2008; Sutherland, 2019). Here, the conceptual tool of 'contact zones' aims to capture both the social and the bio-physical processes that contribute to the way in which people, fire, and the landscape intersect and encounter one another.

Much of the research in both climate change adaptation and disaster risk reduction fields uses 'community' as its unit of study (Cinner et al., 2018). There is great interest in communities becoming more 'resilient' to disasters, by undertaking risk mitigation processes and activities, and there is also interest in those disaster risk reduction actions being 'community-led', so as to ensure local engagement, relevance and therefore effectiveness (McLennan, 2018).

The confidence, and frequency, with which the term 'community' is used, hides several complexities that have deep implications for how 'community' initiatives are designed and implemented in practice. Firstly, it hides how poorly and inconsistently the term is defined in disaster risk reduction literature and policy (Räsänen et al., 2020, p. 1). Secondly, its consistent use glosses over the way in which the term is used to imply a homogeneity—of attitudes, needs, identities and relationships between human community members—that in no way reflects the much more complex reality of people living within a shared location (Mulligan, 2017).

Within this emergency management context, this perceived homogeneity of communities also extends to silently prescribing the relationships that the human inhabitants of an area 'should' have with the more-than-human elements that they share the landscape with. Emergency management demands that more-than-human elements relationships, specifically with elements such as vegetation, topography, weather, and fire, should be viewed using the lens of risk—as the threat they pose to humans. These human-centric conceptualisations of the world mean that the needs and vulnerabilities of the surrounding 'environment' within the same space as people is 'othered'—therefore of lesser priority. This 'othering' aligns with Western thinking that people are somehow separate from, and not reliant on, having healthy, diverse, and functioning ecosystems to survive (Maller, 2018).

However, fire in the landscape, and governance put in place to manage it, is a much more complex story than the one told in community fire safety messaging developed by the emergency sector. Fire is, at its most basic, a chemical reaction and has no inherent morality or meaning. It is not just a destructive force. Fire is often used as a tool—prescribed fire—to reduce risk to human settlements and to promote biodiversity—although doing so requires negotiation of human processes, institutions, time, and knowledges (Sutherland, 2019). Fire itself is impacted by factors such as vegetation, topography, and weather conditions (Price et al., 2021). Any one fire can have both positive and negative impacts—on ecologies, on the built environment, on human and more-than-human lives—and even on the weather in some extreme situations with the creation of storms—seen in the form of pyrocumulonimbus clouds (Bowman et al., 2011). Encounters between fire, the landscape, and the landscape's inhabitants involve both social and bio-physical processes. Controlling fire—using it as a tool as well as suppressing it—is a complex exercise in navigating bureaucratic processes, using regulations to try to control people's behaviour, as well as understanding a range of uncontrollable agents such as weather.

Conceptualising 'communities' as more-than-human contact zones helps expand out the perspectives of who and what matters in a community. It helps examine these encounters between humans and humans; as well as between humans and more-than-human agents. This enables critical examination of how fire and landscape knowledge has been produced and applied in the past; of how and when encounters between agents within communities are identified as emergencies, what roles agents are expected to have in emergencies to 'restore normality', or at least prevent those encounters becoming so disastrous that they overwhelm the communities' abilities to cope.

Identifying, and Responding to, Emergencies in Communities

Australia's Macquarie Dictionary defines 'emergency' as 'an unforeseen occurrence; a sudden and urgent occasion for action' (Macquarie Dictionary, 2020). However, what constitutes an emergency is not clear-cut, and neither is it a neutrally applied term (Anderson et al., 2019). Its use implies two things. Firstly, that a serious, potentially overwhelming, situation exists. Secondly, that intervention is needed to improve that situation, or perhaps merely stop it from worsening.

Emergency management has a legislated status in Australia, with specific roles legally assigned to federal, state, and local government agencies depending on the nature and scale of the emergency (Commonwealth Government, 2020). This management has, at its heart, the four 'phases' of emergencies that were first formalised in United States (Neal, 1997)—Prevention, Preparation, Response, and Recovery (PPRR). These four phases are often presented as a risk management process that can be followed for improved outcomes when an emergency does impact a community (Commonwealth Government, 2018).

This approach assumes that both identifying and responding to emergencies is a neutral and rational process—and that all actors, given the right information, will agree on what constitutes the key risks, and will act to prevent these emergencies from escalating and having disastrous impacts on communities. Where there is a demonstrated failure of communities to be resilient to hazards such as bushfire and floods, it is often represented as a failure of actors to follow the risk-based decision-making process adequately (Parker, 2020).

Bushfires are an integral part of this emergency management landscape. The Australian Institute for Disaster Resilience (2020) defines bushfire simply as being an 'unplanned vegetation fire'. Unplanned, or uncontrolled bushfires within communities, like other 'natural hazards' such as floods or storms, are understood as 'classic' emergencies. As emergencies, bushfires are represented as being crises separate from everyday existence. At the same time, they are seen as being an inevitable part of the 'natural' Australian bush and therefore, like the bush, fighting bushfires has become part of Australian national identity (Griffiths, 2009).

Everyday language also reflects this 'war on wildfire'—the fire services are referred to as combat agencies, and it is not unusual to hear about fire fighters battling flames at the fire front. Increasingly there is militarised equipment such as planes involved in fire activities (Tedim et al., 2020). Extensive governance and community mobilisation is in place to prevent and suppress uncontrolled bushfires at state and local government level in Australia (i.e., NSW Government, 2017). This extends to creating, training, and equipping networks of community volunteer firefighting brigades to suppress fire (i.e., NSW Government, 2020). In turn, all brigades are given the tasks of educating their neighbours about fires, and even monitoring and regulating their usage of it (NSW Rural Fire Service, 2017). Thus, the knowledge that bushfires are a 'problem' to be 'solved' is deeply embedded in our legal systems, policies, language, and practices (Howitt, 2014).

Given the undoubted need to limit the violence that bushfires can bring upon people, what is gained from deconstructing this pervasive narrative? One benefit would be to expand out the scale and complexity of emergencies that are recognised, and then addressed. These could include anthropogenic climate change which is already resulting in long-term droughts and higher bushfire risk (CSIRO & Bureau of Meteorology, 2018). These could include Australia's high rate of land clearance and habitat destruction, and resulting animal extinction rate (Cooke et al., 2016). These could also include the myriad socio-political complexities that have marginalised people and made them 'vulnerable' to the impact of rapid-onset emergencies such as bushfires—the 'disaster risk creators' (Alexander, 2016). Like bushfires, these are all existential issues within the contact zone of communities. Ones that have arisen from decisions and actions over the longer term. These could be recognised as slow-onset emergencies, protracted disasters, that need to be addressed.

Australia's current strategic emergency management approach has been criticised for not actively addressing these enviro-socio-political complexities (i.e., Bosomworth et al., 2016). Interestingly, instead, these social and environmental injustices are acknowledged in the current emergency management narrative in terms of marginalised people and environments being more 'vulnerable' than others to short-term disastrous events such as bushfires (i.e., Handmer, 2003). This can be seen in recent attempts to quantify community resilience or vulnerability, such as in the Australian Exposure Index Platform, or the Australian Natural Disaster Resilience Index, which include markers such as population estimates of socio-economic status, of people that identify as Aboriginal and Torres Strait Islanders, or areas of environmental concern (Geoscience Australia, 2020; Parsons et al., 2016).

Instead of ongoing commitment to addressing these vulnerabilities, what is offered in the current emergency management framework is a narrative that communities can 'build back better' after disasters. Disasters such as bushfires are offered as 'windows of opportunity' for communities to learn, and to use the recovery phase in the PPRR cycle to make themselves less vulnerable to fires in the future, or to adapt to more wildfires in the future (i.e., Mockrin et al., 2018; Schumann et al., 2019). Anderson et al. (2019) argue that this is 'government through emergency', which chooses what an emergency is, and how it is responded to, in a way that continues to consolidate its own power. They write that this 'becomes a problem as it is harnessed to the goal of perpetuating liberal order, and emergency and disaster management are transformed into forms of institutionalized anti-action' (p. 625).

Conceptualising communities as more-than-human contact zones will allow for examination of how and when encounters between agents within communities are identified as emergencies, what roles agents are expected to have in emergencies to 'restore normality', or at least prevent those encounters becoming so disastrous that they overwhelm the communities' abilities to cope. It will also allow for analysis of the entangled relationships between humans and more-than-human agents within that emergency management narrative.

Disconnections—A Discourse of Barriers to Adaptation

Research Method

The aim of the research reported in this chapter was to identify how people currently working with Australian communities to improve their fire knowledge and practices believe Australians can better take an adaptive approach to fire, or to coexist with fire.

The research strategy in this stage was to conduct detailed semi-structured interviews with 16 subject matter experts. Each person interviewed was engaged in developing and running programmes to develop community knowledge and practice about fire, with differing types of expertise and community engagement approach. The intention was to hear from people who had expertise in both engaging communities with the topic of fire, and also with trying to affect change in people's knowledge, and through that, actions.

Research participants were initially recruited through the Interstate Fire Alliance, which is an Australian network of fire ecology practitioners, researchers, and managers which has as its aim to improve fire management on private land. Other participants were recruited through personal networks of emergency service and ecological organisations.

The sixteen people interviewed represent a combination of emergency management and ecological/land management perspectives. They represented programmes from across states and territories of eastern Australia. They work with a range of organisations: government, as well as not-for-profit and volunteer bodies. The community engagement approach of interviewees varied between community education (short-term courses or workshops for residents of communities interested in fire-related topics); community development (long-term programmes designed to facilitate and develop wider community knowledge and practices); policy (government policy development for community risk reduction); and on an ad hoc basis (community groups gathered when necessary—in this case to protect local properties during bush fires).

The interviews took a semi-structured form. All research participants were asked to identify what problems they thought bushfire prone communities in Australia are likely to have to solve into the future, in relation to their surrounding landscapes, built environment and fire.

Secondly, they were then asked to imagine a best-case scenario of communities taking an adaptive approach to fire 20 years from now. This period of time was selected as being long enough to achieve great change, but it is also well within the span of one person's career, and thus it represents change that they might see over their career. Finally, interviewees were asked how these futures could be achieved, using the adaptive capacity domains put forward by Cinner and Barnes (2019). These were: assets; flexibility; organisation; learning; socio-cognitive constructs; and agency.

Findings

All research participants were asked to identify what problems they thought bushfire prone communities in Australia are likely to have to solve into the future, in relation to their surrounding landscapes, built environment and fire. This interview question was intended to let the subject matter experts frame the problems that they anticipated could be addressed by communities taking an adaptive approach to fire. All interviewees agreed that there would be increased risk of fire into the future, all of them saw fire as a land management tool, and the majority of them identified climate change as being a key driver of increased bushfire risk.

When asked to imagine fire-adaptive futures for communities, interviewees did not frame fire as the main problem to be solved. Instead, they focused on social issues—imagining communities that were well-connected and working collaboratively with each other, as well as across levels of government, with good access resources and knowledge. For instance, one interviewee employed in state government fire management, outlined their vision this way:

> *The best guess I could have, is that each community has one or more groups that are looking at fire as an issue … and they're sort of getting information about their local sites, talking about it with each other, coming up with plans, I suppose, what they want to do about it, both the response side of it and the mitigation/resilience side of it. They're well connected to … the various levels of governments in terms of accessing grants and services that might be beyond an individual or a small group of landholders. And that those services are actually in place so that there are programs to help people who need to do fuel management and can't for whatever reason. (Government Sector P5, 2020, interview 27 November)*

A sub-group of interviewees believed that taking an adaptive approach offered an opportunity to challenge the people/'nature' dichotomy prevalent in Western thinking. They argued that learning to live in the Australian landscape meant dealing with bushfire risk, but also learning the positive role that fire could play in some ecosystems. As one interviewee explained:

> *I think if we accept that we're part of the environment and fire is a natural part of environment, that sort of go towards some way of changing people's thinking. (NGO P3, 2020, interview 9 December)*

Their emphasis on connectedness and collaboration reflected an adaptive narrative. And the barriers they identified—which they believed need to be overcome for communities to be able to coexist with fire—took the form of 'disconnections'. These 'disconnections' were often the result of power imbalances, and allowed the continuation of the protracted disasters that contributed to, and arose from, the fires of 2019–2020.

Disconnections as Barriers to an Adaptive Approach to Fire

Disconnections Between People and 'Place'

The majority of interviewees believed that people in bushfire prone areas were fundamentally disconnected from 'place'. The way this was expressed ranged from people being disconnected from the physical environment, or 'nature' generally, to people 'not understanding' that fire was an inevitable part of the Australian landscape.

This disconnection was expressed by some interviewees not just as a problem with fire in the landscape, but as an existential issue. One interviewee explained it this way:

> *In terms of a lack of knowledge or a lack of understanding about how the environment functions and how it supports us is a long-term problem for us as a species. I think we're sort of hurtling towards a cliff, and some people know we [are], and some people don't, and yet we still continue in that direction. (NGO P1, 2020, interview, 24 November)*

A sub-group of interviewees focussed mostly on adapting to increased bushfires in their future-visioning, and envisioned fire-adaptation as primarily scaling up emergency response to bushfires. This sub-group argued that people need an understanding of their surrounding landscape and ecologies, in order to be able to understand the risk that it might pose in the event of a bushfire.

> *Previous to the idea of knowing your risk is to simply know where you are … that I live on a hill, that I live in a valley, that I live on the beach, or … that I live in amongst sclerophyll forest, or I live amongst coastal heath or whatever it is, you know, just to understand. "Where I live" is a really basic thing. (Fire Service P2, 2020, interview, 23 November)*

Disconnections Between People Within Communities

The second disconnection identified by interviewees was between people living within bushfire-prone areas, or within communities. Connected communities were seen by interviewees to be more self-reliant, or more resilient, than those that were not connected. One interviewee expressed it this way:

> *There's no doubt in my mind that connected communities that work well and operate strongly in peacetime function better in recovery. (Government Sector P2, 2020, interview 3 December)*

Knowledge of who was in a community, of their vulnerabilities and their capabilities, was seen to be critical in helping one another when bushfire was a threat. An interviewee from a fire service explained it like this:

> *[your elderly neighbours could] go, 'you know, our plan is to always to leave when there's a high fire danger day'. And they ring you up … and say, 'Hey, we're leaving today … just so you know, we won't be here. We're going to go and stay at our daughter's place. In a safe spot.' You know, even that itself would make you safer. And the whole community safer. (Fire Service P2, 2020, interview, 23 November)*

Social disconnections were also seen as problematic as they prevented landscape level (rather than just individual property) actions being planned and implemented. One interviewee, from an NGO with an ecological focus, argued that governments were guilty of playing identity politics to actively divide communities. They made the case that politicians were doing this to sell political decisions that might be controversial for some of their constituents.

> *I do think governments use landholders and farmers as a pawn when they're talking about legislation ... There's that desire to kind of pigeonhole people and set them off against each other in terms of the greenies versus the farmers, or it's jobs versus the environment. I think that narrative is very negative. I don't think it's going to get us anywhere. It increases division. It doesn't say, "Actually, you can cooperatively manage. You can manage for different objectives. You can manage for different outcomes". (NGO P1, 2020, interview 24 November)*

This leads to the complex world of disconnections between people and government.

Disconnections Between Communities and Government

The third disconnection that interviewees identified was between communities and government. These disconnections were believed to be between communities and governance processes; between communities and the provision of government services (including disconnections between government agencies providing those services); and between people's needs and political decisions. These disconnections were sometimes expressed as 'bureaucracy', 'siloing', and 'playing politics'. They were represented as power imbalances that resulted in restricting communities' knowledge and agency to participate in processes that would contribute to their resilience and safety.

Lack of government action to address climate change—to recognise it as an emergency—was represented by most interviewees as a major disconnection between government decision making and people's safety and needs.

> *The other thing too, that is always a white elephant, [is] the climate change as well. So, there's no integrity at the moment with particularly the federal government ... particularly when they're ... making up numbers to the Paris Climate Agreement. So that's one of the key drivers that's increasing severe fire weather across the nation, and it's then putting our landholders at the frontline of risk. (NGO P4, 2020, interview, 7 December)*

Some interviewees believed that a lack of collaboration between fire governance processes continues and community members actively disempowered people living in bushfire-prone areas. An interviewee who works in fire management in the government sector, believed community members were excluded from their area's strategic bushfire risk management governance processes and actions.

> *I feel like there's an imbalance of implementation but also of empowerment. So, we've taken the on-ground action [of hazard reduction burning] away from community, and almost indicating to the community that, 'it's okay, us as government will do these actions to then reduce your risk', rather than collaboratively working together and empowering each other to do that work. (Government Sector P6, 2020, interview, 27 November)*

Access to services before, during, and after, emergencies—including power, telecommunications, and financial—was seen as problematic. One of the essential services identified was insurance. One interviewee doubted whether people living in bushfire-prone areas would be able to insure their homes against bushfire at all going into the future.

> *People are increasingly not going to be able to get insurance. Insurance premiums are becoming – I mean, if you live in certain parts of Far North Queensland now, you can't get flood insurance. So, I think we are increasingly going to see this, but then I say things like, "People are one disaster away from poverty in some instances." (Government Sector P1, 2020, interview, 4 December)*

These three categories of disconnections—between people, place and government—touch on issues of power, knowledge, values, and resources that reach well beyond a community's ability to respond to bushfires. Addressing these various disconnections was framed by interviewees as an essential step for communities to take an adaptive approach to fire.

No Common Adaptive 'Connections' Narrative

However, the narratives of what comprised adaptive 'connections' differed. These differences seemed to be based on differing narratives of who, within human and more-than-human communities, should be empowered to act, when, and how. This was particularly apparent in discussions about which humans within communities should control—both suppressing or using—fire.

The sub-group of interviewees who focussed mostly on adapting to increased bushfires in their future-visioning, did not tend to question the dominant emergency management narratives, or the roles that those narratives assigned to actors. There was, for example, a narrative of human community members as passive victims, waiting to be saved:

> *We need to change the mindset where people have always relied on fire prevention and then when fires do occur, a big, red fire truck's turning up to save their house ... There's been a pretty heavy reliance on fire authorities to save your house. I think Australian people need to realise if you live in a bushfire-prone area, saving your house is probably going to be your own responsibility. (Fire Service P4, 2020, interview, 11 December)*

Vulnerability to bushfires within communities were sometimes represented by interviewees as being caused by people not 'stepping up' to 'take responsibility' to prevent and/or prepare for bushfires. Alternatively, bushfire vulnerability was believed to be caused by people not having access to, or choosing not to act upon, information about fire risk.

> *[home owners need to understand] in a little more detail what the risk is, and that's a challenge because up until a few years ago, a lot of that that knowledge was either held within agencies or not well documented. At best, we had bushfire-prone areas for building regulation purposes and fire ban districts, but I don't think that risk issue has been very well communicated to the community. (Government Sector P5, 2020, interview 27 November)*

There were some interviewees within this sub-group, who believed that taking an adaptive approach to fires meant to challenging the dominant narratives about who should be empowered to take on which roles to scale up emergency management to bushfires. They argued that the existing top-down relationships encouraged by the dominant narratives restricted the potential gains that might come from working collaboratively. One interviewee, a long-time volunteer fire-fighter, and captain of their local fire brigade for seven years, expressed this as a cultural and gendered problem.

> *That's the RFS [New South Wales Rural Fire Service] culture … the RFS has actually got to think about that. Because it's all about testosterone and reactive and being heroes. It's not about 'let's prepare a community'. (Fire Service P3, 2020, interview, 25 November)*

Several interviewees thought that the tightly regulated use of fire on the landscape was problematic. For example, a community facilitator for community-based bushfire management argued that farmers are not able to burn as easily as they used to, due to cumbersome government regulations, and that this increased bushfire risk.

> *[I would like to see] farmers once again being able to do cool burning or mosaic burning down roadside vegetation strips without having to go through six miles of red ribbon. That just chokes things now. They used to do a much better job of looking after their land and being able to reduce the size and the heat of fires, but ever since the bureaucracies took over, they've lost that ability to do that. (Government Sector P2, 2020, interview 3 December)*

Discussion and Conclusion

Addressing disconnections, or rethinking structural relationships to better coexist with fire, has been a mainstay of arguments made by many social science researchers both in Australia and globally (i.e., Bowman et al., 2011; Howitt, 2014; Moritz et al., 2014; Tedim et al., 2020). If Australia were to discard its 'war on wildfire' and instead seek to take an adaptive approach to fire—learn to coexist with it and address the 'slow emergencies' that make communities vulnerable to fire—it would not be a small change. Howitt (2014), in his essay 'Coexisting with Fire? A Commentary on the Scale Politics of Adaptation' argues that to coexist with fire, there needs to be more than mere acceptance that fires will continue to pose a risk to people and their built environment. "*It needs tackling in terms of recognition, respect, adaptation and a shift in how we think about the terms on which human societies coexist with things*" (Howitt, 2014, p. 61).

There were three categories of disconnections identified in this research—between people, place, and government—that interviewees identified as being barriers to an adaptive approach to fire in Australia. The interviewees were practitioners currently working to improve community knowledge and experience of fire, and they all seemed to envision fire-adaptive communities being collaborative and connected. This narrative, which embraces diversity in fire knowledges and in decision-making, conforms with that seen in climate change adaptation literature. It perhaps offers opportunities to consider inclusion and welfare of agents within communities who have been marginalised (i.e., Bosomworth et al., 2014; Roos et al., 2016). However, the 'connections' that interviewees envisioned between the actors within future fire-adaptive communities, and the roles assigned to them, seemed often deeply informed by the dominant 'war on wildfire' narrative. Only interviewees with a strong ecological or social focus envisioned fire-adaptive communities making better connections to marginalised environments or people to address slow emergencies that are out in the landscape, as well as bushfires. Many interviewees were primarily concerned about how to adapt to a future with increased bushfire risk, and how to scale out emergency management (PPRR) to community members. Some interviewees saw this as community members empowering themselves by 'stepping up' to 'take responsibility'. Some saw it as making more knowledge available. Others saw this as government agencies (especially emergency services) sharing some of their power.

Asymmetrical power relationships play a role in recognising whether particular situations within the contact zones of more-than-human communities are emergencies, and are therefore addressed and remedied (Anderson et al., 2019). Acknowledging this enables analysis of how narratives that might challenge that dominant narrative of 'war on wildfire' are drawn into the dominant emergency narrative, and only discussed in relation to supporting the 'war on wildfire' approach. This has certainly occurred as slow-onset (protracted) emergencies result in marginalised agents within communities, and those marginalised agents are only discussed in emergency management as being 'vulnerable' to rapid-onset emergencies. It has also occurred in the use of prescribed fire in Australia, where fire is primarily used to reduce risk of bushfires, rather than to benefit ecosystems, or to benefit Aboriginal connections to Country (Howitt, 2014). It seems logical that adaptive fire narratives could easily be hijacked by 'war on wildfire' narratives. Certainly, community practitioners who have experienced events like the 2019–2020 bushfires, might be forgiven for focussing on how to adapt to a future with more extreme bushfires.

References

Alexander, D. E. (2016). The game changes: "Disaster Prevention and Management" after a quarter of a century. *Disaster Prevention and Management, 25*(1), 2–10. https://doi.org/10.1108/DPM-11-2015-0262

Almstedt, Å., & Reed, M. G. (2013). Introducing a framework for good and adaptive governance: An application to fire management planning in Canada's boreal forest. *The Forestry Chronicle, 89*(5), 664–674. https://pubs.cif-ifc.org/doi/pdf/10.5558/tfc2013-119

Anderson, B., Grove, K., Rickards, L., & Kearnes, M. (2019). Slow emergencies: Temporality and the racialized biopolitics of emergency governance. *Progress in Human Geography, 44*(4), 621–639. https://doi.org/10.1177/0309132519849263

Australian Institute for Disaster Resilience. (2020). *Australian Disaster Resilience Glossary*. https://knowledge.aidr.org.au/glossary/

Bosomworth, K., Handmer, J., & Dovers, S. (2014). Learning from analyses of policy frames and informal institutions in the fire management sector of Victoria, Australia. In B. C. Glavovic & G. P. Smith (Eds.), *Adapting to climate change: Lessons from natural hazards planning* (pp. 269–289). Springer Science+Business Media.

Bosomworth, K., Owen, C. & Curnin, S. (2016). Addressing challenges for future strategic level emergency management: Reframing, networking, and capacity-building. *Disasters, 41*(2). https://doi.org/10.1111/disa.12196

Bowman, D. M. J. S., Balch, J., Artaxo, P., Bond, W. J., Cochrane, M. A., D'Antonio, C. M., ... Swetnam, T. W. (2011). The human dimension of fire regimes on Earth. *Journal of Biogeography, 38*(12), 2223–2236. https://doi.org/10.1111/j.1365-2699.2011.02595.x

Cinner, J. E., & Barnes, M. L. (2019). Social dimensions of resilience in social-ecological systems. *One Earth, 1*(1), 51–56. https://doi.org/10.1016/j.oneear.2019.08.003

Cinner, J., Adger, W., Allison, E., Barnes, M., Brown, K., Cohen, P., ... Morrison, T. (2018). Building adaptive capacity to climate change in tropical coastal communities. *Nature Climate Change*. https://doi.org/10.1038/s41558-017-0065-x

Commonwealth Government. (2018). *National disaster risk reduction framework*. https://www.homeaffairs.gov.au/emergency/files/national-disaster-risk-reduction-framework.pdf

Commonwealth Government. (2020). *Issues paper: Constitutional framework for the declaration of a state of national emergency*. https://naturaldisaster.royalcommission.gov.au/system/files/2020-05/Issues%20Paper%20-%20Constitutional%20Framework%20for%20the%20Declaration%20of%20a%20State%20of%20National%20Emergency_0.pdf

Cooke, B., West, S., & Boonstra, W. J. (2016). Dwelling in the biosphere: Exploring an embodied human–environment connection in resilience thinking. *Sustainability Science, 11*(5), 831–843. https://doi.org/10.1007/s11625-016-0367-3

CSIRO & Bureau of Meteorology. (2018). *State of the climate 2018*. Australian government. https://www.csiro.au/en/Research/OandA/Areas/Assessing-our-climate/State-of-the-Climate-2018/Australias-changing-climate

Geoscience Australia. (2020). *Australia exposure information platform*. https://www.aeip.ga.gov.au/

Griffiths, T. (2009). *We have still not lived long enough. Inside Story*. https://insidestory.org.au/we-have-still-not-lived-long-enough/

Handmer, J. (2003). We are all vulnerable. *The Australian Journal of Emergency Management, 18*(3), 55–60.

Haraway, D. J. (2008). *When species meet*. University of Minnesota Press.

Howitt, R. (2014). Coexisting with fire? A commentary on the scale politics of adaptation. *Geographical Research, 52*(1), 61–64. https://doi.org/10.1111/1745-5871.12036

Larkin, B. (2019). Pride and prejudice: LGBTIQ community responses to disaster events worldwide. *Australian Journal of Emergency Management, 34*(4), 60–66.

Macquarie Dictionary. (2020). https://www.macquariedictionary.com.au/

Maller, C. (2018). *Healthy urban environments: More-than-human theories* (1st ed.). https://doi.org/10.4324/9781315620534

McLennan, B. J. (2018). Conditions for effective coproduction in community-led disaster risk management. *Voluntas (Manchester, England), 31*(2), 1–17. https://doi.org/10.1007/s11266-018-9957-2

Mockrin, M. H., Fishler, H. K., & Stewart, S. I. (2018). Does wildfire open a policy window? Local government and community adaptation after fire in the United States. *Environmental Management, 62*(2), 210–228. https://doi.org/10.1007/s00267-018-1030-9

Moritz, M. A., Batllori, E., Bradstock, R. A., Gill, A. M., Handmer, J., Hessburg, P. F., ... Syphard, A. D. (2014). Learning to coexist with wildfire. *Nature, 515*(7525), 58 66. https://doi.org/10.1038/nature13946

Mulligan, M. (2017). Rethinking the meaning of 'community' in community-based disaster management. In G. Marsh, I. Ahmed, M. Mulligan, J. Donovan, & S. Barton (Eds.), *Community engagement in post-disaster recovery*. Taylor & Francis Group.

Neal, D. M. (1997). Reconsidering the phases of disaster. *International Journal of Mass Emergencies and Disasters, 15*(2), 239–264.

NSW Government. (2017). *Emergency risk management framework*. https://www.emergency.nsw.gov.au/Documents/publications/ERM-framework.PDF

NSW Government. (2020). *Be part of something special*. https://www.rfs.nsw.gov.au/volunteer/volunteering-with-the-rfs/Join-the-NSW-RFS
NSW Rural Fire Service. (2017). *Community Engagement Strategic Directions 2017–2021*.
Parker, D. J. (2020). Disaster resilience—a challenged science. *Environmental Hazards, 19*(1), 1–9. https://doi.org/10.1080/17477891.2019.1694857
Parkinson, D., & Zara, C. (2013). The hidden disaster: Violence in the aftermath of natural disaster. *The Australian Journal of Emergency Management*, 28(2).
Parsons, M., Morley, P., Marshall, G., Hastings, P., Glavac, S., Stayner, R., … Reeve, I. (2016). The Australian natural disaster index: Conceptual framework and indicator approach (2016.157). https://www.bnhcrc.com.au/publications/biblio/bnh-2585
Pelling, M. (2010). *Adaptation to climate change: From resilience to transformation*. Taylor and Francis.
Pratt, M. L. (1992). *Imperial eyes: Travel writing and transculturation*. Routledge.
Pratt, M. L. (2019). Afterword. *Environment and Planning E: Nature and Space, 2*(4), 799–806. https://doi.org/10.1177/2514848618802535
Price, O. F., Whittaker, J., Gibbons, P., & Bradstock, R. (2021). Comprehensive examination of the determinants of damage to houses in two wildfires in eastern Australia in 2013. *Fire*, *4*(3), 44. https://www.mdpi.com/2571-6255/4/3/44
Räsänen, A., Lein, H., Bird, D., & Setten, G. (2020). Conceptualizing community in disaster risk management. *International Journal of Disaster Risk Reduction, 45*, 101485. https://doi.org/10.1016/j.ijdrr.2020.101485
Roos, C. I., Scott, A. C., Belcher, C. M., Chaloner, W. G., Aylen, J., Bird, R. B., … the Fire and Mankind Discussion Group. (2016). Living on a flammable planet: Interdisciplinary, cross-scalar and varied cultural lessons, prospects and challenges. *Philosophical Transactions of the Royal Society B: Biological Sciences, 371*(1696), 20150469. https://doi.org/10.1098/rstb.2015.0469
Ruane, S. (2019). Applying the principles of adaptive governance to bushfire management: A case study from the South West of Australia. *Journal of Environmental Planning and Management, 63*(7), 1–26. https://doi.org/10.1080/09640568.2019.1648243
Schumann, R. L., Mockrin, M., Syphard, A., Whittaker, J., Price, O., Gaither, C. J., … Butsic, V. (2019). Wildfire recovery as a "hot moment" for creating fire-adapted communities. *International Journal of Disaster Risk Reduction, 13*. https://doi.org/10.1016/j.ijdrr.2019.101354
Sutherland, C. R. (2019). Encountering the burn: Prescribed burns as contact zones. *Environment and Planning E: Nature and Space, 2*(4), 781–798. https://doi.org/10.1177/2514848619871047

Tedim, F., McCaffrey, S., Leone, V., Delogu, G. M., Castelnou, M., McGee, T. K., & Aranha, J. (2020). What can we do differently about the extreme wildfire problem: An overview. In F. Tedim, V. Leone, & T. K. McGee (Eds.), *Extreme wildfire events and disasters: Root causes and new management strategies*. Elsevier.

Williamson, B., Markham, F., & Weir, J. (2020). *Aboriginal peoples and the response to the 2019–2020 bushfires (CAEPR 134/2020)*. https://caepr.cass.anu.edu.au/sites/default/files/docs/2020/3/CAEPR_WP_no_134_2020_Williamson_Markham_Weir.pdf

PART V

Managing Disaster Complexity

CHAPTER 16

National Policy Frameworks for Compound, Cascading and Protracted Disasters: Learning from Other Policy Sectors

Nadeem Samnakay and Stephen Dovers

Introduction

Like other countries, Australia is experiencing greater non-singular disasters resulting in *compound*, being concurrent or overlapping disasters; *cascading*, being sequential and possibly reinforcing disruptions; and *protracted* impacts, where the 'long tail' of consequences following disaster events are layered, more persistent and become more difficult to address. This has been evident over the period 2018–2021 where many

N. Samnakay (✉) · S. Dovers
Australian National University, Canberra, ACT 2600, Australia
e-mail: nadeem.samnakay@anu.edu.au

S. Dovers
e-mail: stephen.dovers@anu.edu.au

A. Lukasiewicz and T. O'Donnell (eds.), *Complex Disasters*, Disaster Risk, Resilience, Reconstruction and Recovery,
https://doi.org/10.1007/978-981-19-2428-6_16

communities in Eastern Australia have sequentially experienced record-breaking droughts, unprecedented bushfires (Binskin et al., 2020) and most recently, widespread flooding.

Triggered in-part by the 2019–20 bushfires, the Australian government established the Royal Commission into National Natural Disaster Arrangements (hereafter the Royal Commission). Its terms of reference were to investigate the responsibilities and coordination mechanisms between Australian, state, territory and local governments; arrangements for improving disaster resilience; mitigation actions and changes to legal frameworks for the Commonwealth (i.e., Australian national) government. (Binskin et al., 2020).

The Royal Commission had as its underlying premise that given compounding, cascading and protracted (i.e., 'complex') disasters, there should be a greater role for the Commonwealth government to show national leadership in disaster risk reduction (DRR). Constitutionally, the great bulk of DRR responsibilities rest with six state and two territories (hereafter: state) governments and with local governments, and not with the Commonwealth government. Of over 1300 recommendations from 55 substantial post-event inquiries held in Australia between 2009 and 2017, only 0.2% related to the role of the Commonwealth government (Cole et al., 2018).

Rising community expectations for improved national DRR coordination were brought into stark focus when the Prime Minister, Scott Morrison, was dismissive of Commonwealth intervention. When asked about the appropriateness of being overseas on holiday while large parts of the nation were consumed in unprecedented bushfires, he stated that "*… I don't hold a hose, mate, and I don't sit in a control room*" (Prime Minister of Australia Website, 2019).

The increasing community expectations for national leadership in disaster response begs the question about the adequacy of national DRR policies. In this paper, we critique national policy settings from the perspective of national policy processes and policy design.

Our focus on one country, Australia, allows a level of detail not possible if multiple countries' policies were to be examined. While the direct comparison between countries is difficult, at a general level we believe the policy success criteria used here allows a finer resolution of analysis and thus of broad insights that may be useful elsewhere, particularly in other federal systems.

Best-Practice Strategic Policy Frameworks

The Royal Commission inquiry in large part focused on operational contexts in relation to disaster arrangements, with only superficial investigation into related matters of broader policy process and design. It recognised the importance of strategic policy, and stated a difficulty it faced which underlines the topic of this paper:

> *National frameworks and strategies generally establish sensible principles. It has, however, been difficult for us to determine the extent to which these principles have been, or will be, translated into tangible outcomes.* (Binskin et al., 2020, p. 505)

However, it did not evaluate the appropriateness of national policy frameworks in the context of best-practice policy making, or against any other standard. Here, we evaluate Australia's national DRR framework policy settings through the lens of best-practice policy processes, drawing lessons from other policy sectors and the public policy literature. Significant bodies of instructive work examining strategic policy and institutional settings exist in the public policy literature, with an emerging and relevant set of insights exploring the attributes of more successful strategic policies in policy domains such as natural resource management, sustainable development and drought. In revealing DRR policy strengths and weaknesses, we hope to inform policy debate to better meet future challenges arising from complex disasters.

National framework policies, also referred to as strategic policies, are typically broad and principles-based, providing national level policy directions and goals. Such policies are particularly important in federal systems, where many responsibilities lie with sub-national (i.e., state, provincial) governments, necessitating the need for coordination and consistency through intergovernmental agreements. These framework policies are often supported by subordinate legislation, policies, programs and plans at national, state and/or regional scales. They typically apply to problems that are complex in nature, intersecting with multiple sectors requiring systemic and integrated approaches to policy design and implementation (Samnakay, 2021).

Disaster policies share many of these characteristics (see for example Binskin et al., 2020; COAG, 2011; DoHA, 2018; Handmer & Dovers, 2013) and for this reason, the application of strategic policy evaluation frameworks is well-suited to evaluating Australia's national DRR policy settings. This is especially so in the context of complex disasters which are multi-jurisdictional, requiring integration with forest and land use policy, urban planning, water resource management, agricultural policy, energy policy and health policy amongst others.

National Disaster Strategic Policies

Two national natural disaster framework policies are currently in effect and serve as the subject for evaluation in this paper: the 2011 National Strategy for Disaster Resilience (NSDR) and the 2018 National Disaster Risk Reduction Framework (NDRRF). Within the constraints of a single chapter, the emphasis here is on the design elements of these two policies, not on how implementation has occurred since the policies were adopted, or the adequacy of subsidiary programs or guidelines.

The NSDR (COAG, 2011), endorsed by the Council of Australian Governments (COAG, a body comprising the Prime Minister, state premiers and territory chief ministers, with local government as observers) recognised that a national, coordinated and cooperative effort is needed to enhance Australia's capacity to withstand and recover from emergencies and disasters. The NDRRF (DoHA, 2018) is an accompanying framework to the NSDR prepared by the Commonwealth Government, guiding national, whole-of-society efforts to proactively reduce disaster risk. It is an attempt to give substance to the aspirations of the NSDR by outlining "… *a coordinated approach to reducing disaster risk*" (DoHA, 2018, p. 4).

Together, the two aim to provide a broad but common and agreed policy framework, enabling more coordinated and consistent approaches across and within governments, while being adaptable to allow flexibility in implementation in different contexts. The following sections describe our approach to and results from exploring whether this aim is likely to be achieved.

Analytical Approach

We apply Samnakay's (2021) strategic policy framework to evaluate the NSDR and NDRRF. The framework presents good-practice policy making criteria informed by three sources. The first is a distillation of policy design success attributes identified by Marsh and McConnell (2010), Howlett (2009) and Pittock et al. (2015), selected from the wider public policy literature. The second source is a review of a wide body of published literature which analyses a number of Australian strategic national policies. Notable amongst these include the National Strategy for Ecologically Sustainable Development (ESD), (Commonwealth of Australia, 1992; Hamilton & Throsby, 1998) the National Forest Policy Statement (NFPS) and Regional Forest Agreements (RFA) (Environment Australia, 1992; Kanowski, 2017), National Water initiative (NWI); (COAG, 2004) and various iterations of Landcare (LC), a longstanding government-supported, community-based program (Robins & Kanowski, 2011). National Competition Policy—(NCP, see legacy website at ncp.ncc.gov.au) has also been included as an example of comprehensive policy processes being applied and enduring for several decades. The third source is a systematic analysis of seventeen national environment and sustainability policy audits and evaluations conducted by the Australian National Audit Office and the Productivity Commission. These two independent public agencies represent the country's most substantive sources of public policy and administration analysis and critique.

The framework presents sixteen policy design criteria which are the primary attributes for analysis in this paper. A further forty-four implementation criteria are presented in the framework but these are not scrutinised here except where some notable findings help reinforce the evaluation (see Samnakay, 2021 for detail on the criteria). The analysis here is a summary: the aim is to demonstrate the utility of taking a finer resolution view of a policy and the process that created it, enabling exploration of specific causes of later policy effectiveness. Where a limitation is raised over one criterion for DRR policy, the comparison cases indicate possible measures that might be considered.

The results of applying these criteria are presented in a summary, comparison fashion in Table 16.1, and in more detail in Table 16.2. These show that different policies have different strengths and weaknesses and indicate the potential to identify opportunities for policy improvement from such a comparative analysis, which is discussed further in the following section.

In Table 16.1, we acknowledge that a simple 'met' 'unmet' 'partially met' assessment masks some complexity and contestability. The intention here is to show that disaster policy has room for improvement and that better policy processes are possible in national settings despite constitutional power settings—political will accepting. Note that all but one of the other policies assessed shared similarly mixed results as the NSDR/NDRRF: it is not uncommon for national scale policies

Table 16.1 Australia's strategic policy frameworks for disasters assessed against attributes of successful strategic policies and compared with other policies

Policy success attributes	*NSDR/NDRRF*	*NCP*	*LC*	*ESD*	*NWI*	*RFA*
Commonwealth has accepted role in policy leadership	P	✓	✓	P	P	X
Objectives are measurable, well-articulated and clear	X	✓	X	X	✓	X
Policy outcomes clearly evident	X	✓	P	X	✓	P
Periodic policy reviews inform policy adaptation	P	✓	P	X	P	P
Policy has broad acceptance from political and other stakeholders	X	✓	✓	X	P	X
Policy support has endured decadal timeframes	X	✓	✓	X	✓	P
Governance, legislative and institutional arrangements clear across tiers of government	X	✓	P	X	✓	✓
Financial resources are apparent and enduring	P	✓	✓	X	✓	P
Other policy instruments support implementation (e.g. penalties, tax incentives etc.)	X	✓	P	X	X	P

Policy success attributes	*NSDR/NDRRF*	*NCP*	*LC*	*ESD*	*NWI*	*RFA*
Risks of policy failure recognised and compliance mechanisms strong	X	✓	P	X	X	X

Source Authors' analysis
Key NSDR/NDRRF, National Strategy for Disaster Resilience/National Disaster Risk Reduction Framework; NCP, National Competition Policy; LC, Landcare; ESD, National Strategy for Ecologically Sustainable Development; NWI, National Water Initiative; RFA, Regional Forest Agreements; ✓, Criteria mostly met; X, Criteria mostly not met; P, Criteria partially met

Table 15.2 Evaluation of NSDR and NDRFF against good-practice strategic policy design criteria

Evaluation criteria	*Assessment of the NSDR and NDRFF*	*Examples of good policy practices from other framework policies*
1. Policy initiation Is the case for a national policy broadly accepted as a means to resolve problems traditionally vested in the States?	Yes. COAG gave strong endorsement for national approaches to disaster management coordination. Community sentiment also advocating for strong national leadership in large-scale disasters	RFAs were driven by political imperative to reduce conflict/uncertainty over forest use; NCP driven by the necessity to drive national economic reforms to encourage productivity
2. Problem Definition Is the case for a national policy articulated through clarity of the problem being solved and a strong narrative for change in disaster management?	Partially. The NSDR and NDRFF mention that climate change will exacerbate complex disasters. However, current policy emphasis is on recovery and adaptation, not risk management. Recent cross-jurisdictional disasters and findings from the Royal Commission strongly advocate for changes to national coordination but implementation of recommendations remains unfinished	NWI creation driven by drought, water shortages and environmental degradation and poor prior policy success; NCP driven by an 'efficiency' gospel

Evaluation criteria	*Assessment of the NSDR and NDRFF*	*Examples of good policy practices from other framework policies*
3. Consultation and buy-in Is policy design informed by a comprehensive and extensive stakeholder consultation process?	This is unclear for NSDR given available descriptions of the process. NDRRF states that over 80 organisations were represented in its design process without identifying the types of organisations or how their views were incorporated	ESD utilised a model consultation process well suited to disaster policy formulation, consisting of sectoral and cross-sectoral Working Groups under three independent Chairs. Stakeholder consultations were central in formation of RFAs, via high-level consultation and structured, regional-scale forums. Natural Heritage Trust (NHT) and various Landcare programs reflect national policy leadership with bottom-up build also well suited to disaster policy
4. Policy agency Have representative parties including portfolio agencies, Prime Minister and Cabinet (PM&C), finance and treasury departments at federal and state levels been actively engaged and involved in policy design?	Not evident at inception of NSDR in 2011. However recent events, including droughts, floods and COVID-19 have seen Department of Prime Minister and Cabinet take a lead role in post-disaster recovery. Examples include the 2021 establishment of the National Recovery and Resilience Agency. Little evidence that other portfolio agencies were actively involved in policy design at inception of NSDR	RFA via the regional forest forums as a structured engagement mechanism. NCP strongly supported by central agencies, with strong implementation agencies established to drive compliance. NWI in initial phases had very strong policy agency which diminished as severe drought eased, and with a change of government

(continued)

Table 16.2 (continued)

Evaluation criteria	*Assessment of the NSDR and NDRFF*	*Examples of good policy practices from other framework policies*
5. Policy objectives and outcomes Are policy objectives and principles coherent and achievable?	No. The NSDR lists 47 highly aspirational outcomes against 7 actions with no clear identification of responsibilities and how implementation will be assessed. NDRFF identifies 25 strategies for action against 4 priorities with only a high-level articulation of who has implementation responsibility	The states were signatory to the NWI agreeing to implement a schedule of thirty-nine key actions with stipulated delivery dates and responsibilities tied to financial incentives. Implementation responsibilities were clearly articulated, and the National Water Commission (NWC, disestablished in 2014) was established to report biannually on implementation, recommending policy options for subordinate policies, plans and programs. Large investments in research and development were also supported to give effect to reforms in how water is managed nationally. RFA process was subject to a 20-year review. NCP had an implementation schedule, with significant funds tied to milestones
6. Implementation Timeframes Are timeframes specified for policy implementation processes, outputs and outcomes?	No for NSDR. NDRFF identifies 5-year outcomes at a high-level only. Both documents acknowledge that disaster resilience is a long-term objective	

Evaluation criteria	*Assessment of the NSDR and NDRFF*	*Examples of good policy practices from other framework policies*
7. Governance and Institutional Arrangements Are national coordinating mechanisms clear, either through nominated lead agencies or new institutional arrangements (e.g. Commissions/Authorities or new functions within existing agencies)?	The 2021 Federal Budget announced commitments to the National Recovery and Resilience Agency (NRRA) but its functions are not fully determined. The Royal Commission identifies numerous agencies who have a national role but noting poor integration amongst these. It recommended that an authoritative disaster advisory body be established at national and state levels to consolidate strategic advice and operations	Establishment of lead coordinating bodies (e.g. commissions) have been successfully established in the case of NWI, NCP (Productivity Commission and National Competition Council as influential entities). ESD had no effective implementation lead in the long term
Are State–Federal roles and responsibilities clarified?	No. Both the 2019–20 bushfire response and COVID-19 pandemic have highlighted tensions and inconsistencies between state and federal responsibilities. It is not clear how states' functions will integrate with functions of the NRRA	Very clear with NCP; yes re the NWI at least in the Murray–Darling Basin (MDB) as a key focus; one aim of the RFA was to be able to define roles, then for the Australian government to leave policy to the states
Are there mechanisms for vertical integration between Federal and State responsibilities (e.g. Ministerial forums, senior officials' committees etc.)?	Currently not at strategic level. The abolition of COAG and replacement by 'National Cabinet' has left a void. Ministerial and Senior Official committees are no longer formalised. National Cabinet lacks coordination mechanisms as it only includes Premiers, Chief Ministers and the Prime Minister	NWI/MDB Plan as functional examples, even though messy and difficult—an example that shows it is possible to integrate but always open to ongoing tension, etc. RFA process was somewhat less seamless in integration noting that RFAs did not equate to wholescale forest reforms intimated in NFPS. ESD lacked such measures

(continued)

Table 15.2 (continued)

Evaluation criteria	*Assessment of the NSDR and NDRFF*	*Examples of good policy practices from other framework policies*
Are there mechanisms for horizontal integration across State or Federal portfolio agencies?	No. Roles and responsibilities are unclear. Ambiguity of the role of Defence has been a notable theme and in COVID-19, responsibility for quarantine has been distributed to states despite it being a Commonwealth responsibility constitutionally. This is a recurring issue at state level identified in many post-disaster inquiries (Cole et al., 2018), but little attention to this at Commonwealth level until the Royal Commission	ESD engaged most federal portfolios but failed to continue engagement. NCP legislative review covered many government portfolios
How will non-government stakeholders (private sector and NGOs) remain engaged in policy implementation and evaluation?	Unclear. No mechanisms evident for other parties to engage in policy design or implementation. Most attention has been at the state level where the most relevant levers and opportunities for engagement exist	Landcare as a bottom-up build can provide learning opportunities for disaster policy implementation given cross-sectoral implications. The ESD process recommended continuation of the foundational ESD Working Groups to ensure continued whole-of-society engagement, but government did not accept this

Evaluation criteria	*Assessment of the NSDR and NDRFF*	*Examples of good policy practices from other framework policies*
8. Policy Tools Is implementation supported by a broad mix of policy tools (regulatory frameworks, market mechanisms, education, incentives, stewardship schemes) to achieve the systemic nature of addressing the problem?	No. Policies are largely communicating broad aspirational goals and principles but lack specific mechanisms and tools for implementation. Most implementation effort is focussed on grants for disaster recovery and not for risk minimisation, avoidance or planning	Australian fisheries management (state and Commonwealth responsibilities) adopts a broad mix of regulatory, compliance, education and quota-based mechanisms accompanied with ongoing research and development funds. RFA used planning, forestry standards and protected areas to implement outcomes. NWI used of statutory plans, institutional reforms, water markets, and improved hydro-climatic information. ESD process made numerous policy recommendations, but little follow up

(continued)

Table 16.2 (continued)

Evaluation criteria	*Assessment of the NSDR and NDRFF*	*Examples of good policy practices from other framework policies*
9. Funding Arrangements Are funding arrangements stipulated to support implementation?	No ongoing funds are committed for implementation of either of the policies. Commonwealth funds are allocated ad-hoc for subordinate policies such as for flood and drought recovery and $2 billion was allocated to the Bushfire Recovery Fund. The 2021 Federal Budget commits A$600 million to the NRRA and A$210 million for a new, consolidated Australian Climate Service initiative	NWI was established with an initial allocation of A$2 billion. The Future Drought Fund (no longer disaster policy) had an initial allocation of A$3.9 billion as a perpetual fund for building drought resilience and expending A$100 million per year on programs. NHT and Landcare have had ongoing funding allocations at the programmatic level. NCP used tranche payments tied to state reforms as a supreme model of ensuring implementation

Evaluation criteria	*Assessment of the NSDR and NDRFF*	*Examples of good policy practices from other framework policies*
10. Monitoring and Reporting Are reporting and review requirements and responsibilities clearly stated?	No stipulated review timeframe for NSDR. NDRRF is to be reviewed at the end of its five-year life cycle (2023)	RFAs had a twenty-year time frame with five-yearly reviews and an end-of-cycle review (albeit not well-publicised or rigorous). NWI audit process world class but discontinued. It included biannual assessments of progress and clear identification of progress or failure in delivering reforms Murray-Darling Basin Plan under the NWI umbrella is increasingly gaining stronger compliance measures Risk assessments of not achieving the policy objectives are almost always poorly done. Most national reform policies degrade over time due to inadequate consideration of what success would look like (See for example Samnakay, 2020)
11. Compliance and Audit Assessments Are compliance and audit functions stipulated?	No. The absence of a national coordinating agency and inadequate portfolio of policy instruments presents weaknesses with respect to achievement of policy objectives. Most scrutiny and audit is via post-event inquiries, almost all at state level until recent national Royal Commission	

(continued)

Table 15.2 (continued)

Evaluation criteria	*Assessment of the NSDR and NDRFF*	*Examples of good policy practices from other framework policies*
12. Risk Assessment Are the risks of not achieving the objectives identified?	Partially. Both documents present a strong case for enhanced disaster coordination and management. What is to be lost or gained from an absence of national policy is not clearly stated given that states have their own (numerous and overlapping) disaster management policies independent of Commonwealth policies	

Source Authors' analysis

Key NSDR/NDRRF, National Strategy for Disaster Resilience/National Disaster Risk Reduction Framework; NCP, National Competition Policy; LC, Landcare; ESD, (National Strategy for) Ecologically Sustainable Development; NWI, National Water Initiative; RFA, Regional Forest Agreements to have deficiencies. In contrast, the National Competition Policy was an exception in meeting all criteria for successful, enduring and impactful policy.

Table 16.2 describes the criteria more fully, and offers summary assessments of the NDRS/NDRRF against these and comments on commonalities and contrasts across the other framework policies. Direct transfer of approaches or lessons is not possible between policies developed in different circumstances, however, general lessons can be inferred against specific criteria. Importantly, Table 16.2 serves as evidence that the criteria and examples from other policy sectors show viable policy processes and institutional options to address specific weaknesses in another sector (in this case, DRR).

Results and Discussion

We aggregate the summary results in Tables 16.1 and 16.2 to assess disaster policy against the following questions:

i. Is there sufficient policy agency and buy-in at the national level to empower a national response?
ii. Are policy objectives and implementation pathways appropriate given the complex nature of the problem? And
iii. Are mechanisms in place to inform policy effectiveness?

In the following, we note that many national strategic policies, including those referred to above, have deficiencies and may still have positive effects. The intention is not to single out the NSDR and NDRRF, but to emphasise the potential to target particular features of policy process and design that may be useful as a focus for further improvement of disaster policy generally, and particularly in light of the heightened pressures arising from increased complex disasters.

Is There Sufficient Policy Agency and Buy-In at the National Level to Empower a National Response?

Our assessment finds that the Australian government's commitment to implementing the NSDR and NDRRF are largely reactive. Action is driven after an event, which Raikes et al. (2019) define as 'focusing or hazard events' with an emphasis on recovery rather than proactive leadership. National responses have been inconsistent with the stated, more proactive intent of framework policies.

For example, in 2019, the federal Department of the Prime Minister and Cabinet established the North Queensland Flood Response Agency in response to damage to pastoral properties in a portion of northern Queensland. Around the same time, drought conditions were intensifying in several eastern states which elicited an Australian government recovery response coordinated by the renamed National Drought and North Queensland Flood Response Agency. Under the National Drought Agreement (COAG, 2018), all governments have agreed to classify droughts as a risk which has to be managed as a normal cycle of Australia's climate rather than as natural disasters, and thus a reliance on recovery assistance can be seen as being contrary to agreed policy. Conservatively, we estimate that some AUS$6 billion dollars of in-drought assistance, which includes Commonwealth funds, has been offered to farming communities by Australian Governments since 2015 (DAWE, 2020) despite the policy objective of self-reliance.

In contrast, with large-scale trans-jurisdictional disasters which should trigger national policy responses as per the NSDR and NDRFF, the responses have been found wanting. The Royal Commission's findings in the context of having clear, robust and accountable national arrangements are captured by the statement that.

> *Better national coordination is required to enable* ***significant reduction*** *in disaster risks and* ***impacts*** *in the future. Australia is facing* ***increasingly frequent*** *and* ***intense*** *natural disasters, a significant number of which are likely to be* ***compounding****. Governments will need to prepare for more* ***large-scale, multijurisdictional crises****. Clear, robust and accountable arrangements for national coordination will greatly assist with addressing these future challenges.* (Binskin et al., 2020, p. 94 emphasis added)

We highlight in the above quote, key criteria for accountable national action. These are in relation to significantly reducing disaster risk and impact wherein disasters will be increasing in frequency, intensity and compounding, with large-scale and multi-jurisdictional impact.

The North Queensland flood and eastern states drought policy responses do not meet these criteria but resulted in Commonwealth response. Yet, the 2019–20 bushfires which were truly national and large-scale initially elicited a poor federal response until mounting community pressure forced an Australian government recovery response, including deployment of both regular and reserve Australian Defence Force capabilities.

Contrary to national frameworks, the policy positioning has frequently been governed by constitutional powers. The Morrison government (2019–2022) has often deflected national disaster policy responses by claiming that these are state responsibilities such as in the case of bushfire response (Wootton, 2020), climate change and energy policy (Hewson, 2020), quarantine for international travellers during COVID-19 (Tingle, 2021) and for cladding standards (AAP, 2019). The Productivity Commission (2014) notes that engineering standards, such as for cyclone tolerance (and more recently with respect to bushfire building standards), are important considerations for national disaster response policy.

Fast forward ten years from the 2011 NSDR and Commonwealth policy inertia may be at a turning point. The functions of the National Drought and North Queensland Flood Response and Recovery Agency have now been combined with the National Bushfire Recovery Agency to form the National Recovery and Resilience Agency (NRRA). From July 2021 the NRRA is to incorporate the disaster risk reduction and recovery functions from the Department of Home Affairs and will take on responsibility for the management of the Drought Policy's Rural Financial Counselling Program from the Department of Agriculture, Water and the Environment (National Recovery and Resilience Agency, n.d.). It remains to be seen whether this apparently more focused arrangement will result in a strong policy agency focused on prevention or whether the Commonwealth's role will continue to be focused on post-event recovery response of large-scale events as singular crises to be managed. The latter is likely to be insufficient in the case of increasing complex disasters.

Are Policy Objectives and Implementation Pathways Appropriate Given the Complex Nature of the Problem?

The policy objectives of both the NSDR and the NDRFF are highly aspirational, resulting in inadequate direction setting for sharing powers and responsibilities in disaster management (Lukasiewicz et al., 2017). The NSDR lacks detail in identifying responsible parties (the *Who*) with no specificity as to priority actions and their delivery timeframes (the *When*).

In acknowledging the complex and interconnected nature of disaster risk resilience, the NDRFF lists twenty-five strategies for action against four priority themes to be implemented over the period 2019–2023. The NDRFF is a laudable addition to the overly principles-based objectives of the NSDR but still lacks sufficient detail in relation to who is responsible for implementation and oversight. Good practice strategic policy design requires that sufficient consideration be given to guiding question of *who, why, what, where* and *when*, to aid implementation and evaluation (Samnakay, 2021).

The complexity of disaster policy on account of its need for cross-sectoral integration is acknowledged in the NDRFF. It provides an indicative list of twenty-five policy domains intersecting with disaster policy (DoHA, 2018, Fig. 3) including health, energy, telecommunications, natural resource management, standards and codes, critical infrastructure and community development amongst others.

This complexity presents the compelling justification for a national coordinating body as recommended by the Royal Commission. In Australia's federal model, effective trans-jurisdictional (vertical) and cross agency (horizontal) integration mechanisms are attributes of good-practice strategic policy design but are lacking in disaster policy. The abolition of the Council of Australian Governments (COAG) by the Prime Minister in May 2020 presents a significant vulnerability for vertical and horizontal integration across state-federal responsibilities. The reporting chains between departmental senior officials, their respective Ministers and to Government leaders, while seen as administratively bureaucratic, are necessary governance arrangements for complex policy design and implementation. This view is supported by the Royal Commission which finds that while the 'National Cabinet' that has replaced COAG has been a good model for managing COVID-19 response, it is not the preferable arrangement for natural disaster coordination, lacking as it does structured interactions between governments in the finer detail of policy formulation and coordination, and recommends a model more akin to the now disbanded COAG.

The focus of the NRRA is currently on drought, flood and bushfire recovery but in the absence of COAG-like integrating mechanisms, it will be increasingly important for the NRRA to establish effective state-federal integrating mechanisms given looming complex disasters. While the emphasis on post-disaster recovery grants is understandable, disaster policy lacks a broad toolkit of policy instruments. Risk avoidance in the face of complex disasters should manifest in reforms to cross-sectoral policies which use diverse policy instruments. For example, legislative changes to land zone planning, tightening of building codes and changes in community behaviour through bottom-up community-based initiatives present opportunities to implement risk management approaches to disaster resilience rather than crises-driven reactionary interventions (Duckworth, 2021; Raikes et al., 2019) which currently prevail.

This is also evidenced by the funding commitments shown in Table 16.2 which are responses to 'focusing events' to deal with crises rather than commitments to underpin the implementation of the NDRS and NDRRF objectives.

Are Mechanisms in Place to Inform Policy Effectiveness?

The inadequacy of policy agency and the limitations in policy design such as through aspirational objectives or limited instrument choice selection have direct implications for assessing policy effectiveness. The NSDR does not stipulate any review periods, while the NDRRF has a nominated review period in 2023.

The question of who will conduct reviews and what elements are to be reviewed present ongoing challenges for institutionalising national policy frameworks. Establishing the NRRA outside of the Department of Home Affairs is an important first step in reforming disaster policy coordination. The Department of Home Affairs, largely a policing and enforcement agency, arguably is not the best lead agency to drive disaster policy, in-part given the absence of national disaster legislation for it to enforce, and in-part to not having a *modus-operandi* of community-centred '*soft skill*' approaches to achieving policy objectives. Soft skills in this context include critical disaster management features such as engaging stakeholders, leveraging investments, institutionalising coordination and sharing knowledge (Albris et al., 2020).

The absence of clear delineations of *who, why, what, where* framing questions in national policy present ongoing challenges for monitoring policy effectiveness, evaluating policy strengths and weaknesses, reporting on progress and consequently on determining where policy adaptation is required. Establishing an independent oversight and coordination agency such as the NRRA presents an opportunity to improve objective setting and monitoring of disaster policies through more focused alignment of policies and practices but its functions are as yet unclear (Barnes, 2021).

Examples from other policy sectors (see NWI and NCP above) might be informative.

At this point, we note that the wielding of some such soft skills, but also intergovernmental coordination and development of detailed doctrine, has been led for many years nationally by the Australasian Council of Fire and Emergency Authorities (afac.com.au), the peak body representing state level emergency and land management authorities and by the information hub of the Australian Institute for Disaster Resilience (aidr.org.au), as well as by NGOs and state agencies. It is unclear what relationship or demarcation of roles is intended between AFAC, these other bodies and the new NRRA.

What Needs to Change in National Policy Frameworks?

The above analysis points to some key opportunities for policy reform at the national level in Australia. These apply to the current situation, and even more so to a future where complex disasters are more common:

1. There needs to be greater political acceptance of a federal role in disaster policy coordination, with an emphasis on avoidance and mitigation, not just post-event recovery. We note the importance of political will needing to be bipartisan to ensure policy agency endures over long timeframes commensurate with the scale and scope of the policy problem.
2. Federal coordination necessitates that agreements or collaborative arrangements with the states must be accompanied by clarification of objectives, responsibilities and timelines for delivering on policy reforms. In addition to establishing an accountable national coordinating agency, there needs to be an acceptance that vertical and horizontal coordination mechanisms are an essential component of working in a federal model where power sharing and financial resource imbalances can be negotiated to ensure that the whole is greater than the sum of the parts. If this is not the case, there is no need for a national policy.
3. The consideration of more granular deliverables in agreement with the states permits a more measurable assessment to be made of progress and to hold implementing agencies accountable. Accompanied by a broad mix of policy instruments such as improved planning guidelines and education programs, incentives for 'behavior change'

(including in mitigating harm from climate change) will need to be considered.

As examples from the other policy sectors show, improvements to national strategic policy processes are not beyond the realm of possibility: there are models or indicators of better approaches. We note that there is very little literature analysing national strategic policies generally, and specifically for disasters. We conclude by looking at some international examples of disaster policy in federated nations. Our aim here is to point to other situations that reflect some improvements in national policy, thus encouraging further comparative analysis of national level policies.

- The federal Disaster Mitigation Act of 2000 in USA establishes a program of technical and financial assistance to states and local governments to assist in the implementation of pre-disaster hazard mitigation measures accompanied by a National Pre-disaster Mitigation Fund to support public–private disaster hazard mitigation partnerships (Library of Congress, n.d.).
- Germany's constitution empowers states with the responsibility for disaster management similar to Australia. Germany's Federal Office of Civil Protection and Disaster Assistance (BBK) carries out specific tasks of the Federation with regard to civil protection, such as risk management, warning of the population, information and resource management, chemical, biological, radiological and nuclear defence and health protection, protection of critical infrastructure and cultural property, research and international cooperation. The BBK's experts develop strategies, conduct crisis management exercises and raise awareness amongst the public to enhance self-protection. The Federal Agency for Technical Relief (THW) is a governmental non-profit organisation. As a technical and operational agency, its tasks include technical relief and assistance in emergencies, within Germany and abroad (European Civil Protection and Humanitarian Aid Operations, n.d.).
- On behalf of the Government of Canada, the Government Operations Centre (GOC) leads and supports response coordination of events affecting the national interest. The GOC provides an all-hazards integrated federal emergency response to events (potential or actual, natural or human-induced, accidental or intentional)

> of national interest. It provides 24/7 monitoring and reporting, national-level situational awareness, warning products and integrated risk assessments, as well as national-level planning and whole-of-government response management. During periods of heightened response, the GOC is augmented by staff from other government departments/agencies (OGD) and non-governmental organizations (NGO) who physically work in the GOC and connect to it virtually (Public Safety Canada, n.d.).

Further identification and analysis of international examples and approaches would yield additional insights and establish a wider suite of ideas and policy reform options, capable of being adapted and discussed in the context of specific countries. The application of a set of criteria such as used here would provide direction and structuring to such an exercise.

Conclusion

Under Australia's constitutional arrangements, the primary responsibility for disaster management rests with the states. However, given the threat of increasing and compounding disasters, in large part due to climate change, there is increasing recognition that national leadership and collaboration is required to manage large scale, multi-jurisdictional and compounding disasters. Current national strategic policies have considerable deficiencies in policy design and policy processes. Drawing upon other policy sectors and the public policy literature, key areas of policy weakness in national disaster policy have been highlighted. Well-designed national policies have the ability to deliver outcomes that are in the national interest, beyond the separate actions of states. National and international examples suggest numerous opportunities for continued improvement in Australia's strategic disaster policy formulation. There is also much scope for research attention to be paid to international comparative analysis of national level disaster policies to draw out lessons of good practice in policy design and implementation.

References

AAP. (2019). *Scott Morrison says cladding issues a state responsibility*. The Sydney Morning Herald. https://www.smh.com.au/politics/federal/cladding-crisis-scott-morrison-says-cladding-issues-a-state-responsibility-20190719-p528p0.html

Albris, K., Cedervall, K., & Raju, E. (2020). Strengthening governance for disaster prevention: The enhancing risk management capabilities guidelines. *International Journal of Disaster Risk Reduction, 47*. https://doi.org/10.1016/j.ijdrr.2020.101647

Barnes, P. (2021). *The government has pledged over $800m to fight natural disasters. It could be revolutionary — if done right*. The Conversation. https://theconversation.com/the-government-has-pledged-over-800m-to-fight-natural-disasters-it-could-be-revolutionary-if-done-right-160348

Binskin, M., Bennett, A., & Macintosh, A. (2020). *Royal Commission into National Natural Disaster Arrangements*. Commonwealth of Australia.

COAG. (2004). *Intergovernmental agreement on a national water initiative* (pp. 1–30). Council of Australian Governments. http://www.nwc.gov.au/__data/assets/pdf_file/0008/24749/Intergovernmental-Agreement-on-a-national-

COAG. (2011). *National strategy for disaster resilience* (Issue February). Council of Australian Governments.

COAG. (2018). *National Drought Agreement*. Council of Australian Governments.

Cole, L., Dovers, S., Gough, M., & Eburn, M. (2018). Can major post-event inquiries and reviews contribute to lessons management? *Australian Journal of Emergency Management, 33*(2), 34–39.

Commonwealth of Australia. (1992). *National Strategy for Ecologically Sustainable Development*. Prepared by the Ecologically Sustainable Development Steering Committee.

DAWE. (2020). *Interim National Drought Agreement Report*. Department of Agriculture, Water and the Environment.

DoHA. (2018). *National disaster risk reduction framework*. Department of Home Affairs.

Duckworth, M. (2021). *"Top down" disaster resilience doesn't work. The National Recovery and Resilience Agency must have community at its heart*. The Conversation. https://theconversation.com/top-down-disaster-resilience-doesnt-work-the-national-recovery-and-resilience-agency-must-have-community-at-its-heart-160363

Environment Australia. (1992). *National Forest Policy Statement: A New Focus for Australia's Forests*. Environment Australia.

European Civil Protection and Humanitarian Aid Operations. (n.d.). *Germany: Overview of the National Disaster Management System*. Retrieved May 30,

2021, from https://ec.europa.eu/echo/what/civil-protection/disaster-management/germany_en

Hamilton, C., & Throsby, D. (Eds.). (1998). *The ESD process: evaluating a policy experiment*. Academy of the Social Sciences.

Handmer, J. W., & Dovers, S. (2013). *Handbook of disaster policies and institutions: improving emergency management and climate change adaptation* (2nd ed.). Routledge.

Hewson, J. (2020). *The Morrison government has abrogated responsibility for the climate crisis to the states*. The Guardian Australia. https://www.theguardian.com/commentisfree/2020/nov/25/the-morrison-government-has-abrogated-responsibility-for-the-climate-crisis-to-the-states

Howlett, M. (2009). Governance modes, policy regimes and operational plans: A multi-level nested model of policy instrument choice and policy design. *Policy Sciences, 42*(1), 73–89. https://doi.org/10.1007/s11077-009-9079-1

Kanowski, P. J. (2017). Australia's forests: Contested past, tenure-driven present, uncertain future. *Forest Policy and Economics, 77*, 56–68. https://doi.org/10.1016/j.forpol.2015.06.001

Library of Congress. (n.d.). *Disaster Mitigation Act of 2000 (USA)*. Retrieved May 30, 2021, from https://www.congress.gov/bill/106th-congress/house-bill/707

Lukasiewicz, A., Dovers, S., & Eburn, M. (2017). Shared responsibility: The who, what and how. *Environmental Hazards, 16*(4), 291–313. https://doi.org/10.1080/17477891.2017.1298510

Marsh, D., & McConnell, A. (2010). Towards a framework for establishing policy success. *Public Administration, 88*(2), 564–583. https://doi.org/10.1111/j.1467-9299.2009.01803.x

National Recovery and Resilience Agency. (n.d.). *About the National Recovery and Resilience Agency*. Retrieved May 22, 2021, from https://recovery.gov.au/about

Pittock, J., Hussey, K., & Dovers, S. (2015). Ecologically sustainable development in broader retrospect and prospect: Evaluating national framework policies against climate adaptation imperatives. *Australasian Journal of Environmental Management, 22*(1), 62–76. https://doi.org/10.1080/14486563.2014.999725

Prime Minister of Australia Website. (2019). *Radio Interview with John Stanley, 2GB*. https://www.pm.gov.au/media/radio-interview-john-stanley-2gb

Productivity Commission. (2014). *Natural Disaster Funding Arrangements* (Vol. 2, Issue 74, p. 598). Productivity Commission.

Public Safety Canada. (n.d.). *Government Operations Centre (Canada)*. Retrieved May 30, 2021, from https://www.publicsafety.gc.ca/cnt/mrgnc-mngmnt/rspndng-mrgnc-vnts/gvrnmnt-prtns-cntr-en.aspx

Raikes, J., Smith, T. F., Jacobson, C., & Baldwin, C. (2019). Pre-disaster planning and preparedness for floods and droughts: A systematic review. *International Journal of Disaster Risk Reduction, 38*. https://doi.org/10.1016/j.ijdrr.2019.101207

Robins, L., & Kanowski, P. (2011). 'Crying for Our Country': Eight ways in which 'Caring for our Country' has Undermined Australia's Regional Model for Natural Resource Management. *Australasian Journal of Environmental Management, 18*(2), 88–108. https://doi.org/10.1080/14486563.2011.566158

Samnakay, N. (2020). Understanding design and implementation attributes for strategic policies: The case of Australia's national environment policies. *Policy Studies*, *0*(0), 1–23. https://doi.org/10.1080/01442872.2020.1760232

Samnakay, N. (2021). A framework for analysing and informing Australia's National strategic natural resource management policies. *Australasian Journal of Environmental Management, 28*(3), 267–286. https://doi.org/10.1080/14486563.2021.1959427

Tingle, L. (2021). *Who is responsible for quarantine in Australia?* Australian Broadcasting Commission. https://www.abc.net.au/news/2021-01-19/who-is-responsible-for-quarantine-in-australia/13070108

Wootton, H. (2020). *States and federal governments should work together on fire crisis*. Financial Review. https://www.afr.com/politics/federal/states-and-federal-governments-should-work-together-on-fire-crisis-20200112-p53qrw

CHAPTER 17

Definition and Explanation of Community Disaster Fatigue

Valerie Ingham, Mir Rabiul Islam, John Hicks, Anna Lukasiewicz, and Christopher Kim

Introduction

In the normal course of events, communities are dynamic, with ongoing adaptation to change, they constantly address new issues. Healthy communities endeavour to ensure that their standard of living continues

V. Ingham (✉)
Australian Graduate School of Policing & Security, Charles Sturt University, Bathurst, Australia
e-mail: vingham@csu.edu.au

M. R. Islam
School of Psychology, Charles Sturt University, Bathurst, Australia
e-mail: rislam@csu.edu.au

J. Hicks
School of Business, Charles Sturt University, Bathurst, Australia
e-mail: jhicks@csu.edu.au

A. Lukasiewicz and T. O'Donnell (eds.), *Complex Disasters*, Disaster Risk, Resilience, Reconstruction and Recovery,
https://doi.org/10.1007/978-981-19-2428-6_17

to improve and that the benefits of this growth are increasingly spread throughout the community. Disaster events can interrupt this process and even reverse it. However, for the most part, communities can respond and fight back against such disruption to re-establish themselves on a progressive path. But what if the characteristics of community resilience are so severely damaged by a disaster event (or series of events) that progress ceases and the quality of life within a community is negatively affected? Could the result be the failure of the community to take advantage of the resources and opportunities at its disposal? Is it possible that community drive and decision making, once exhibited in a network of effective community groups—supportive, social and commercial—goes missing as people become inward-looking, reclusive, and fail to lend their talent and expertise to the community's needs? Community members could also believe that they no longer have anything to offer that could make a difference by being overwhelmed by the decisions that confront them. In other words, are these manifestations evidence of a community suffering from collective disaster fatigue?

A complex disaster has many varied definitions (see Chapter 1) but may be characterised by either multiple and interacting drivers of risk and/or cascading or compounded impacts. For Australia in 2020 the application of the term was not difficult. The Black Summer bushfires eventuated after years of drought and were quickly followed by the COVID-19 pandemic. Unusually, over 2019/2020 Australian summer, various regions in the eastern states flooded. Detailed coverage of all three events is found in Cunningham (2020). Many communities had not completed their rebuild after the Black Summer fires and subsequent floods when COVID-19 took the nation to a lockdown standstill in March 2020.

A. Lukasiewicz
Institute for Climate, Energy & Disaster Solutions, Australian National University, Canberra, Australia
e-mail: anna.lukasiewicz@anu.edu.au

C. Kim
Industry Policy and Program, Department of Defence, Canberra, Australia

Although not directly part of a research project, the authors, each with a substantial research background in disaster and emergency management and familiar with many of the locations impacted and the frontline workers involved, observed, on a casual basis, how tired people were prior to returning to work in 2021. We observed the Australian university sector, which in 2020 grappled with moving face to face classroom teaching to online platforms in the midst of staff cuts brought on by the consequences of the pandemic. In addition, in our areas of residence—particularly for one team member living in the Blue Mountains—we observed that professional people who had been involved in and impacted by a number of disasters in our locality and who had been participants in our previous disaster-based research were commencing their working year in an already exhausted and debilitated state. We have sought ways to describe and explain the unfolding exhaustion and, in this chapter, contend that a situation of complex disasters may induce 'community disaster fatigue'.

Defining and exploring community disaster fatigue has significant consequences. For example, there could be major implications for national, state, and local government officials and emergency service workers tasked with addressing a new impending disaster, when there exists in the community people whose houses have not been restored from the previous disaster, whose businesses or livelihoods are still to recover, whose neighbours are still in need, whose children are only just back at school, whose social life is on hold or badly disrupted and so on. Dealing with a new disaster is going to be difficult when the community is already in a state of disaster fatigue.

The concept of community disaster fatigue has implications for how communities perceive, react, and adapt to risk. This can impact not only the way in which they make (or do not make) decisions but the decisions themselves. For example, complex disasters create conflicts in decision making. A community severely impacted by fire has to take decisions with respect to providing residents with access to provisions and building materials and the restoration of basic needs. However, in a COVID environment, the decisions that might normally be taken are compromised by public health regulations requiring social distancing, lockdown, and closure of businesses. Not knowing how to act, the community may choose not to act, and as the issues mount, inaction becomes the default as community disaster fatigue sets in.

This chapter has two principal aims: first, to synthesise related concepts and findings from existing studies, and second, to build an overarching conceptual framework of community disaster fatigue for guiding subsequent development.

Approach

This chapter reports a snapshot review of literature which first aims to identify the determinants of community disaster fatigue (exhibited, primarily, by the failure of a community to make decisions that will return it to a thriving state following a series of disasters). We start with the use of a simple diagram (Fig. 17.1).

The three circles represent three key themes derived from the disaster management literature, which generally arranges itself according to the psychosocial, economic, and physical environment (sometimes further

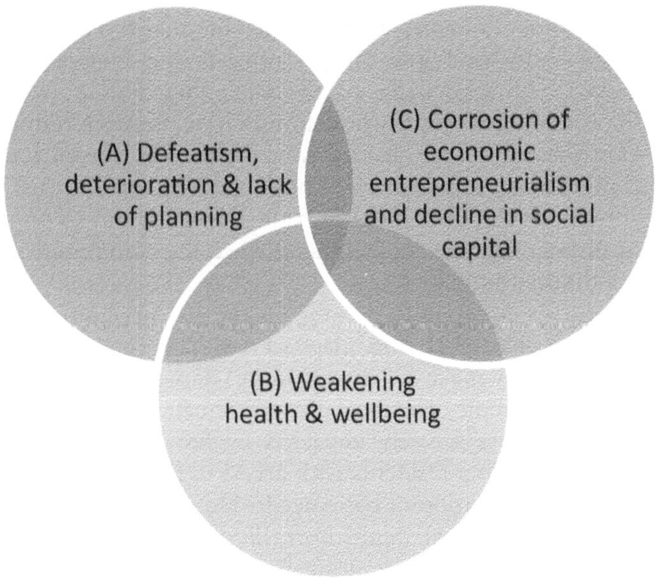

Fig. 17.1 Components of community disaster fatigue (*Source* Authors' construct)

delineated as natural and man-made) as constructs to measure a community's resilience. However, community resilience is diametrically opposed to the concept of community disaster fatigue. Rather than looking at a confident, aspiring community and on the rebound, we are considering communities that are defeated, languishing, and slipping further into an oblivion generated by a series of disaster events. We have therefore condensed the elements of the disaster management literature into three components of community disaster fatigue: defeatism, deterioration, and lack of planning (circle A); weakening mental health and wellbeing (circle B), and corrosion of economic entrepreneurialism and decline in social capital (circle C). When all three components are combined in a community (where all three circles overlap), that community is at its lowest ebb in terms of dynamism—exhibiting only lethargy and apathy in all areas of community activity from which it is unlikely to emerge without help. Less severe levels of community disaster fatigue can also be recognised in the diagram where only two of the three circles intersect or when just one component is present. From the perspective of these three components, we surveyed relevant literature relating to complex disasters to identify indicators of each component that, when identified and aggregated following a series of disaster events, could act as an indicator of the component's contribution to community disaster fatigue. A summary of these findings is reported in Table 17.1, and we expand upon the table's construction in the following three sections before discussing our definition and description of community disaster fatigue, drawing our conclusions, and providing suggestions for further research.

Defeatism, Deterioration, and Lack of Planning

The component of community disaster fatigue 'Defeatism, deterioration, and lack of planning' (see Fig. 17.1) is situated within the vulnerability-resilience narrative, where resilience is often spoken of in terms of 'time taken' to recover, depending on what criteria are prioritised (Norris et al., 2008; Rus et al., 2018; Sudmeier-Rieux, 2014). However, when the community forms the view that the responders have established the wrong priorities, difficulties will arise. Public–private partnerships (Busch & Givens, 2013) will be non-existent or weak, and the community is going to fail to adapt in a manner that will give rise to resilience (Kapucu et al., 2014). Under such circumstances, appropriate disaster risk governance (Lassa, 2015) and planning will fail to emerge.

Table 17.1 Collective indicators of disaster fatigue

Lens through which disaster fatigue was viewed	*Collective indicators of disaster fatigue*
Defeatism, deterioration, and lack of planning (A)	Dissatisfaction with disaster risk governance in terms of recovery progress, resulting in group protest causing disruption Lack of response by the community Increasing levels of unemployment Communities lacking focus or purpose Declining ability and opportunities for community-based initiatives; reduced participation in regional and State level decision making Increased demands on police and other public officials
Weakening mental health and wellbeing (B)	*On a collective scale known as public health* Increased demand for various mental health and community support services Mistrust of other communities fed by social distancing e.g. not wanting tourists to visit as they may bring COVID-19 Reduction (either temporary or permanent) in organised social activities Widespread non-participation in organised activities Collective trauma and helplessness (Chang, 2017; McFarlane & Norris, 2006) Collective complicated grief (Gesi et al., 2020) Extra burden on social and community support services
Corrosion of economic entrepreneurialism and decline in social capital (C)	Waning production in local businesses Dissolution or a prolonged hiatus of previously active community groups Lack of safe housing (Gearhart et al., 2018) Lack of community networks and social support (Gearhart et al., 2018) Sub optimal decision making due to poor information flow and time pressure Increased political polarisation

Source Authors' construct

Rouhanizadeh and Kermanshachi (2021) examined post-disaster recovery barriers (after hurricanes). They found that for experts, "*undefined roles and responsibilities in the process of recovery*" was the highest weighted barrier, whilst the public ranked "*insufficient built infrastructures*" (p. 2). The results table constructed by Rouhanizadeh and Kermanshachi registers the levels of each identified barrier to recovery, with one column for 'experts' and the other for 'public'. Of relevance to complex disasters and community disaster fatigue, the occurrence of multiple disasters is rated as of medium concern for experts but of great concern for individual community members (p. 2). Further, in the context of multiple disasters, experts are failing to register the same level of concern for issues that concern community members in the disaster recovery process. Therefore, this chapter aims to increase the awareness of the 'experts' as to what is distressing and contributing to disaster fatigue in public. The scope of this chapter does not extend to 'what to do about community disaster fatigue', as this will be a subsequent research project.

There has been increased international research focusing on disaster risk governance (for example, Alam & Ray-Bennett, 2021; Cheek, 2020; Sakic Trogrlic et al., 2021; Shmueli et al., 2020). A common theme is community disappointment at the lack of opportunity for their engagement and consultation by the politically appointed recovery body (Ingham & Redshaw, 2017). The community must be involved, and a failure to secure their involvement will disrupt planning efforts and contribute to a growth of defeatism throughout the community with a consequent deterioration in outcomes. Community resilience will depend on the community's ability to respond to changes in the physical, social, and psychological aspects of community life during post-disaster recovery (Paton, 2006). As Quarantelli (1998) explained, a disaster is a social event in that it disrupts the community social system. After the onset of an extreme natural catastrophe, returning to a previous state becomes the least likely possible scenario. Community members will experience collective stress when they come to understand that the formal response mechanisms have failed to take their perceived needs into account (Barton, 2005, p. 126). Thus, community resilience will extend to members' capacity to respond autonomously and collectively to hazardous situations (Norris et al., 2008; Paton, 2006). Such a view is well incorporated in recent studies on community disaster resilience (Carvalhaes et al., 2021; Wadi et al., 2020). Most studies have attempted to identify a set of adaptive capacities (Erdiaw-Kwasie et al., 2019; Norris et al., 2008).

In several studies, community resilience is viewed as a set of community capacities. Magis (2010) stressed that community resilience is primarily about community members' capacities to adapt to change. Norris et al. (2008) echoed this, suggesting that a set of networked capacities constructs community resilience (p. 135). Community resilience emerges where the effects of the community capacity lie on community members' ability to identify and solve social problems and accumulate this capacity as community resources and knowledge (Goodman et al., 1998, pp. 259–60). Brown and Kulig (1996, p. 43) define community resilience as "*the capacity of community members to engage in projects of coordinated action within the context of their community despite events and structures that constrain such projects.*" This notion presents several key elements of community resilience. First, community resilience is reflected in the community's ability to engage in deliberate actions for solving the challenges they experience (Magis, 2010, p. 407). In addition, coordinated actions require collaboration amongst individuals in the community (Kulig et al., 2008, p. 761). Finally, in the face of challenges, the community is able to exert its resilience to recover from an acute disaster shock (Norris et al., 2008, p. 130).

The literature suggests that the seeds for the emergence of community disaster fatigue can be sown long before a disaster presents because of a failure of communication between official responders and the community. In preparing for a crisis, a community must identify the threats to its existence and develop a plan to meet these threats (Parnell & Crandall, 2017). However, there may be a lack of information in the planning stage, not the least of which is the nature of the crisis that will manifest. Planning for fire in a fire-prone area may be high on the agenda of official responders and therefore follow a conventional strategy. Still, little thought may be given to the potential for flood, famine, or disease. The planners may be familiar with the planning needs of the first of these, but far less so with the needs of the others. Thus, whilst the plans put in place may be adequate and rationale for one disaster, they may be completely inadequate if other types of disaster present one after the other, such as in a complex disaster situation. A community may be convinced that they have planned well for disaster—and indeed, they may have planned well for a specific type of disaster. Such a situation may reflect what Parnell and Crandall (2017) refer to as an optimism bias—where the community has underestimated the probability of another disaster arising on the heels of the last one. In this unexpected situation, where the community is still in recovery, a conflict of priorities may arise between meeting the needs of the previous event (for example, a lack of housing or food) whilst trying to meet the needs of the current event (for example, social distancing and complying with lockdown restrictions due to a pandemic).

Finally, when a community and/or its leaders do not adequately plan for a crisis– usually because of an expectation that an event is unlikely (Finucane et al., 2020; Parnell & Crandall, 2017), one can expect that the experience of multiple events and a multiple variety of events will have an even greater impact.

Weakening Mental Health and Wellbeing

The weakening mental health and wellbeing of a community can be evidenced in several areas. For example, the rise in instances of substance abuse (Crompton et al., 2018) and suicide (Jafari et al., 2020; Gesi et al., 2020) severely impacts a community's ability to respond to a disaster or series of disasters as it effectively eliminates potential contributors both in terms of those directly impacted and those who are required to divert their energies to meet the needs of those directly impacted.

The emergence of vicarious trauma, also known as compassion fatigue (Pearlman & Saakvitne, 1995), is also an issue. In its original context, vicarious trauma was understood as the traumatic experiences of counsellors due to continuous over engagement with traumatised clients. Extrapolated to the general population, vicarious trauma is also known as compassion fatigue (Beaglehole et al., 2018; Kristensen et al., 2009). We may be already seeing the signs of both vicarious trauma and compassion fatigue within communities in Canada, the UK, and Australia due to the constant media focus on issues such as COVID-19 and climate change (Clayton, 2020).

A wide range of research exists on compassion fatigue in career and volunteer emergency workers, community service and social workers, nurses, and frontline occupations exposed to tragedy and disaster (Cocker & Joss, 2016). In a recent study, the consequences of compassion fatigue to nurses were identified as exhaustion, difficulties in maintaining the optimal level at work, repercussions on family and private life, apprehension, increased stress level, and the desire to quit the profession (Pérez-García et al., 2021).

Essential public health responses such as social distancing can also add to a decline in mental wellbeing, notwithstanding the essential nature of their application. Social distancing creates a mistrust at the exact time when members of a community need to gather to redefine their community persona, trust in others (Naser et al., 2019), and sense of community success and achievement (Beaglehole et al., 2018). Elcheroth and Drury (2020) ask, "*how long can we expect people to 'sacrifice daily freedoms' when we are experiencing economic hardship, and across the world, the chasm of social inequity exposed by the pandemic becomes more and more publicly visible?*" (p. 1).

It is challenging to assess the cost of the effects of psychosocial trauma experienced by a community (Deloitte Access Economics, 2016; Gurtner & King, 2021). The mental anguish of the large-scale loss of capacity and the resulting trauma is not highly visible in the initial stages of recovery (Chang, 2017; McFarlane & Norris, 2006). It is far easier to identify the effects at the individual household level. For example, we know that in the aftermath of a disaster, there may be an increase in domestic violence and relationship breakdowns (Parkinson & Zara, 2013; Rezaeian, 2013; Thurston et al., 2021), poor school and work performance (Duffy, 2014; Gibbs et al., 2019), however, on a national level in the disaster recovery phase, a deficiency of limited safe housing, lack of community networks and social support can contribute to personal distress and create a sense of disempowerment (Gearhart et al., 2018). There may also be a collective reduction in work capacity in areas such as decision making and reaction time (Qin & Jiang, 2011).

Corrosion of Economic Entrepreneurialism and Decline in Social Capital

How have the complex disasters of fire, flood, and the COVID-19 pandemic influenced economic and social structures and community and business engagement? Collapsing economic systems place increased stress on those working in industries catering for discretionary spending (Cavallo & Noy, 2009). An example would be people working in the travel and hospitability industries.

Behavioural economics addresses some of the issues important in the context of complex disasters– especially how people might change their behaviour following a series of unexpected adverse events (Cameron & Manisha, 2012). For example, people might like living amongst large trees (and think that they are prepared) until they experience a fire– then they may want to move. Another aspect is that people facing a bushfire may be prepared to stay and defend their properties until they have to do it. Once experienced, however, it becomes an event that they never want to face again (McCaughey et al., 2018). Thus, their perception of costs and benefits change. With complex disasters, this realignment of thought and actions may be ongoing. As pointed out by Samson (2014), decision making is context-dependent and as the context changes, so will the decisions and the way decisions are made. Further, he argues that although individuals make decisions, they are part of a society that shapes their way of thinking. If that society changes (perhaps because of complex disasters), the individual's decisions, conforming to that change, may also shift.

Samson (2014) describes behavioural economics as the development of theories based on psychological experimentation of how people make decisions. Earlier in his work, when referring to Sunstein and Thaler (2008), he notes that good decisions result when the decision-makers have experience of the issues, good information, and receive prompt feedback on any issues raised. When information and knowledge are limited, the quality of decision making will suffer. Consequently, a behavioural economics perspective suggests that decisions are not always optimal as we do not have all the information and, indeed, we are often incapable of processing what information we do have. Samson (2014, pp. 1–3) notes that given this uncertainty, and the different contexts (and with complex disasters, changing contexts) in which we may have to make decisions, we may not carefully identify the optimal choice. Instead, it is argued, we rely on several shortcuts, which can introduce bias into our decision making and result in poor decisions being made. The examples provided include: (1) An excessive reliance on our memory of events from the past which may be distorted and/or may not be relevant to the events we are currently confronting; (2) An affect heuristic which leads us to make decisions based on our feelings rather than reflecting on and assessing the situation before making a decision; (3) We may also base our decisions on the most obvious or readily available information. Because this information is salient, it is given an emphasis that it may not deserve.

This brief synopsis begins to raise some issues of relevance in the current context. Shortcuts in decision making will arise when information is not flowing freely when the information, we have is difficult to process, or we have time constraints that make the processing of information impossible. Each of these factors can be generated by a disaster event.

Therefore, the disaster event (or a series of disaster events) can give rise to the circumstances in which shortcuts in decision making are taken—indeed they may be necessary. As Parnell and Crandall (2017) point out, individuals have prescribed bounds in which to make decisions. That is, they cannot thoroughly examine every angle relevant to a decision, nor do they have all the information needed to decide or understand all the information they do have. Thus, they are forced to make decisions that are not perfect. They must base their decisions on heuristics, or rules of thumb that have served them well in the past. The hope is that the decisions they make are, in the circumstances, satisfactory (see Simon, 1956). Parnell and Crandall (2017) concede, however, that successful decisions based on heuristics assume a predictable environment in which decisions are routine. When the circumstances are unfamiliar, the reliance on heuristics is less effective. Disasters often, although not always, place us in unfamiliar circumstances. Complex disasters sharply increase the unfamiliarity of what we are experiencing. Faced with the unfamiliar, the application of heuristic solutions may become increasingly irrelevant to the point where decision-making becomes less effective or ceases and disaster fatigue sets in. This situation is exemplified by a lack of action in the face of seemingly overwhelming events. Indeed, we need not just focus on the event or events. Constantly changing circumstances will also impact decisions we make with respect to both preparation and recovery. A reliance on heuristics will be quite limiting.

One way of approaching the economics of community disaster fatigue is to look at how the perception of the costs (of the event) and benefits (of taking action to prevent or fight the event) change and how this change is translated into action (or inaction) and how the theory of behavioural economics might explain this. This leads to looking at how the perceptions (both before and after) correspond to reality. Before a disaster, people may tend to be overly optimistic and therefore plan to take actions that are not practical. After a disaster, it may be that they become overly pessimistic and, as a result, forego taking actions that might be of value.

A policy could be targeted at modifying perceptions so that, in both cases, they more closely correspond to reality and, as a result, actions are more efficient and beneficial. Such policies could be targeted, for example, at increasing the flow of (accurate) information.

COVID-19, in particular, has the potential to change economies through the possible disappearance of industries or, at least, the substantial restructuring of industries, for example, the cruise ship industry, hospitality, professional sports, airlines, public transport, entertainment, housing construction—indeed anything that brings large groups of people together or facilitates the spread of COVID-19 in an economy and society that does not have the appropriate medical facilities in place or in which vaccinations are not the panacea they were first trumpeted to be. Community disaster fatigue has the potential to slow down or prevent changes in economics and society that are optimal and necessary for moving forward. For example, the continuing push to 'open up' and get back to 'normal'—an outcome that could bring its own disaster without an adequate distribution of a vaccine (or the distribution of a vaccine that is not adequate)—or even a failure to strengthen sufficiently societies' public health response capacity.

These factors, when they interact, may have a much more significant impact when social capital is compromised. It has been argued that the allocation of funding that does not consider public preferences may cause unnecessary disruption and delay (Omidvar et al., 2011, p. 1399). To what extent do such problems arise from community disaster fatigue? Public officials are not immune. The role of politics is also important in terms of the services the public expects (Omidvar et al., 2011, p. 1399). "*Appropriate performance of the reconstruction authorities, including all of its fractions, namely cooperativeness, behavior, pace of work cycles, sense of responsibility, and their working experience may all increase the participation level extensively*" (Omidvar et al., 2011, p. 1410). But in all cases, continuing disasters may compromise public officials' ability to make appropriate decisions, thus compromising their contribution to community welfare.

Discussion: Defining and Describing Community Disaster Fatigue

On a global collective scale, right down to communities and individual households, people want to get back to normal, get on with their lives, however a whole new reconstruction, a reinvention of a community's fabric may be necessary when a community experiences a situation of complex disasters. It will not be possible for some communities to 'get back to normal'. In resilience arenas the term, 'the new normal' is used in recognition of the fact that it is not actually possible to go back to the pre-disaster state and that what will be regarded as normal into the foreseeable future will involve, at times, considerable change and adaptation. (Goromaru et al., 2021).

What COVID-19 has done on a global scale is to reveal the cracks in attempting to go back to previous ways of participating in a community and pursuing the previous priorities as if a community transformation had not taken place. Whether a nation decided to follow herd immunity such as Sweden or enforced lockdowns as China, makes little difference to the result, which is the resilience, mental health, and socio-economic situation of a community coming to grips with the loss and trauma of family, friends, livelihoods and a way of life.

As previously indicated, Table 17.1 is designed to provide an initial checklist for leaders and people in positions of community decision making, enabling them to recognise the onset of community disaster fatigue. We review our discussion here.

Resnik and Fins quote Hirschberger (2018) in the construction of their definition of collective trauma: "*Collective trauma refers to the 'psychological reactions to a traumatic event that affect an entire society'*" (Resnik & Fins, 2021, p. 3). For Resnik and Fins (2021, p. 3), referring to Park (2013), "*It threatens the fundamental fabric of identity, undermines community, and creates a crisis of meaning.*" From our perspective, social distancing and associated COVID-19 safety requirements have broken the fabric of interpersonal and community trust to the extent that there is community mistrust (for example, locals disliking tourists as they may bring COVID-19 with them). This disconnection of the individual from the collective makes them vulnerable. To counter this, it may be that leaders of the collective should concentrate on re-creating a sense of community and a sense of belonging.

The importance of the psychosocial health of a community in preparing for and recovering from a large-scale disaster, which in 2021 was a pandemic, is paramount to a nation's mental health recovery from disaster. For years to come, disaster-related stress and trauma will continue to be exhibited on a mass scale in national psyches (Dawel et al., 2020). It will require a national economic effort to address the long-term fallout to psychological health.

Mental health already looms as an emerging issue in Western nations, as collapsing economic systems put increased stress on those who work in industries catering for discretionary spending, such as travel and hospitability. Considered 'non-essential' these workers and their organisations struggled with the COVID-19 situation. The importance of cleaners, hospital personnel, and food-producing industries was swiftly elevated. This is a considerable rearrangement of the employment hierarchy, as a new delineator of essential vs non-essential workers comes into play.

In addition, community disaster fatigue might have a wide range of cultural and political impacts, such as underprivileged, vulnerable, and marginalised people entering into an era of social disadvantage, as well as gender and social exclusion (Resurrección et al., 2019). Further, the consequences of these impacts could last for many years.

Community disaster fatigue is initially evidenced by the disconnection of community members from former points of collective activity. Community disaster fatigue is indicated on a collective level by the breakdown or absence of community resilience. A healthy and robust community exhibits collective strength in the areas of mental health and wellbeing and through socio economic indicators. When these points of community collaboration and cohesion are weakened or broken, for example, through a complex disaster situation, there may be a lessening of community resilience. For example, Gearhart et al. (2018) demonstrate a sharp rise in the incidence of domestic violence after a disaster. The individual needs arising from relationship breakdown then place an extra burden on social and community support services. We posit that the increase in demand for such services should serve as an indicator of community disaster fatigue.

Conclusion: The New Normal

In this chapter, we extended Aridi's (2020) definition of disaster fatigue as "*a form of emotional exhaustion that can reshape how people make choices*" to encompass a more complex description of disaster fatigue as evidenced on a community level. We define community disaster fatigue as the failure of a community to make decisions that will return it to a thriving state following a series of disasters.

Community disaster fatigue is a complex issue, and the areas of resilience, mental health, and socio-economics are multi-faceted and interrelated, as depicted in Fig. 17.1. Communities are complex and dynamic, displaying varying levels of adaptiveness (Faulkner & Sword-Daniels, 2021, p. 3). The way the various elements or components within the system morph and change, and adapt is complex and 'non-linear' (Faulkner & Sword-Daniels, 2021, p. 3). This discussion of community disaster fatigue is designed to assist with the identification of the phenomenon in order to determine and design adaptation that is beneficial and contributes to the collective wellbeing, economy, and society in general—as opposed to a collective situation of disaster fatigue in which the components of a complex system 'wrongly' adapt and lead to inactivity, public health issues, and dysfunctional institutions.

In this chapter, we establish the issue of community disaster fatigue specifically to raise awareness of how difficulties in coping with the economic, physical, political, social, and emotional impacts of multiple disaster events are expressed collectively on a community level. We need to look at how leaders make decisions and the additional strain put on decision making by complex disasters. In the face of multiple disasters, it will always be impossible to identify the optimal policy, but we must get as close to optimal as possible. Of necessity, we will be forced to adopt heuristic shortcuts in our decision-making processes. But we must recognise that such an approach is fraught with problems that may make the situation worse.

What people look for coming out of a complex disaster situation is leadership. On a community level, this means resourcing community groups. Decision-makers must be provided with enhanced tools which will cover a range of potential events, and strategies have to be put in place to educate communities in areas where education, in the past, was not regarded as necessary.

For Further Research

Are we still 'all in this together'? A failure to focus on collective trauma will result in an immense negative impact upon a community's social, economic, and cultural profile, and therefore this is an area that requires concerted attention. The consequences of community disaster fatigue and potential policy remedies require research. Potential research questions could include: How to engage a disaster fatigued public to participate and collaborate in post-disaster decision making and recovery? What incentives would increase and promote community resilience when the indicators of community disaster fatigue begin to surface? How do experiences manifest in different countries?

References

Alam, E., & Ray-Bennett, N. S. (2021). Disaster risk governance for district-level landslide risk management in Bangladesh. *International Journal of Disaster Risk Reduction, 59*(1), 1–13.

Aridi, R. (2020, November 24). For scientists studying 'disaster fatigue,' this has been a year like no other, *Science Magazine*. Retrieved 29 June from https://www.sciencemag.org/news/2020/11/scientists-studying-disaster-fatigue-has-been-year-no-other

Barton, A. H. (2005). Disaster and collective stress. In R. W. Perry & E. L. Quarantelli (Eds.), *What is a disaster?* (pp. 125–152). Xlibris Corporation.

Beaglehole, B., Mulder, R., Frampton, C., Boden, J., Newton-Howes, G., & Bell, C. (2018). Psychological distress and psychiatric disorder after natural disasters: Systematic review and meta-analysis. *The British Journal of Psychiatry, 213*(6), 716–722.

Brown, D., & Kulig, J. C. (1996). The concept of resiliency: Theoretical lessons from community research. *Health and Canadian Society, 4*(1), 29–52.

Busch, N., & Givens, A. (2013). Achieving resilience in disaster management: The role of public-private partnerships. *Journal of Strategic Security, 6*(2), 1–19.

Cameron, L., & Manisha, S. (2012). Risk taking behaviour in the wake of natural disasters, Institute for the Study of Labor (IZA), Discussion Paper No. 6756, accessed 1 July 2021, http://ftp.iza.org/dp6756.pdf

Carvalhaes, T. M., Chester, M. V., Reddy, A. T., & Allenby, B. R. (2021). An overview and synthesis of disaster resilience indices from a complexity perspective. *International Journal of Disaster Risk Reduction, 57*, 102165.

Cavallo, E., & Noy, I. (2009). The economics of natural disasters: A survey, IDB Working Paper Series, No. IDB-WP-124, Inter-American Development Bank (IDB), Washington, DC, accessed 1 July 2021, https://www.econstor.eu/bitstream/10419/89151/1/IDB-WP-124.pdf

Chang, K. (2017). Living with vulnerability and resiliency: The psychological experience of collective trauma. *Acta Psychopathologica, 3*(53). https://doi.org/10.4172/2469-6676.100125

Cheek, W. (2020). The Paradox of community involvement: Rebuilding Minamisasanriku. *Disaster Prevention and Management: An International Journal, 29*(6), 893–907.

Clayton, S. (2020). Climate anxiety: Psychological responses to climate change. *Journal of Anxiety Disorders, 74*(102263), 1–7.

Cocker, F., & Joss, N. (2016). Compassion fatigue among healthcare, emergency and community service workers: A systematic review. *International Journal of Environmental Research and Public Health, 13*(6), 1–18.

Crompton, D., Young, R., Shakespeare-Finch, J., & Raphael, B. (2018). Responding to disasters: More than economic and infrastructure interventions. *Insights on the Depression and Anxiety, 2*(1), 14–28.

Cunningham, S. (Ed.). (2020). *Fire flood. Plague: Australian Writers Respond to 2020*. Vintage.

Dawel, A., Shou, Y., Smithson, M., Cherbuin, N., Banfield, M., Calear, A. L., Farrer, L. M., Gray, D., Gulliver, A., Housen T., McCallum, S. M., Morse, A. R., Murray, K., Newman, E., Harris, R. M., & Batterham, P. J. (2020). The effect of COVID-19 on mental health and wellbeing in a representative sample of Australian Adults. *Frontiers in Psychiatry, 11*. https://doi.org/10.3389/fpsyt.2020.579985

Deloitte Access Economics. (2016). *The economic cost of the social impact of natural disasters*. Retrieved May 20021 from http://australianbusinessroundtable.com.au/our-research/social-costs-report

Duffy, N. (2014). Opportunities for disaster resilience learning in the Australian curriculum. *Australian Journal of Emergency Management, 29*(1), 12–16.

Elcheroth, G., & Drury, J. (2020). Collective resilience in times of crisis: Lessons from the literature for socially effective responses to the pandemic. *British Journal of Social Psychology, 59*(3), 703–713.

Erdiaw-Kwasie, M. O., Abunyewah, M., & Yamoah, O. (2019). 'After the disaster comes destination thoughts': A review and conceptualization of consolidative disaster adaptive capacity model. *International Journal of Disaster Risk Reduction, 35*, 1–9.

Faulkner, L., & Sword-Daniels, V. (2021, March 18). *Improving resilience measurement: Learning to adapt*. Practice Paper 01. ITAD. Retrieved 22 April 2021, from https://www.itad.com/knowledge-product/improving-resilience-measurement/

Finucane, M. L., Acosta, J., Wicker, A., & Whipkey, K. (2020). Short-term solutions to a long-term challenge: Rethinking disaster recovery planning to reduce vulnerabilities and inequities. *International Journal of Environmental Research and Public Health, 17*(2), 482.

Gearhart, S., Perez-Patron, M., Hammond, T. A., Goldberg, D. W., Klein, A., & Horney, J. A. (2018). The impact of natural disasters on domestic violence: An analysis of reports of simple assault in Florida (1999–2007). *Violence and Gender, 5*(2), 87–92.

Gesi, C., Carmassi, C., Cerveri, G., Carpita, B., Cremone, I. M., & Dell'Osso, L. (2020). Complicated grief: What to expect after the coronavirus pandemic. *Frontiers in Psychiatry, 11*, 489.

Gibbs, L., Nursey, J., Cook, J., Ireton, G., Alkemade, N., Roberts, M., Gallagher, H. C., Bryant, R., Block, K., Molyneaux, R., & Forbes, D. (2019). Delayed disaster impacts on academic performance of primary school children. *Child Development, 90*(4), 1402–1412.

Goodman, R. M., Speers, M. A., McIreoy, K., Fawcett, S. B., Kegler, M. C., Parker, E., Rathgerb Smith, S., Sterling, T. D., & Wallerstein, N. (1998). *Health Education & Behavior, 25*(3), 258–278.

Goromaru, H., Kokogawa, T., Ueda, Y., & Fukaya, S. (2021). Study of new normal business continuity to improve resilience against uncertain threat. *Journal of Disaster Research, 16*(1), 31–39.

Gurtner, Y., & King, D. (2021). Socio-economic vulnerabilities to natural disasters and social justice. In: Chaiechi, T. (Ed.), *Economic effects of natural disasters: Theoretical foundations, methods, and tools*. Academic Press, an imprint of Elsevier, London, UK, 493–509.

Hirschberger, G. (2018). Collective trauma and the social construction of meaning. *Frontiers in Psychology, 9*(1441), 1–14.

Ingham, V., & Redshaw, S. (2017). Connecting community organisations for disaster preparedness. *International Journal of Safety and Security Engineering, 7*(1), 52–64. https://doi.org/10.2495/SAFE-V7-N1-52-64

Jafari, H., Heidari, M., Heidari, S., & Sayfouri, N. (2020). Risk Factors for suicidal behaviours after natural disasters: A systematic review. *The Malaysian Journal of Medical Sciences, 27*(3), 20–33.

Kapucu, N., Hawkins, C. V., & Rivera, F. I. (2014). Disaster preparedness and resilience for rural communities. *Risk, Hazards & Crisis in Public Policy, 4*(4), 215–233.

Kristensen, P. Weisæth, L., & Heir, T. (2009). Psychiatric disorders among disaster bereaved: An interview study of individuals directly or not directly exposed to the 2004 tsunami. *Depression and Anxiety*, 1127–1133.

Kulig, J., Edge, D., & Joyce, B. (2008). Understanding community resiliency in rural communities through multimethod research. *Journal of Rural and Community Development, 3*(3), 77–94.

Lassa, J.A. 2015. Post disaster governance, complexity and network theory: Evidence from Aceh, Indonesia after the Indian Ocean Tsunami 2004. *PLoS Currents, 7*, ecurrents, https://doi.org/10.1371/4f7972ecec1b6

Magis, K. (2010). Community resilience: An indicator of social sustainability. *Society & Natural Resources, 23*(5), 401–416.

McCaughey, J., Daly, P., Mundir, I., Mahdl, S., & Patt, A. (2018). Socio-economic consequences of post-disaster reconstruction in hazard-exposed areas. *Nature Sustainability, 1*, 38–43.

McFarlane, A. C., & Norris, F. H. (2006). Definitions and concepts in disaster research. In F. H. Norris, S. Galea, M. J. Friedman, & P. J. Watson (Eds.), *Methods for disaster mental health research* (pp. 3–19). Guilford Press.

Naser, A. Y., Al-Hadithi, H. T., Dahmash, E. Z., Alwafi, H., Alwan, S. S., & Abdullah, Z. A. (2020). The effect of the 2019 coronavirus disease outbreak on social relationships: A cross-sectional study in Jordan. *International Journal of Social Psychiatry*. https://doi.org/10.1177/0020764020966631

Norris, F. H., Stevens, S. P., Pfefferbaum, B., Wyche, K. F., & Pfefferbaum, R. L. (2008). Community resilience as a metaphor, theory, set of capacities, and strategy for disaster readiness. *American Journal of Community Psychology, 41*(1–2), 127–150.

Omidvar, B. H., Zafari, H., & Khakour, M. (2011). Evaluation of public participation in reconstruction of Bam, Iran, after the 2003 earthquake. *Natural Hazards, 59*(3), 1397–1412.

Park, C. L. (2013). Trauma and meaning making: Converging conceptualizations and emerging evidence. In J. A. Hicks & C. Routledge (Eds.), *The experience of meaning in life* (pp. 61–76). Springer.

Parkinson, D., & Zara, C. (2013). The hidden disaster: Domestic violence in the aftermath of natural disaster. *Australian Journal of Emergency Management, 28*(2), 28–35.

Parnell, J., & Crendall, W. (2017). The contribution of behavioral economics to crisis management decision-making. *Journal of Management and Organization, 26*, 585–600.

Paton, D. (2006). Disaster resilience: Building capacity to co-exist with natural hazards and their consequences. In D. Paton & D. Johnson (Eds.), *Disaster Resilience: An integrated approach, Charles C* (pp. 3–10). Thomas.

Pearlman, L. A., & Saakvitne, K. W. (1995). Treating therapists with vicarious traumatization and secondary traumatic stress disorders. In: C. R. Figley (Ed.), *Brunner/Mazel psychological stress series, No. 23. Compassion fatigue: Coping with secondary traumatic stress disorder in those who treat the traumatized* (pp. 150–177). Brunner/Mazel.

Pérez-García, E., Ortega-Galán, Á. M., Ibáñez-Masero, O., Ramos-Pichardo, J. D., Fernández-Leyva, A., & Ruiz-Fernández, M. D. (2021). Qualitative study on the causes and consequences of compassion fatigue from the perspective of nurses. *International Journal of Mental Health Nursing, 30*(2), 469–478.

Qin, X., & Jiang, Y. (2011). The impact of natural disaster on absenteeism, job satisfaction, and job performance of survival employees: An empirical study of the survivors in Wenchuan Earthquake. *Frontiers of Business Research in China, 5*, 219–242.

Quarantelli, E. L. (Ed.). (1998). *What is a disaster?* (pp. 31–55). A dozen perspectives on the question.

Resnik, K. S., & Fins, J. J. (2021). *Professionalism and resilience after COVID-19* (pp. 1–5). Springer.

Resurrección, B. P., Goodrich, C. G., Song, Y., Bastola, A., Prakash, A., Joshi, D., Liebrand, J. & Shah, S. A. (2019). In the shadows of the Himalayan mountains: Persistent gender and social exclusion in development. In: Wester, P., Mishra, A., Mukerji, A., Shrestha, A. B. (Eds.) *The Hindu Kush Himalaya assessment: Mountains, climate change, sustainability and people* (pp. 491–516). Springer Nature.

Rezaeian, M. (2013). The association between natural disasters and violence: A systematic review of the literature and a call for more epidemiological studies. *Journal of Research in Medical Sciences, 18*(12), 1103–1107.

Rouhanizadeh, B., & Kermanshachi, S. (2021). Barriers to an effective post-recovery process: A comparative analysis of the Public's and experts' perspectives. *International Journal of Disaster Risk Reduction, 57*(102181), 1–12.

Rus, K., Kilar, V., & Koren, D. (2018). Resilience assessment of complex urban systems to natural disasters: A new literature review assessment of complex urban systems to natural disasters. *International Journal of Disaster Risk Reduction, 31*, 311–330.

Sakic Trogrlic, R., Duncan, M., Wright, G., van den Homberg, M., Adeloye, A., Mwale, F., & McQuistan, C. (2021). External stakeholders' attitudes towards and engagement with local knowledge in disaster risk reduction: Are we only paying lip service? *International Journal of Disaster Risk Reduction, 58*, 102196.

Samson, A. (2014). An introduction to behavioral economics. In: A. Samson (Ed.), *Behavioral economics guide*. Retrieved 28 April 2021 from https://www.behavioraleconomics.com/be-guide/the-behavioral-economics-guide-2014/

Shmueli, D. F., Ozawa, C. P. & Kaufman, S. (2020). Collaborative planning principles for disaster preparedness. *International Journal of Disaster Risk Reduction, 52*, https://doi.org/10.1016/j.ijdrr.2020.101981

Simon, H. A. (1956). Rational choice and the structure of the environment. *Psychological Review, 63*(2), 129–138.

Sudmeier-Rieux, K. I. (2014). Disaster resilience—An emerging paradigm of danger or of hope? *Disaster Prevention and Management, 23*(1), 67–80.

Sunstein, C. R., & Thaler, R. H. (2008). Privatizing marriage. *The Monist, 91*(3/4), 377–387.

Thurston, A. M., Stöckl, H., & Ranganathan, M. (2021). Natural hazards, disasters and violence against women and girls: A global mixed-methods systematic review. https://doi.org/10.1136/bmjgh-2020-004377

Wadi, M. M., Nordin, N. I., Roslan, N. S., Celina, T., & Yusoff, M. S. B. (2020). Reframing resilience concept: Insights from a meta-synthesis of 21 resilience scales. *Education in Medicine Journal, 12*(2), 3–22.

CHAPTER 18

Enabling a Collaborative Research Environment to Meet Complexities of Compound, Cascading, and Protracted Disasters

Alistair Dawson, Allison Rifai, Claudia Baldwin, and Matthew Dyer

Introduction

Inherent in effective disaster management is multi-level and multi-stakeholder cooperation and collaboration. Compound cascading and protracted disasters are not only a challenge to predict and prepare for but add complexity to response and recovery… and research. Integration

A. Dawson · A. Rifai (✉)
Office of the Inspector-General of Emergency Management, Brisbane, QLD, Australia
e-mail: Allison.Rifai@igem.qld.gov.au

A. Dawson
e-mail: Alistair.Dawson@igem.qld.gov.au

A. Lukasiewicz and T. O'Donnell (eds.), *Complex Disasters*, Disaster Risk, Resilience, Reconstruction and Recovery,
https://doi.org/10.1007/978-981-19-2428-6_18

of local knowledge with additional scientific and technical knowledge can improve disaster risk reduction and climate change adaptation (IPCC, 2012). Enabling evidence-based research on this topic requires insight and careful planning, as evidence needs to be based on measurable indicators of cause and effect where possible. One of the things that research institutions can contribute is the development of meaningful indicators, monitoring, and independent evaluation.

This chapter shares case studies of existing disaster management research programs with a view to identifying success factors for industry-government-university collaboration. The chapter identifies implications for research on the management of complex disasters. We draw on Rybnicek and Konigsgruber's (2019) conceptual model outlining institutional, relationship, output, and framework factors which influence the success of industry-university collaborations. Through case studies of local and state entities, industry sectors, and university collaborations, this chapter examines the indicators enabling a collaborative research environment. Whilst we document evidence from Queensland case studies, broader applications emerge.

Why Partnerships and Collaboration Are Important to Disaster and Emergency Management

Australia is regularly impacted by complex events that cause serious disruption. Preparing for, responding to, and recovering from such events is beyond the ability of any single agency. Disaster management demands local level capability supported through entities that have expertise, experience, and or authority to assist with a comprehensive, all-hazards all agencies approach to disaster management.

C. Baldwin
Sustainability Research Centre, University of the Sunshine Coast, Maroochydore, QLD, Australia
e-mail: CBaldwin@usc.edu.au

M. Dyer
Queensland Fire and Emergency Services, Brisbane, QLD, Australia
e-mail: matthew.dyer@qfes.qld.gov.au

Australia has adopted a comprehensive and integrated approach for the effective management of emergencies and disasters. Each Australian state and territory have legislation that establishes how disasters will be managed (Elphick, 2020). Whilst the term 'emergency' is used interchangeably with 'disaster' at the Australian Government level and in several jurisdictions, legislative arrangements ultimately recognise that what is of most concern is the impact on community, economy, and environment (State of Queensland, 2021).

Queensland is arguably Australia's most disaster-prone state (Commonwealth of Australia, 2014) and perhaps not surprisingly, Queensland's legislation explicitly adopts the term 'disaster', differentiating a disaster from an incident or emergency (State of Queensland, 2019). This is consistent with research that argues disasters are more complex than emergencies and incidents. Disasters require extraordinary, coordinated decision making across many organisations and sectors (Auf der Heide, 1989; Etkin & Davis, 2007). The complexity of disaster events is exacerbated by changes in climate and vulnerability (IPCC, 2012). The sector has come to anticipate cascading events and now also finds itself preparing for repeating unprecedented events.

Collaboration in Queensland is enabled through Queensland's *Disaster Management Act* (2003) (State of Queensland, 2020), which "*...provides a framework in which all levels of government, government owned corporations, non-government organisations, partners and stakeholders can work collaboratively to ensure disaster management across the state*" (State of Queensland, 2021: 12). Indeed, Queensland's disaster management arrangements require collaboration to work optimally and speak to a sharing of responsibility to enhance community resilience. Disaster management stakeholders must consider how they can best collaborate, consult, and communicate (QFES, 2018). Whilst this is a requirement under Queensland's arrangements, it runs deeper than just compulsion by virtue of the fact that local people, with local knowledge and relationships—formal and informal—are front-and-centre when it comes to disaster prevention, preparedness, response, and recovery.

The sharing of data, information, and knowledge has been identified through disaster management reviews as important to creating effective disaster management (State of Queensland [Inspector-General of Emergency Management (IGEM)], 2017, 2019a, 2019b). Cross-sectoral collaboration is necessary to avoid a lack of consistency in stakeholders' ability to create and use data and avoid disconnected efforts to make improvements in specific areas. Striking a balance between mitigating risks and realising the benefits of data sharing is grounded upon collaboration between data owners and end users.

Collaboration between the disaster management and research sector has long been an important means of effective disaster planning and improving disaster prevention, preparedness, response, and recovery (Quarantelli, 1988). The role of the research sector in disaster management remains vital and has been recognised as such through the inclusion of Disaster Management Research as a discrete chapter within the Queensland State Disaster Management Plan (State of Queensland, 2018). The complex nature of disaster management and the multidisciplinary nature of the disaster management sector—with practitioners from diverse professional backgrounds—sees a broadening opportunity for collaboration amongst sectors to position the sector to meet the challenges of the future. Collaboration and partnerships within and across sectors also create synergies and value for money propositions that are otherwise unattainable.

What Makes University-Industry Research Partnerships and Collaborations Work?

The value of research to the emergency and disaster management sectors has been well established (Frykmer et al., 2021; Owen, 2018; Owen et al., 2020; Pardee et al., 2018; Taba et al., 2020). Collaborations amongst practitioners and academic researchers from a range of disciplines can help address complex problems created from the intersection of social, economic, environmental, political, and hazardous environments (Gilligan, 2019; Hall et al., 2018; Leroy et al., 2017; Stibb et al., 2018; Sapat, 2018; Sherman-Morris et al., 2021).

Reduction and management of complex disasters depend on our improved understanding of causal connections and mechanisms where one event precipitates another. Collaborative approaches mean that a range of perspectives can provide insight into and data about the causal connections. For example, an extreme event that affects electricity will affect the water supply, and lead to possible contamination and health impacts. Disruption of telecommunication or transportation networks may lead to an untimely evacuation response. Such situations put additional pressure on the full range of emergency response agencies. Likewise continuing to place sewage treatment facilities in flood-prone areas increases the risk of cascading disaster, causing contamination of water supplies and the environment. A system-wide approach is needed. Hence the need for intersectoral factors to be identified.

Generating knowledge and research is one part of the puzzle, and translating research into practice is another. Around these two elements are the relationships, networks, and collaborations which help to make both knowledge creation and implementation happen. *The fundamental core of good partnerships is their ability to bring together diverse resources in ways that can together achieve more: more impact, greater sustainability, increased value to all* (Stibbe et al., 2018: 6). Collaborations can range from being transactional in scope to transformational partnerships generating co-knowledge and system change (Stibb et al., 2018; The Partnering Initiative, 2021). Both literature and practice illustrate that collaborations between researchers and industry can be done well or poorly (Butcher et al., 2019), and despite the value placed on collaborations, they can be difficult to define and evaluate (Australian Research Alliance for Children and Youth [ARACY], 2013). A review of the literature identifies a range of factors that can either help and/or hinder collaboration between academic researchers and industry/government sectors (Ankrah & AL-Tabbaa, 2015; Antonella Maiello et al., 2013; ARACY, 2013; Barnes et al., 2002; Fothergill, 2000; Hall et al., 2018; Ishiwatari, 2019; Leroy et al., 2017; Sapat, 2018).

Rybnicek and Konigsgruber (2019) developed a conceptual model to organise and categorise the factors which they found influenced the success of industry/university collaborations:

(a) Institutional Factors—including resource availability, structure of entity/university and scale/size of entity, willingness to change/ flexibility throughout the collaboration
(b) Relationship Factors—including communications, commitment, trust, culture, expectations, conflicts, team experience, range of disciplines involved, partner selection
(c) Output Factors—including clarity around objectives, goal compatibility, systems, and processes for knowledge and technology transfer
(d) Framework Factors—including governing environment and processes, contracts, intellectual property rights, geography (proximity to partners).

The model also identifies a series of potential moderators which can further impact the factors—scale, discipline, level, and phase. The factors and moderators combine to illustrate what can help to make collaborations between universities and industries work.

Queensland Collaboration Case Studies—Applying the Conceptual Model

The Rybnicek and Konigsgruber model was used as a framework to identify the factors which influenced five examples of disaster/emergency management and university collaborations in Queensland. The collaborations themselves emerged directly from experiences and learnings from extraordinary disaster events, planning insights following exercises, or through strategic interventions to reduce climate and disaster risk. Increasingly there is a requirement to deliver outcomes that are tangible and demonstrable to community and stakeholders across the disaster management sector: research projects have to 'mean something' and create value for the people that live in the communities within which research is situated. Also, government and funding bodies have concerns that public value is being achieved, never more so than now as economies recover from the pandemic.

As alluded to earlier, Queensland is well-familiar with significant natural disasters. The impetus for the following case studies include:

- A need to recognise and enhance the bringing-together of expertise, experience, and technical acumen to solve complex problems.
- A better-evidenced understanding of the impacts of climate change on the environment and heat (State of Queensland, 2019c).
- Hard-won learnings about decision making in disasters, as a consequence of Ex Tropical Cyclone Oswald in 2013, a compounding event that included torrential rain, five tornados within one local government area alone that caused damage to hundreds of homes and depleted local disaster response capacity, a storm tide that exceeded highest astronomical tide and impacted coastal catchments and a flood of record through a major town, causing record evacuations, long-term isolation and displacement of whole suburbs and massive economic cost (State of Queensland, 2013).
- Identification of planning gaps in regard to resupply operations.

- Adaptation planning that takes account of potential compounding, cascading and protracted events can reduce the risk which would otherwise lead to more severe outcomes.

The Queensland Disaster Management Research Framework

Acknowledging the importance of facilitating partnerships and promoting the use of research in decision making, the Office of the Inspector-General of Emergency Management Queensland has taken on a knowledge broker role connecting policymakers, practitioners, and researchers in sector-wide collaborative disaster management research. The Queensland Disaster Management Research Framework (the Framework) supports collaboration between the tertiary and disaster management sectors and promotes the use of research by disaster managers. The purpose of the Framework is to:

a. Develop a coordinated approach to undertaking, managing, and sharing research
b. Support the development of strategic research priorities for the disaster management sector
c. Promote engagement between government and the tertiary sector
d. Promote transitions between research and practice.

The Framework focuses on the development of relationships between researchers and decision makers/industry sectors. It adopts a peer driven, collaborative engagement approach, enabled by IGEM through Section 16C(g) and (i) of the Act. The key features of the Framework are knowledge brokering; the Research Advisory Panel; a researcher register; mapping and gap analysis; helping translate research into practice.

Heat Adaptation Planning—Green Infrastructure for Mitigating Heat Stress in Aged Care Facilities

This project addresses the critical issue of the role of urban greenery in reducing heat stress for seniors given increasingly longer and hotter heat waves into the future. Unique aspects of the project involve installing heat/humidity sensors and conducting multi-functional heat and universal design audits at case study sites to determine priority areas for intervention. Residents and staff were consulted about how they use and would like to use the facility's outdoor environment and to provide input into the cost-effective design options for reducing heat. The outcome is a model heat adaptation plan for Tropical and Sub-tropical aged care facilities. Benefits of the partnership meant that researchers learned of the need to retain warm areas for winter use, as older people also feel the cold. Making such centres more heat proof will reduce the need to move older people to air-conditioned or congested public centres putting them at higher risk of illness.

In the Heat of the Moment: Understanding the Impact of Emotional Preparedness on Decision Making During Natural Disasters

The 'In the Heat of the Moment' project explored the notion of emotional preparedness. The project drew on the findings of earlier bushfire research (Every et al., 2015) which highlighted that preparation efforts to date had primarily focused on physical preparation. Whilst many residents felt able to physically prepare for events, emotional responses during the event undermined best-laid plans. Data were gathered from residents, employees of State and Local Government agencies, and community support service providers in two local government areas recently impacted by a flooding event and explored what psychological and emotional preparedness meant to them, what prevents people from engaging with this type of risk and how best to communicate it. Importantly, this research has changed the practices of partner organisations. Multi-modal approaches, which draw on the findings and insights from the research have been adopted, and more targeted resources used to support the local community (Every et al., 2018).

Enabling Action for Local Food Resilience and Contingency in Cairns

Food is a connector to multiple resiliencies including positive outcomes for health, wellbeing, and economic stability. Disaster agencies know that if communities share responsibility for local food resilience, they recover more effectively from the impacts of disasters, from extreme weather events through to pandemics. However, in Australia, our current Local Disaster Management Plans (LDMPs) do not yet include food arrangements to deal within the PPRR cycle of disaster 'Prevention, Preparedness, Response and Recovery'. Our research team is responding to this gap by investigating ways to address short-term 'acute' food supply shortage and long-term 'chronic' food disadvantage.

Through enabling formalised food resilience arrangements within the Cairns LDMP the team are demonstrating how local food procurement procedures can inform emergency management operations and empower communities to share responsibility for their own food resilience (Reis et al., 2021). The team is also enabling an online Local Food Resilience and Contingency Hub, as a living 'source of truth' to inform local food resilience and contingency arrangements, and to encourage shorter food supply chain development alongside a supermarket and other conventional food systems.

Sector Adaptation Plans—Development and Implementation

Under the Queensland Climate Adaptation Strategy, the Department of Environment and Science (DES) supported the development of Sector Adaptation Plans (SAPs) for seven important sectors and systems on which we depend. The development of each SAP was led by sector stakeholders to ensure they reflected the needs and adaptation priorities of each sector or system. SAPs were developed between 2017 and 2020 for the following sectors: built environment and infrastructure, agriculture, tourism, human health and wellbeing, emergency management, biodiversity and ecosystems, and small and medium enterprises. Completed SAPs were made publicly available on the Government website and actively promoted by stakeholders through their networks and media. SAP development was guided by a SAP Development Model designed by the University of Queensland; this model is grounded in collaboration. SAP developers were encouraged to, and actively sought, wide representation of other relevant stakeholders, including researchers, through extensive consultation. Once all SAPs were developed, DES developed the SAP+ co-investment program to support the implementation of priority measures identified in the SAPs. This program adopted a collaborative design process to ensure proposed projects addressed multiple objectives, including the consideration of complex cross-sectoral issues. The program is currently funding seven collaborative adaptation projects.

The Emergency Management Sector Adaptation Plan for climate change (EM-SAP) was launched in September 2018 under the broader Queensland Climate Change Response which outlines the commitments and actions that will be taken by the Queensland Government to transition to a low carbon, clean growth economy and to adapt to the impacts of a changing climate. The EM-SAP, developed by QFES in partnership with the Department of Environment and Science (DES), and the National Climate Change Adaptation Research Facility (NCCARF, Griffith University) seeks to foster 'An adaptive emergency management sector that is fully engaged with the risks and opportunities of a changing climate, building resilience together with the communities of Queensland'. The plan identifies eight sector-specific priorities and 32 actions, existing climate adaptation activities, and current knowledge gaps and barriers to adaptation for the emergency management sector. It was subsequently adopted by the State Disaster Coordination Group.

Findings

The case studies indicated that the four groups of factors identified in the model have both positive and negative impacts on disaster/emergency management and university research collaborations. Relational factors were the most significant, followed by institutional, framework, and output factors. The addition of COVID-19 also impacted the influence of relationship factors, highlighting the importance of in-person contact when establishing teams.

Relationship Factors

Open and regular **communications** played a key role in helping to build collaboration across the five case studies. Whilst in-person communication is always preferable, COVID-19 limited the ability to travel and interact face-to-face. Webinars and teleconferences replaced direct contact, with the technology also generating firewall challenges with different participants using different platforms. Two case studies reported that work arounds were eventually developed, but this took additional project time. The opportunities available to use online forms of communication were not always recognised early and were not always utilised as broadly as they could be to enable broader stakeholder engagement.

Team experience was important, with case studies reporting a mix of research and practical/sector experience helping address the complex problems. Importantly, a willingness to share experience across the team and to learn from other members of the collaboration helped develop and sustain relationships. One case study noted that engaging with university researchers enabled the agencies to utilise a broader range of expertise and disciplines than they would have exposure to individually.

Trust, whilst taking time to build, was a core factor in the success of the collaborations. Trust enabled sharing of information, expansion of networks, and created learning opportunities amongst teams. Trust was developed through open and regular communication, the timely delivery of objectives, and conflict resolution. The case studies noted the importance of 'incidental and casual communications' as a cornerstone of building team trust and dynamics, as well as being inclusive. The researchers noted that having a range of experience on the team meant they could choose the best person to communicate with for different purposes.

Entity **culture** and willingness to learn and explore new ideas helped open spaces for collaborations. Another case study reported that the **commitment** to better understanding the topic and implementing practice change provided a 'united course' for the collaboration to focus on. Where research was not a core function of the entity, the desire to build capability in this space helped with engagement. Being able to see the 'bigger picture' within which the problem being addressed sits, opened the door for collaboration outside of traditional networks and boundaries. Ensuring that partners were able to commit to the collaboration, be that through time, resources, or financial considerations, was an ongoing factor. In some cases, initial commitments around project involvement had to be modified, which impacted the timelines of the data collection stages. Case study participants reported ongoing commitment to the project/program implementation with the local entities, beyond the end of the contracted collaboration. The importance of showcasing good practice and outcomes helps build a collaborative culture, with one case study reporting that it assisted with new co-investment in additional stages of the program. Collaboration did not always automatically lead to co-investment, with case study participants noting challenges securing funding to support projects despite goodwill and a willingness to contribute.

Partner Selection was important across the case studies. The importance of pre-existing relationships and agreements along with proven track records in subject-matter areas and research were noted throughout the case studies. Case study participants reported the importance of networks to partner selection, particularly for leadership roles.

The case studies reported examples of project **leadership** and the importance of good leadership in helping develop and drive collaborations. By creating the environments in which teams and relationships could develop, collaboration leaders enabled representatives from entities to be fully engaged in the projects. The importance of 'common ground' for collaboration participants was noted.

Institutional Factors

Resources are an important factor in building and sustaining collaborations. The case studies reported that most partners contributed financially to the project/program, but they also noted the importance of in-kind contributions. The flexibility to recognise in-kind contributions enables a broader range of participants to be part of a collaboration through the provision of data, access to networks, venue support and catering, and focus group coordination. One case study participant did note that not all governance systems and grant requirements recognise the value of in-kind contributions. Underestimating the resources required for travel and engagement can also negatively impact collaboration.

The case study participants reported both positive and negative experiences with **processes**. Identifying a formal representative from each partner entity helped navigate the broad range of different processes involved in establishing and maintaining formal collaborations. One participant noted that there were variable lead times for different internal entity processes and functions which needed to be accommodated during project planning and implementation. This was a notable point of difference between university, government, and not-for-profit agency processes. Partners with clear delegations of authority also helped navigate through formal processes. Clearly established processes for project management, contracts, and procurement, can also help work through any process challenges.

The case studies identified the important role postgraduate and student placements can play in helping develop and sustain collaborations. Processes which enable students to be part of projects on-the-ground, where they are supported by both university and workplace supervision, and organisational protocols such as workplace health and safety and ethics clearance help build relationships as well as providing powerful learning opportunities for the students.

Grounding program models and development in collaboration principles and practices was an important **structural** element identified by the case study participants, along with clearly established roles and responsibilities.

Flexibility and **willingness to change** are important across the life of collaborations. One participant noted that, with changes to timelines, the way stakeholders could be engaged and learnings from initial project stages, the ability to be flexible with the scope and direction of the collaboration is important.

Framework Factors

Case study participants identified the **governance** of information technology systems as a risk to building and sustaining collaborations. One participant noted that firewalls can limit the availability of platforms necessary for online collaboration, causing project delays, additional work, and the development of new processes and systems to compensate. Workarounds were put in place, but these resulted in a lesser experience than would otherwise have been possible (for example, asynchronous rather than synchronous communication). The benefits far outweighed negatives, however, that resulted in time delays and extra work to problem solve and introduce alternate processes and systems. Participants also noted that, where possible, allowing the local disaster management group to provide governance support for the program/project, helped facilitate agency involvement and limited the range of internal processes each participating entity had to work through.

It was important that each participant contributed to the development of **contracts** and grant applications to support the collaboration. The university **environment** provided access to ethics committees and processes, something many entities do not have access to when researching independently. Existing processes and arrangements for data confidentiality also helped build trust in the collaboration. Standard contract processes can help in generating a shared understanding of expectations and requirements for collaborators.

Geography/locational proximity was important when building collaborations, with travel impacted by COVID-19 restrictions. In some cases, this resulted in lower costs due to less physical travel whilst in other cases, costs increased due to the need to attend smaller more distributed meetings. Technology became more important in helping to overcome geographical divides. Where it was possible, the collaborations benefited from embedding researchers into the local communities and entities. Case study participants noted the significance of having local researchers as part of the collaboration, to help overcome any obstacles with community engagement and context, as well as data collection. One case study respondent noted the ***role of regional universities brokering, and leading research of this nature should not be understated.***

Intellectual property can often be a barrier to collaboration, however, one case study participant noted that when the core desire is to 'partner' not just 'research on partners', all members of the collaboration can be equal. This facilitates product ownership and authorship of outputs.

Output Factors

Objectives need to be clearly articulated from the initiation of any collaboration. Case study participants noted the importance of mutually developed and agreed upon objectives as a key influence on the outputs from collaborations.

Knowledge transfer was a key output factor. Participants identified the importance of providing an end product that is of value to all collaboration partners and the need to ensure this is a focus from the initiation of the project/program. The importance of workshops, regular briefings, and summaries at milestone points helps with knowledge transfer. The inclusion of knowledge transfer from the start of the program/project helps to facilitate discussion and collaboration.

Discussion

Compound, cascading and protracted disasters are a part of the future of disaster management. The research sector also sees the challenges future disasters create and has capability to enhance problem solving and decision making. The range of disciplines working in adaptation planning can help work towards reducing the risks associated with compounding and cascading disasters. These sectors have a history of collaborating to achieve improvements and solve difficult problems, but the pathways to this collaboration have remained elusive. This chapter provided five case studies from Queensland that demonstrated effective collaboration and discussed the factors enabling a collaborative research environment.

The case studies confirmed the importance of the factors identified in Rybnicek and Konigsgruber's model and resulted in slight adaptation to reflect Queensland findings. The case studies illustrated that the factors have both positive and negative impacts on building collaborations between academic researchers and the disaster management sector. Relationship factors were the most significant of the four groups, with communications crucial. Within the Framework Factors, governance played a key role in enabling and sustaining collaborations, whilst resources were a critical Institutional Factor. The importance of objectives was reinforced when considering Output Factors for collaborations (Fig. 18.1).

Conclusion

Across Australia, collaboration across sectors including researchers is essential to a comprehensive integrated approach to disaster management. A sound research basis is needed for evidence-based planning. When it's at its best, it brings together a wide array of entities to solve potentially wicked problems. Inter-disciplinary, inter-agency, and inter-sector collaboration is a basis for addressing incompatibilities in culture, lack of trust, lack of awareness, role confusion, and indeed competition (Noran, 2014). In examining four case studies from Queensland using Rybnicek and Konigsgruber's model, relationship and institutional factors are identified as important in the creation of collaborations to address challenges faced by the disaster management sector. Collaboration between the research community and the disaster management sector is vital to achieving disaster resilience outcomes as existing challenges become exacerbated by a changing climate. Additional research is needed to help identify the value of multi-sector experience inputs into our ability to understand and manage the risks from cascading and compound disasters.

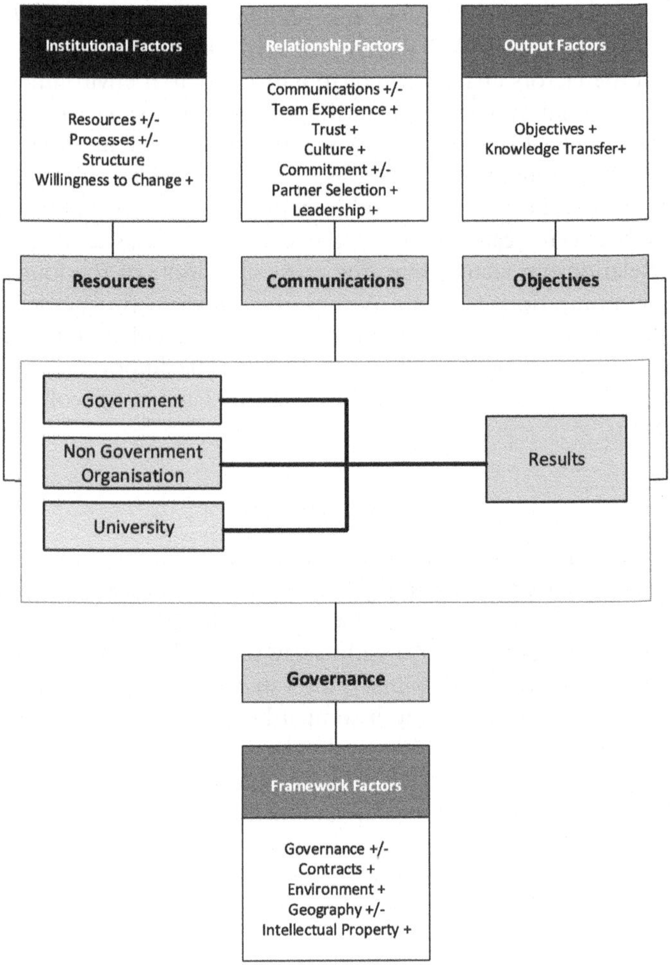

Fig. 18.1 Queensland adaptation of Rybnicek and Konigsgruber's model (*Source* Authors' construct)

References

Ankrah, S., and AL-Tabbaa, O. (2015). Universities–industry collaboration: A systematic review. *Scandinavian Journal of Management*, *31*(3), 387–408.

Antonella Maiello, A., Viegas, C. V., Frey, M., & Ribeiro, J. L. D. (2013). Public managers as catalysts of knowledge co-production? Investigating knowledge dynamics in local environmental policy. *Environmental Science and Policy*, *27*, 141–150, ISSN 1462–9011, https://doi.org/10.1016/j.envsci.2012.12.007

Auf der Heide, Erik. (1989). *Disaster response: Principles of preparation and coordination*. Erik Auf der Heide [online book version]. Retrieved from http://sheltercentre.org/sites/default/files/CVMosby_DisasterResponsePrincipl%20es.pdf

Australian Research Alliance for Children and Youth. (2013). *Key elements of collaboration—Fact sheet 5*. Australian Research Alliance for Children and Youth. Retrieved from https://www.aracy.org.au/publications-resources/area?command=record&id=25

Barnes, T., Pashby, I., & Gibbons, A. (2002). Effective university–industry interaction: A multi-case evaluation of collaborative RandD projects. *European Management Journal*, *20*(3), 272–285.

Butcher, J. R., Gilchrist, D. J., Phillimore, J., & Wanna, J. (2019). Attributes of effective collaboration: Insights from five case studies in Australia and New Zealand. *Policy Design and Practice*, *2*(1), 75–89.

Commonwealth of Australia. (2014). Natural disaster funding arrangements: Productivity commission inquiry report, Canberra, Australia: Author. ISBN 978-1-74037-523-8 (Volume 1). Retrieved from https://www.pc.gov.au/inquiries/completed/disaster-funding/report/disaster-funding-volume1.pdf

Commonwealth of Australia. (2020). *State of the climate 2020*, Canberra, Australia: Author. ISBN Print 978-1-4863-1508-6, Online 978-1-4863-1509-3. Retrieved from https://www.csiro.au/en/research/environmental-impacts/climate-change/State-of-the-Climate

Doyle, E. E. H., Becker, J. S., Neely, D. P., Johnston, D. M., & Pepperell, B. (2015). Knowledge transfer between communities, practitioners, and researchers: A case study for community resilience in Wellington, New Zealand. *Australasian Journal of Disaster and Trauma Studies*, *19*(2), 55–66.

Elphick, K. (2020). *National emergency and disaster response arrangements in Australia: a quick guide*. Commonwealth of Australia. Retrieved from http://www.aph.gov.au/About_Parliament/Parliamentery_Departments/Parliamentary_Library/pubs/rp/rp1920/Quick_Guides/NationalEmergencyDisasterManagement

Etkin, D., & Davis, I. (2007). *The search for principles of disaster management. Working paper* (draft). Unpublished manuscript. Retrieved from: https://view.officeapps.live.com/op/view.aspx?src=https%3A%2F%2Ftraining.fema.gov%2Fhiedu%2Fdocs%2Femprinciples%2Fthe%2520search%2520for%2520principles%2520of%2520disaster%2520management.doc

Every, D., Reynolds, A., Clarkson, L., Bearman, C., Matthews, R., Haigh, L., & Dawson, D. (2015). Capturing community experiences in the 2015 Sampson Flat fire, report for the South Australia Country Fire Service, Bushfire and Natural Hazards CRC, Melbourne.

Every, D., Reynolds, A., Keen-Dyer, H., Dyer, M., Miller-Rose, P., & Trigg, J. (2018). That psychological bulls**t': Surprising findings from community research are improving preparedness communication. *Australia Journal of Emergency Management, 33*(4), 57–61.

Fothergill, A. (2000). Knowledge transfer between researchers and practitioners. *Natural Hazards Review, 1*(2), 91–98.

Frykmer, T., Tehler, H., Uhr, C., & Wester, M. (2021). Advancing the field of disaster response management: Toward a design science approach. *International Journal of Disaster Risk Science, 12*, 220–231.

Gilligan, J. M. (2019). Expertise across disciplines: Establishing common ground in interdisciplinary disaster research teams. *Risk Analysis*. Retrieved from https://onlinelibrary.wiley.com/doi/epdf/10.1111/risa.13407

Hall, K. L., Vogel, A. L., Huang, G. C., Serrano, K. J., Rice, E. L., Tsakraklides, S. P., & Fiore, S. M. (2018). The science of team science: A review of the empirical evidence and research gaps on collaboration in science. *American Psychologist, 73*(4), 532–548.

IPCC. (2012). Field, C. B., V. Barros, T. F. Stocker, D. Qin, D. J. Dokken, K. L. Ebi, M. D. Mastrandrea, K. J. Mach, G.-K. Plattner, S. K. Allen, M. Tignor, and P. M. Midgley (Eds.), *Managing the Risks of Extreme Events and Disasters to Advance Climate Change and Adaptation*. Cambridge University Press.

Isiwatari, M. (2019). Flood risk governance: Establishing collaborative mechanism for integrated approach. *Progress in Disaster Science, 2* (July): 100014, ISSN 2590-0617, https://doi.org/10.1016/j.pdisas.2019.100014

Leroy, L., Rittner, J. L., Johnson, K. E., Gerteis, J., & Miller, T. (2017). Facilitative components of collaborative learning: A review of nine health research networks. *Health Policy, 12*(3), 19–33.

Noran, O. (2014). Collaborative disaster management: An interdisciplinary approach. *Computers in Industry, 65*(6), 1032–1040.

Owen, C. (2018). How emergency service organisations can—and Do—utilise research. *Australian Journal of Emergency Management, 33*(2), 28–33.

Owen, C., Krusel, N., & Bethune, L. (2020). Closing the research-practice gap in emergency service organisations. *Australasian Journal of Disaster and*

Trauma Studies, 24(2), 103–115. http://trauma.massey.ac.nz/issues/2020-2/AJDTS_24_2_full.pdf

Pardee, J. W., Fothergill, A., Weber, L., & Peek, L. (2018). The collective method: Collaborative social science research and scholarly accountability. *Qualitative Research, 18*(6), 671–688.

Quarantelli, E. L. (1988). Assessing disaster preparedness planning. *Regional Development Dialogue, 9*, 48–69.

Queensland Fire and Emergency Services. (2018). *Queensland prevention, preparedness, response and recovery disaster management guideline*. Retrieved from https://www.disaster.qld.gov.au/dmg/Documents/QLD-Disaster-Management-Guideline.pdf

Reis, K., Desha, C., Bailey, M., Liddy, P., & Campbell, S. (2021). Towards local food resilience. Key considerations for building local food resilience and contingency plans: A focus on the Cairns region. Final Report, 26 March. Cities Research Institute in consultation with the Cairns Regional Council.

Rybnicek, R., & Konigsgruber, R. (2019). What makes industry-university collaboration succeed? A systematic review of the literature. *Journal of Business Economics, 89*, 221–250.

Sapat, A. (2018). Lost in translation? Integrating interdisciplinary disaster research with policy praxis. *Risk Analysis. (early View)*. https://doi.org/10.1111/risa.13198

Sherman-Morris, K., Houston, J. B., & Subedi, J. (2021). Theoretical matters: On the need for hazard and disaster theory developed through interdisciplinary research and collaboration. *Risk Analysis, 41*(7), 1059–1065.

State of Queensland. (2013). *Regional economic impacts of the 2013 floods in Queensland: Regional economic impact study august 2013*. Unpublished manuscript.

State of Queensland (Inspector-General of Emergency Management). (2017). *The Cyclone Debbie review: Lessons for delivering value and confidence through trust and empowerment*. Author. Retrieved from https://www.igem.qld.gov.au/sites/default/files/2019-02/Cyclone%20Debbie%20Review%20Rpt1-17-18_PUBLIC_WEB.pdf

State of Queensland. (2018). *Queensland state disaster management plan*. Retrieved from: https://www.disaster.qld.gov.au/cdmp/Documents/Queensland-State-Disaster-Management-Plan.pdf

State of Queensland. (2019). *Queensland disaster management Lexicon*. Retrieved from: https://www.igem.qld.gov.au/sites/default/files/2020-07/2020%20Qld%20Disaster%20Management%20Lexicon%20v2.0.pdf

State of Queensland (Inspector-General of Emergency Management). (2019a). *The 2018 Queensland Bushfires Review: A climate for good neighbours*. Brisbane, Queensland. Retrieved from https://www.igem.qld.gov.au/sites/default/files/2019a-12/IGEM%20Queensland%20Bushfire%20REVIEW%202019a.pdf

State of Queensland (Inspector-General of Emergency Management). (2019b). *2019b Monsoon Trough Rainfall and Flood Review*. Brisbane, Queensland. Retrieved from https://www.igem.qld.gov.au/sites/default/files/2019b-12/IGEM%20MTRF%20Review.pdf

State of Queensland. (2019c). Queensland State Heatwave Assessment. Brisbane, Queensland. Retrieved from https://www.disaster.qld.gov.au/dmp/Documents/QFES-Heatwave-Risk-Assessement.pdf#search=heat

State of Queensland. (2020). *Disaster management act*. Retrieved from https://www.legislation.qld.gov.au/view/pdf/inforce/current/act-2003-091

State of Queensland. (2021). *Queensland disaster management arrangements participant guide*. Retrieved from https://www.disaster.qld.gov.au/dmp/Documents/Queensland-Disaster-Management-Arrangements-Participant-Guide.pdf

Stibbe, D. T., Reid, S., & Gilbert, J. (2018). *Maximising the Impact of Partnerships for the SDGs*. The Partnering Initiative and United Nations Department of Economic and Social Affairs.

Taba, S. T., Mojtahedi, M., & Newton, S. (2020). Disaster risk management approaches in construction and built environment—A research collaboration networks perspective. *International Journal of Disaster Resilience in the Built Environment, 11*(1), 85–99.

The Partnering Initiative. (2021). *The relationship spectrum*. The Partnership Initiative. Retrieved from https://thepartneringinitiative.org/the-relationship-spectrum-2

CHAPTER 19

Complex Disasters as Part of Everyday Life

Tayanah O'Donnell and Anna Lukasiewicz

Introduction

In Chapter 1, we noted at length the definitional challenges with respect to disasters. While we unpacked relevant literature there, we also posited that time is of the essence with the implication that we just need to get on with things. Climate change, rising inequities, and inequality, rapid biodiversity loss: systemic, compound, cascading, protracted issues that affect us now and will affect many future generations. Chapter 2 expands the theoretical underpinnings of compound, cascading, and protracted, and analyses these with reference to two recent, extreme, weather events occurring in Australia: fires and floods. There, with Chantra, we explore

T. O'Donnell
Fenner School of Environment and Society, Australian National University, Canberra, ACT, Australia
e-mail: tayanah.odonnell@anu.edu.au; taodonnell@deloitte.com.au

A. Lukasiewicz (✉)
Institute for Climate, Energy & Disaster Solutions, Australian National University, Canberra, Australian Capital Territory, Australia
e-mail: anna.lukasiewicz@anu.edu.au

A. Lukasiewicz and T. O'Donnell (eds.), *Complex Disasters*, Disaster Risk, Resilience, Reconstruction and Recovery,
https://doi.org/10.1007/978-981-19-2428-6_19

these frames and underscore them with particular attention to their widespread and harmful societal impacts.

Thinking about and treating disasters as complex, non-linear occurrences is critical. This book is not new in this systematic approach, building on Cutter (2018) and Pescaroli and Alexander (2018). We see throughout this book a focus on either part, or all, of our tripartite complex disaster framing (i.e., compound, cascading, protracted).

In Part I, we present a range of chapters that explore all three frames, whether implicitly or explicitly, to illustrate both the interconnectedness of these frames as well as their underpinning hazards such as climate change risks. In Chapter 3, Barnes presents an epistemological history of concepts that are now ubiquitous in disaster management—hazard, risk, and resilience. Barnes highlights an 'evolving threat landscape', necessitating a nationally coordinated multi-threat, risk management strategy, a comprehensive threat taxonomy, and a common knowledge base for terminology and use in regulatory, policy, and decision-making settings. In Chapter 4, Steffen provides a detailed analysis of climate risks faced by Australia. As noted in Chapter 1, following Simpson et al. (2021), we have used the term complex to mean 'the diversity of interactions among sectors and systems' (p. 489) including how this relates to disasters. Disasters do not occur in a vacuum, Steffen's analysis illustrates climate science that helps underpin 'the complexity of interactions among multiple drivers of climate change risk' (Simpson et al., 2021, p. 489). These multiple drivers include economic, social, cultural, political, and technological systems. Critically, these systems operate over temporal and spatial scales. Time and geography continue to be essential considerations for how we both theorise and act on complex disasters. O'Donnell and Thom describe in Chapter 5 the conflicting policies of different levels of government attempting to deal with the complexities posed by coastal development and the inherent, underlaying risks and hazards. They also explore how value systems, sitting as they do alongside the built environment and climate risk impacts to coastlines, are playing out over varying time and spatial scales. They observe that addressing this complexity requires a multi-pronged approach, but that a good starting point is in national policy. There is also a need for long-term support of research into climate change adaptation, a point also made in later chapters.

Conclusions from Part II, which deals with compound risk, revolve around the need for knowledge development and dissemination—with the respective authors highlighting public education, research, and awareness as responses to the compound, cascading, and protracted risks and disasters described in those chapters. Given that the understanding of disasters has evolved in academic and practitioner circles, these conclusions are especially pertinent to translate this understanding into improved practice and implementation.

In Chapter 6, Cook and Harrison describe the Brisbane River floodplain as a living example of compounding risk: including historical and current development, a disregard of risk due to a belief that dams provide flooding immunity and an increase in the risk of flooding due to climate change. They highlight the need for public education to raise the awareness of risk. In Chapter 7, Graham explores the compound risks of sea level rise from the very unique perspective of the individual—and their attachment to multiple at-risk places. She concludes with the need for more research in interactions between multi-place attachment and climate change. The focus on individuals continues in Chapter 8 where Legg details the intersections between an ongoing environmental hazard (chemical contamination), extreme weather events (droughts and floods) and the mental health of community residents.

The conclusions in Part III highlight issues of communication between government officials and citizens affected by disasters, participation and enfranchisement, and the difficulties of integrating informal adaptation options into formal planning systems. All of these issues revolve around governance and more specifically, interactions between formal government spheres and disaster-affected citizens. While issues of participation and disaster governance are not new, they bring novel challenges when applied to compound, cascading, and protracted disasters, as these complex disasters connect to all spheres of 'everyday' governance.

In Chapter 9, Wolff and Ramírez-Lovering discuss the lived experiences of the urban poor in Indonesia living with frequent flooding. The lack of reliable information regarding an upstream dam comes out as a central finding, highlighting the need for communications between government officials and communities affected by a disaster. In their discussion of 'chokepoints' in Chapter 10, Thompson et al. describe governance, enfranchisement, and participation as important determinants of what surveillance and preparedness may or may not be viable the in Indo-Pacific Region. Chapter 11 by Mackenzie provides an interesting study of the role of gardening as an adaptation option, pointing out its benefits as an alternative to formal employment, a source of fresh produce and material for barter for the urban poor. In his conclusion, Mackenzie discusses how the diffuse, opportunistic, and fragmented nature of gardening makes integrating it into formal land use (and disaster management) planning difficult for government officials.

Part IV examines diverse case studies of protracted disasters from the Asia Pacific region. The conclusions from these individual chapters hone in on the need to reformulate the concept of disasters from a sudden-onset, exceptional event to an ongoing part of everyday life. They also tackle the limits of formal government structures in places where central and regional governments do not have resources to lead long-term recovery efforts. Both threads come together: formal government structures may not have the resources or political will to provide ongoing support in either communicationor material assistance to rebuild and government agencies trained in a managerial culture may hold much different views and understandings of disasters compared to the lived experiences of the affected communities. In most of the chapters, formal government agencies, representing, national or provincial governments were physically and, in some cases, culturally removed from affected communities, highlighting once again the need for participation of affected people in formal disaster management planning.

Chapter 12, by Green and Whitney, evaluates the practices of local government information dissemination targeting culturally and linguistically diverse communities in Japan during the COVID-19 pandemic. Their conclusions highlight the lack of national government guidance in messaging to CALD residents in Japan and point out that hazard awareness needs to be integrated into multicultural policy frameworks. This point echoes Mackenzie's preceding chapter about integrating adaptation actions into land use planning. Disaster management and response planning need to be integrated with planning processes for 'everyday' and 'normal' systems. In Chapter 13, Turner-Walker's conclusion continues the importance of communication by detailing how villagers affected by multiple ongoing earthquakes were subjected to multiple sources of information, including traditional knowledge, social media, hoax news, and sensitisation to tsunami events. The case study highlights how the lack of trust and legitimacy of formal government institutions affect local community responses to protracted events. This theme is continued in Chapter 14 by Cajilig, where she describes the informal recovery assistance systems of fisherfolk in the Philippines. There, kinship networks bypass and augment formal government structures as state-led efforts are limited to immediate relief after typhoons and inadequate for long-term recovery. Chapter 15 by D'Arcy turns the spotlight to Australia in an examination of the 'war on wildfire' narrative; the conceptualisation and management approach to wildfire in Australia. An important finding in D'Arcy's research related to the asymmetrical power relationships between government agencies (especially emergency services) and communities at risk of wildfire.

Section 5 looks beyond individual disaster typologies and its three chapters focus on Australia—a relatively wealthy, (mostly) politically stable country, with a well-developed federated system. Two powerful themes emerge from these chapters: the need for national leadership and coordination of complex disasters (echoing the conclusions of Chapters 3 and 5), and the need for collaborative research with partnerships between governments, communities, and industry.

In Chapter 16, Samnakay and Dovers argue for national leadership and collaboration to manage large scale, multi-jurisdictional, and compounding disasters, given the threat that compound, cascading and protracted disasters pose, and the underlying risks that ongoing climate change presents. Disaster fatigue at a collective (community) level is introduced in Chapter 17 by Ingham et al. who highlight the need for further research and the development of tools and frameworks for disasters leaders and decision-makers (both formal and informal). Chapter 18 by Dawson et al. is a showcase of case studies of existing disaster management research programs aimed at identifying success factors for industry-government-university collaboration.

Throughout all chapters, we see that while the challenges, frequency, and severity of disasters is now more complex, what we need to do to address these challenges hasn't really changed. Transparent governance, institutional reform, wider participation, better planning, and preparedness, ensuring relief occurs before recovery, and clear and open communication are all needed as we respond to the impacts of climate risk and biodiversity loss. The ongoing COVID-19 pandemic also makes a frequent appearance throughout this book. While it is the subject of only four chapters, its cascading, protracted (and ever-changing) effects have affected the daily lives of all the contributing authors, and, we'll hazard a guess, most of its immediate readers. The consequences of this pandemic may only be fully appreciated by future generations. Another ongoing hazard and risk, climate change, is slowly being incorporated into daily lives, albeit imperfectly, through the most recent pledges by world leaders to commit to a net zero emissions target by 2050. We hope that the confluence of both these crises highlights the need to re-orient and re-conceptualise disasters not as single, isolated, unexpected events, but as a part of daily life, permeating the background on which other, sudden-onset events take place.

References

Cutter, S. L. (2018). Compound, cascading, or complex disasters: What's in a name? *Environment: Science and policy for sustainable development*, *60*(6), 16–25. https://doi.org/10.1080/00139157.2018.1517518

Pescaroli, G., & Alexander, D. (2018). Understanding compound, interconnected, interacting, and cascading risks: A holistic framework. *Risk Analysis*, *38*(11), 2245–2257. https://doi.org/10.1111/risa.13128

Simpson, N. P., Mach, K. J., Constable, A. J., Hess, J. J., Hogarth, R., Howden, M., ... Trisos, C. H. (2021). A framework for complex climate change risk assessment. *One Earth*, *4*, 489–501. https://doi.org/10.1016/j.oneear.2021.03.005

Index

A. Lukasiewicz and T. O'Donnell (eds.), *Complex Disasters*, Disaster Risk, Resilience, Reconstruction and Recovery, https://doi.org/10.1007/978-981-19-2428-6

D

The manufacturer's authorised representative in the EU is Springer Nature Customer Service Centre GmbH, Europaplatz 3, 69115 Heidelberg, Germany. If you have any concerns regarding our products, please contact ProductSafety@springernature.com

Printed and bound by CPI Group (UK) Ltd, Croydon, CR0 4YY

30/09/2026

02225327-0015